**ATLAS
OF THE
HUMAN SKULL**

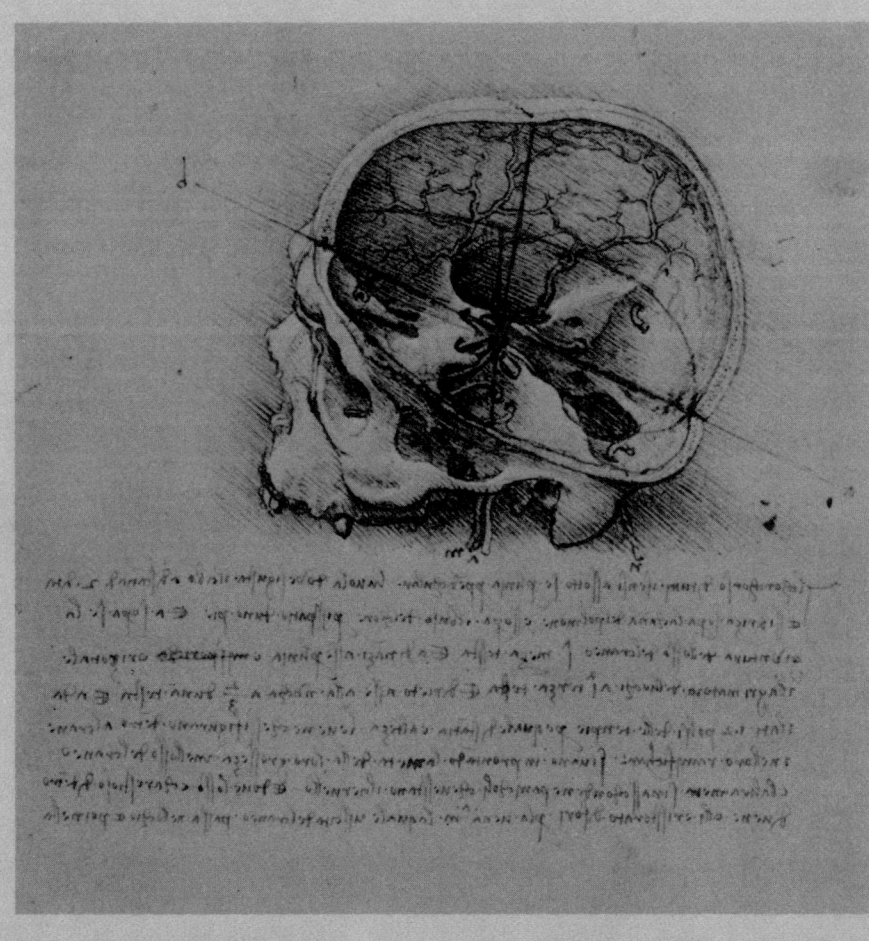

Leonardo da Vinci (1452-1519)
Drawing, Royal Library, Windsor Castle

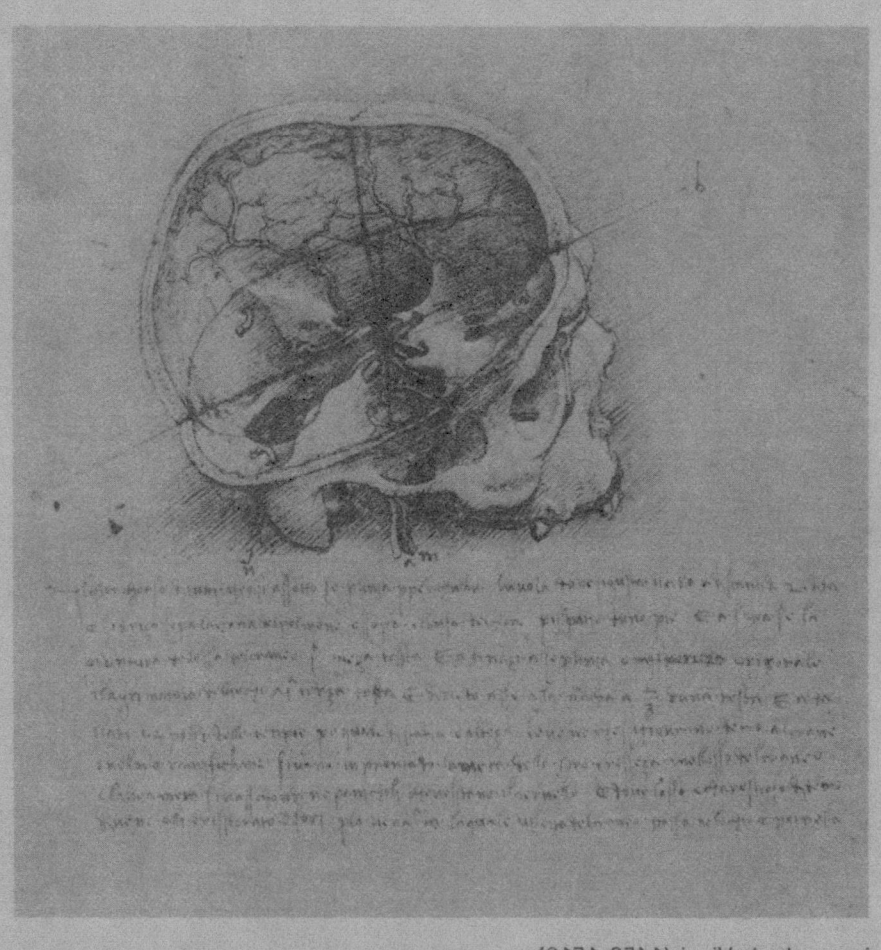

Leonardo da Vinci (1452-1519)
Drawing, Royal Library, Windsor Castle

ATLAS OF THE HUMAN SKULL

Margaret M. Waddington, M.D.

Adjunct Associate Professor
of Clinical Medicine (Neurology)
Dartmouth Medical School
& Dartmouth-Hitchcock Clinical Center,
Hanover, New Hampshire

Illustrations by the Author

Foreword by
R. M. Peardon Donaghy, M.D., F.A.C.S.
Professor of Surgery
Department of Surgery
Division of Neurology
University of Vermont,
College of Medicine
Burlington, Vermont

Academy Books
Rutland

Copyright (c) 1981 by Margaret Miles Waddington

All rights reserved. No part of this book may be reproduced in any form or by any electronic or mechanical meas, including information storage and retrieval systems, without permission in writing from the publisher, except by a reviewer who may quote brief passages in a review.

Library of Congress Catalog Card No. 81-68348
ISBN 0-914960-36-9

First Edition

Printed and bound in
the United States of America
by Sharp Offset Printing Inc.
P.O. Box 757, Rutland, VT 05701

FOREWORD

Writing a foreword for this beautiful and significant volume is a distinct pleasure. The words "significant" and "pleasure" are not chosen lightly. It is significant because this is the most easily understandable exposition I have seen of a very complicated segment of osteology. Not only is the clarity outstanding, but the portrayal by photograph and drawing is done with grace and beauty—a compliment to author and publisher alike.

It is significant because it is a "teaching" book, and because it does not force one to resort to an adequate memory of pages of detail, but rather makes learning a pleasant experience so that the student may say, "Ah, yes! Now that I see the relationship I know why my patient's symptoms progressed as they did." The simulation of structures that traverse cranial foramina is helpful. The labeling of areas of muscle attachment to bone is a welcome return to a feature of many texts of the nineteenth century, and color coding of various bones is a simple, basic yet novel approach that will be genuinely appreciated by the reader.

More than anything, however, the significance to this writer is that he has had the privilege of knowing Margaret Waddington as medical student, resident, and devoted and beloved clinician. Her fertile and facile mind was not content until she had applied it to anatomic research, and then she did not rest until she had developed a medium for teaching others what she had learned and had become a nationally recognized authority along the way.

All this was done without grants or support from government or great institutions, but rather in a small city and at personal sacrifice of time and capital. There is not a surgeon, pathologist, or medical student who does not owe her a debt of gratitude, nor a cynical elder of the profession who cannot find pleasure in the knowledge that the force that drove Hippocrates still functions and the spirit of the perennial student and teacher yet lives, sometimes in the same individual.

R. M. Peardon Donaghy, M.D.

DEDICATED TO FRIENDS AND COLLEAGUES

PREFACE

This atlas is intended primarily for medical students, neuro-scientists, and those surgical specialists requiring understanding of the intricacies of the skull in order to perform surgical procedures in the regions adjacent to it. It seems particularly appropriate to attempt to meet the gap created by the relative paucity of excellent anatomic specimens in our teaching institutions. The skulls and disassociated bones are sometimes badly damaged because of prolonged use. Time often does not permit an immediate grasp of the three dimensional concept underlying the construction of the skull. By using color photography, taking many different views, magnification when indicated in order to bring out details, and assembling disassociated bones in composites to emphasize the relation of one bone to the other, it is hoped that the book will be profitable and enjoyable.

The line drawings are made as a supplement, with no intention to detract from the photographs, but rather to facilitate the learning process and to avoid labeling photographs. For this reason, shading of the illustrations is purposely deleted. Minor inaccuracies in the drawings may be found, although every attempt has been made to keep them at a minimum.

In general, English terminology is used, though in the Glossary the appropriate Nomina Anatomica term is given. The Glossary deals with the important terms used in the atlas, particularly to serve as a quick reference to the content of foramina, fissures or canals. Translation of the common Latin terms is included. The Table of Muscles is limited to those with attachments to the skull. The conventional reference to insertion origin, nerve supply, and action is listed. Neither the Glossary nor Table of Muscles should be interpreted to be an adequate replacement for a detailed textbook of anatomy.

The author takes full responsibility for errors and omissions.

ACKNOWLEDGMENT

I am under great obligation to many colleagues and friends whose suggestions and useful criticism have helped shape the course of this atlas. Dr. R. M. Peardon Donaghy, Dr. Frank D. Lathrop, Dr. Alexander G. Reeves and Dr. Paul I. Yakovlev have all reviewed and appraised the manuscript at one or another stage in its career. The author is pleased to acknowledge the editorial assistance of Professor Theodore Steele and Louise McCoy. Dr. Wilbert F. Chambers early endorsed the project and provided bone models from the Department of Anatomy of the Dartmouth Medical School. Dr. Theodore G. Shattuck introduced me to the *Surgical Anatomy of the Temporal Bone and Ear* by Dr. Barry J. Anson and Dr. James A. Donaldson. Dr. Donaldson and the publisher, W. B. Saunders Company, gave permission to make line drawings from the inner ear photographs taken through a microscope, forming elegant illustrations of this intricate anatomy. Many of the definitions from *Dorland's Illustrated Medical Dictionary*, 25th Edition, have been used in the Glossary with the permission of the editor at the W. B. Saunders Company. I am indebted to Mr. Frank C. Mallory and Mr. Wing Woon who took most of the photography at the Department of Photography, University of Vermont College of Medicine. Some slides were taken by Dr. Philip G. Merriam, Chairman of the Department of Pathology, The Rutland Hospital, Rutland, Vermont, leaving but several for the author to take in completing the atlas.

The frontispiece, drawing by Leonardo da Vinci, is reproduced by gracious permission of Her Majesty The Queen.

I wish to thank Mr. John C. Matthews, Mr. Herman B. Horsman and the staff of Coach House Design, Ltd., whose conversion on the computer printing press of the 35 mm. slides and colored drawings achieved such fine quality. It has been a pleasure to work with Mr. Robert Anderson Sharp and the staff of Academy Books whose perceptiveness and assistance eased the throes of getting this book in print.

My secretaries, Miss Rosalie Gregory and Mrs. Cheryl Young, were invaluable in verifying the nomenclature and typing the manuscript.

I have had good advice. The shortcomings are mine.

CONTENTS

- v   Foreword by Dr. R. M. Peardon Donaghy
- vii   Preface
- viii   Acknowledgment
- xii   Color code

1 Introduction

2 PART I
Skull

2 SECTION 1
Skull from front
- 3 Bones
- 4 Sutures
- 5 Points and foramina
- 6 Muscle attachments

8 SECTION 2
Skull from side
- 9 Bones
- 10 Sutures
- 11 Points and landmarks
- 12 Muscle attachments

14 SECTION 3
Skull from behind
- 15 Bones
- 16 Sutures
- 17 Points and foramina

18 SECTION 4
Skull from below
- 19 Bones and landmarks
- 20 Mandible removed
- 21 Bones and sutures
- 22 Foramina and canals
- 23 Points and landmarks
- 24 Muscle attachments
- 26 Closeup view, foramina and adjacent areas

30 SECTION 5
Skull lines

32 SECTION 6
Skull from above, bones and sutures
- 33 Internal surface, bones and vascular grooves

34 SECTION 7
Base of skull, internal surface
- 35 Bones and vascular grooves
- 36 Foramina and canals
- 37 Cranial fossae

38 SECTION 8
Hypophyseal fossa
- 38 From above
- 39 From side
- 41 From behind

42 SECTION 9
Orbital cavity
- 42 From front, bones and foramina
- 43 Margins, fissures and foramina
- 45 Orbital floor
- 46 Orbital rim and orbital roof
- 49 Orbital walls
- 51 Nasolacrimal canal
- 52 Floor and walls, composite of maxilla and adjacent bones
- 54 From the side, skull with frontal bone resected

56	**SECTION 10** **Nasal cavity**
56	Midsagittal sections
58	Composite of maxilla and adjacent bones
60	From front
63	Oblique view
66	From below
68	From front
70	From behind
72	Sphenopalatine foramen
74	**SECTION 11** **Pterygopalatine fossa**
74	Diagrammatic, from side and from front
75	View through orbital cavity
77	From side
78	Posterior wall
79	Medial wall
80	**SECTION 12** **Infratemporal fossa**
82	**SECTION 13** **Temporomandibular joint**
84	**PART II** **Facial bones**
84	**SECTION 1** **Mandible**
84	From front
85	From behind, and above
86	From side
87	Landmarks and muscle attachments
88	From behind and below
89	Landmarks and muscle attachments
90	From above
91	**SECTION 2** **Maxilla**
91	Outer and oral surface, alveolar process removed to expose roots of teeth
92	From front and behind
93	Nasal side and lateral side, muscle attachments
94	Diagrammatic illustrations
96	**SECTION 3** **Zygomatic**
96	Composite of zygomatic with adjacent bones, temporal fossa
97	From above and below, from front and behind
98	**SECTION 4** **Palatine**
98	From sides
99	From behind and from front
100	**SECTION 5** **Small facial bones**
100	Vomer, inferior nasal conchae, lacrimal and nasal bones
101	**PART III** **Cranial bones**
101	**SECTION 1** **Frontal**
101	From side
102	From front
103	From behind

104 SECTION 2
Ethmoid
104 From side and from behind
105 From above and from below

106 SECTION 3
Sphenoid
106 From front
107 From above
109 From behind
111 From below
112 From side

114 SECTION 4
Occipital
114 From above
116 From behind
117 From below
118 Occipital and sphenoid, from above
119 Occipital and sphenoid, from side

120 SECTION 5
Parietal
120 External surface
121 Internal surface
122 From side

123 SECTION 6
Temporal
123 With sphenoid, middle cranial fossa
124 From side, muscle attachments
125 External auditory canal
126 From below
127 Internal acoustic meatus
128 From behind, internal surface
129 From above, internal surface
130 Semicircular canals exposed
132 Facial canal and auditory ossicles
133 From side, with facial canal exposed
136 From above, cochlear spiral canal and semicircular canal exposed

138 SECTION 7
Auditory ossicles
140 Ligaments and muscles

141 SECTION 8
Tympanic cavity
141 Tympanic membrane, external and internal surface
142 Facial canal exposed
144 Stapes and adjacent nerves
145 Semicircular canals in relation to auditory ossicles

146 PART IV
Infant skull at birth
146 From front and from side
147 From above and from behind, and infant temporal bone
Glossary
Table of Muscles of the Skull
Bibliography
Index

COLOR CODE: SKULL BONES

Facial Bones

1. Mandible (1)
2. Maxilla (1)
3. Zygomatic (2)
4. Lacrimal (2)
5. Nasal (2)
6. Inferior nasal concha (2)
7. Palatine (2)
8. Vomer (1)

—
(13)

Cranial Bones

9. Occipital (1)
10. Parietal (2)
11. Temporal (2)
12. Sphenoid (1)
13. Frontal (1)
14. Ethmoid (1)

—
(8)

INTRODUCTION

In order to study the anatomy as though looking at the real object, all labels are placed on accompanying line drawings made from the original photograph. Colored rods are placed in canals and foramina to emphasize direction. The bones are color coded to emphasize shape, size and boundaries.

To avoid over-labeling of any given illustration, individual drawings are used for the bones, sutures, points on the skull, and muscle attachments. Stippling is employed to identify larger cavities and fossae.

The atlas is basically divided into four sections: the skull, facial bones, cranial bones, and the infant skull at birth. The skull has subsections for each view, as well as for the hypophyseal fossa, the orbital and nasal cavities. These are shown in closeup as part of the skull, followed by a series of composites of disarticulated bones explaining the actual intricacy of construction.

The facial and cranial bones are photographed from each direction and are shown as composites with one or several adjacent bones. The teeth, the bony cochlear and vestibular canals, and the auditory ossicles are treated separately, but with less detail than in specialized texts for dentists or otolaryngologists.

Though illustrations of the infant skull at birth are included in Part IV, the embryology of the skull at various stages of gestation is omitted as being beyond the scope of this atlas.

PART I. SKULL

SECTION 1. Skull from front.

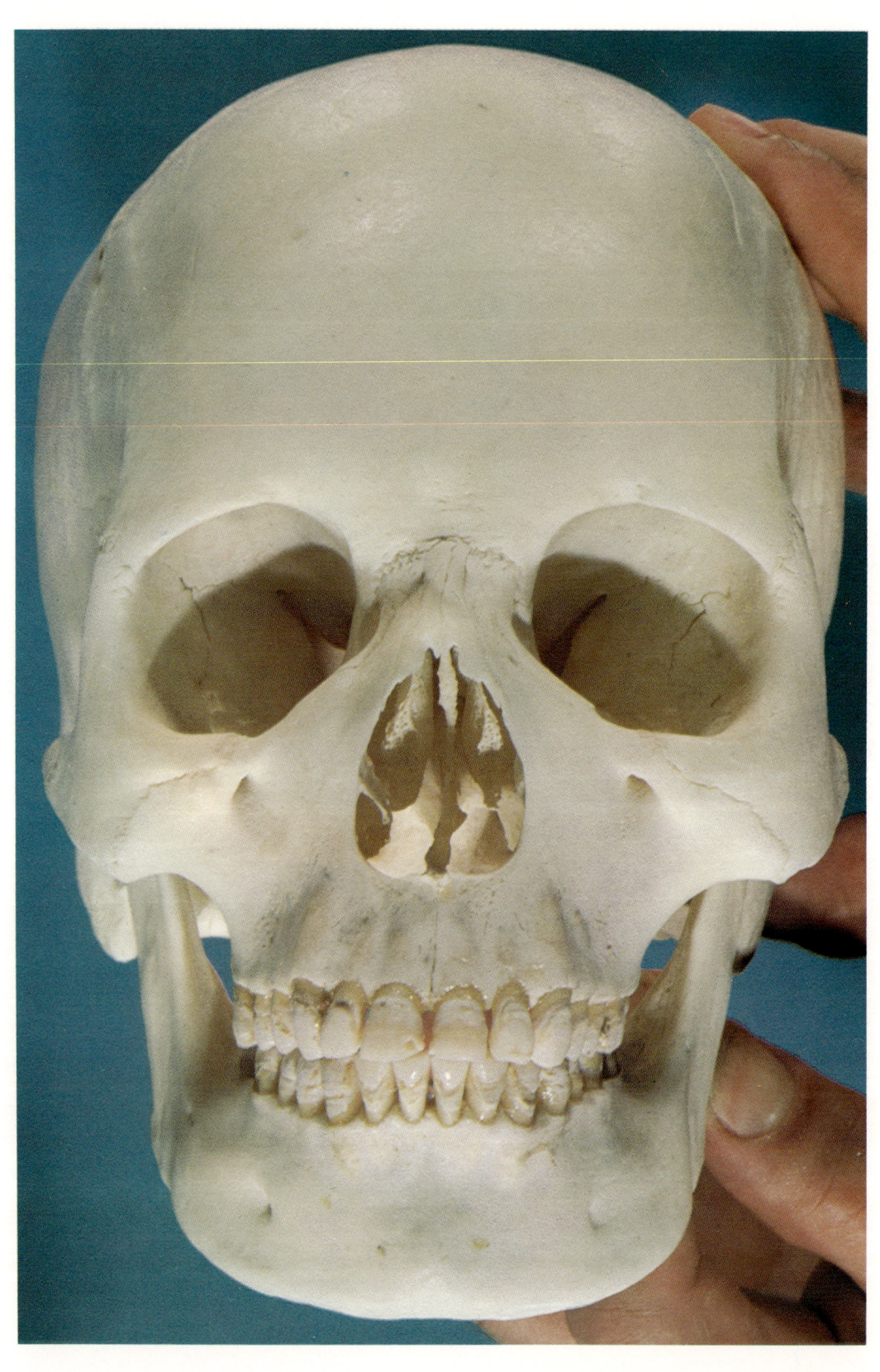

Skull from front.
Bones. (Palatine and occipital not illustrated)

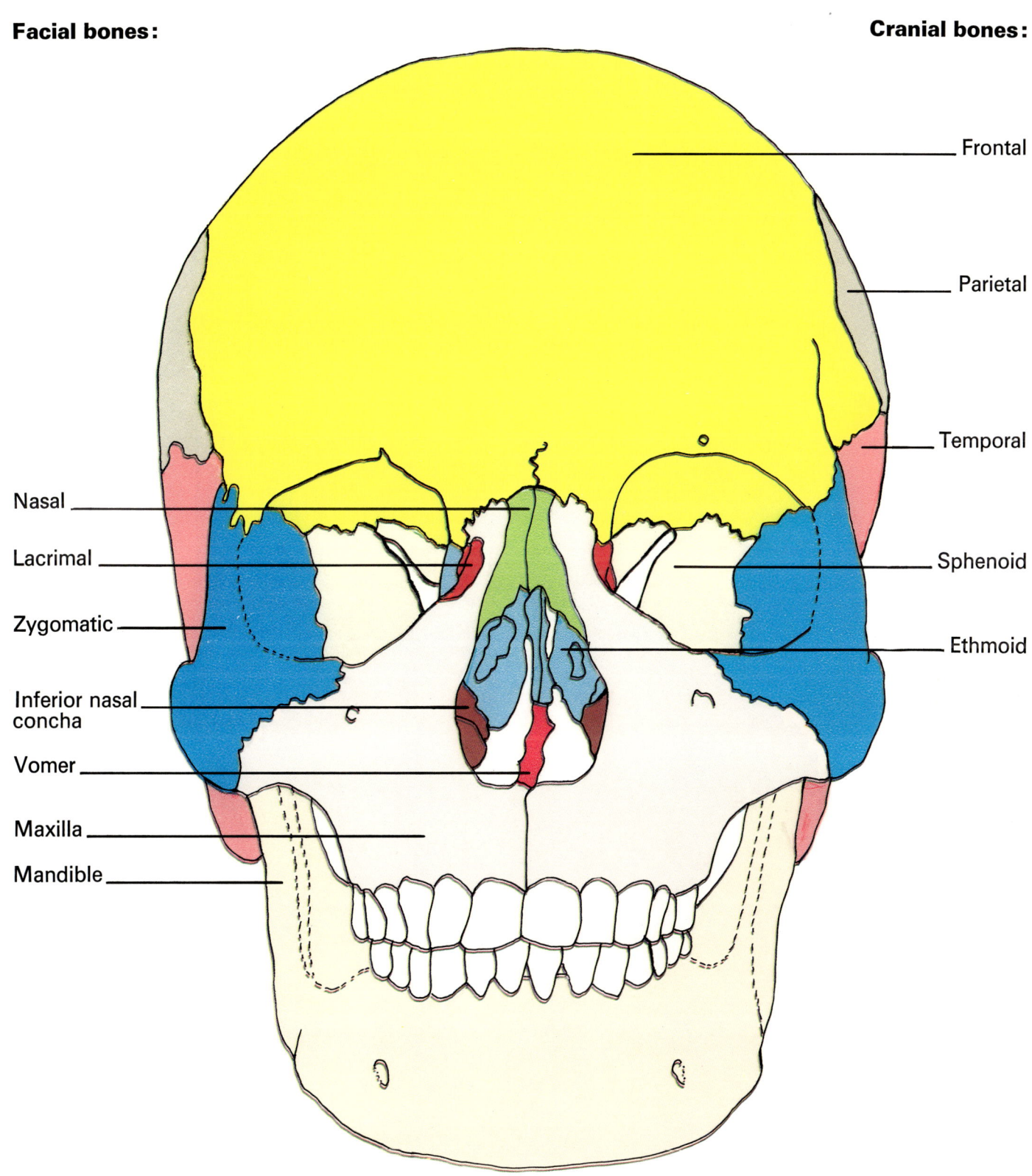

Skull from front.
Sutures.

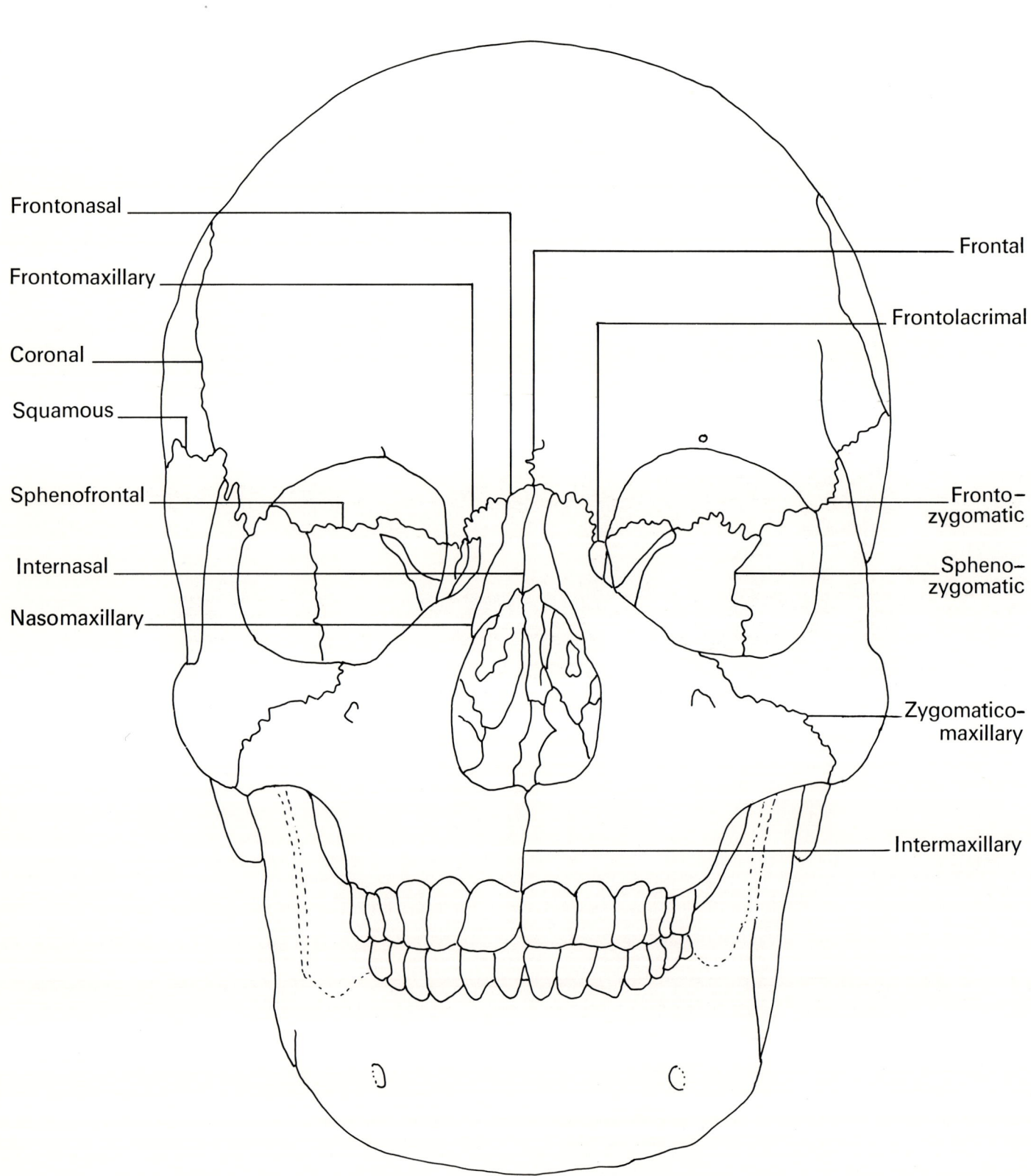

Skull from front.
Points and foramina.

Points:
- Vertex
- Glabella
- Supraorbital
- Nasion
- Infraorbital
- Anterior nasal spine
- Superior alveolar
- Inferior alveolar
- Mental

Foramina:
- Supraorbital
- Fossa for lacrimal sac
- Superior orbital fissure
- Infraorbital
- Nasal cavity
- Mental

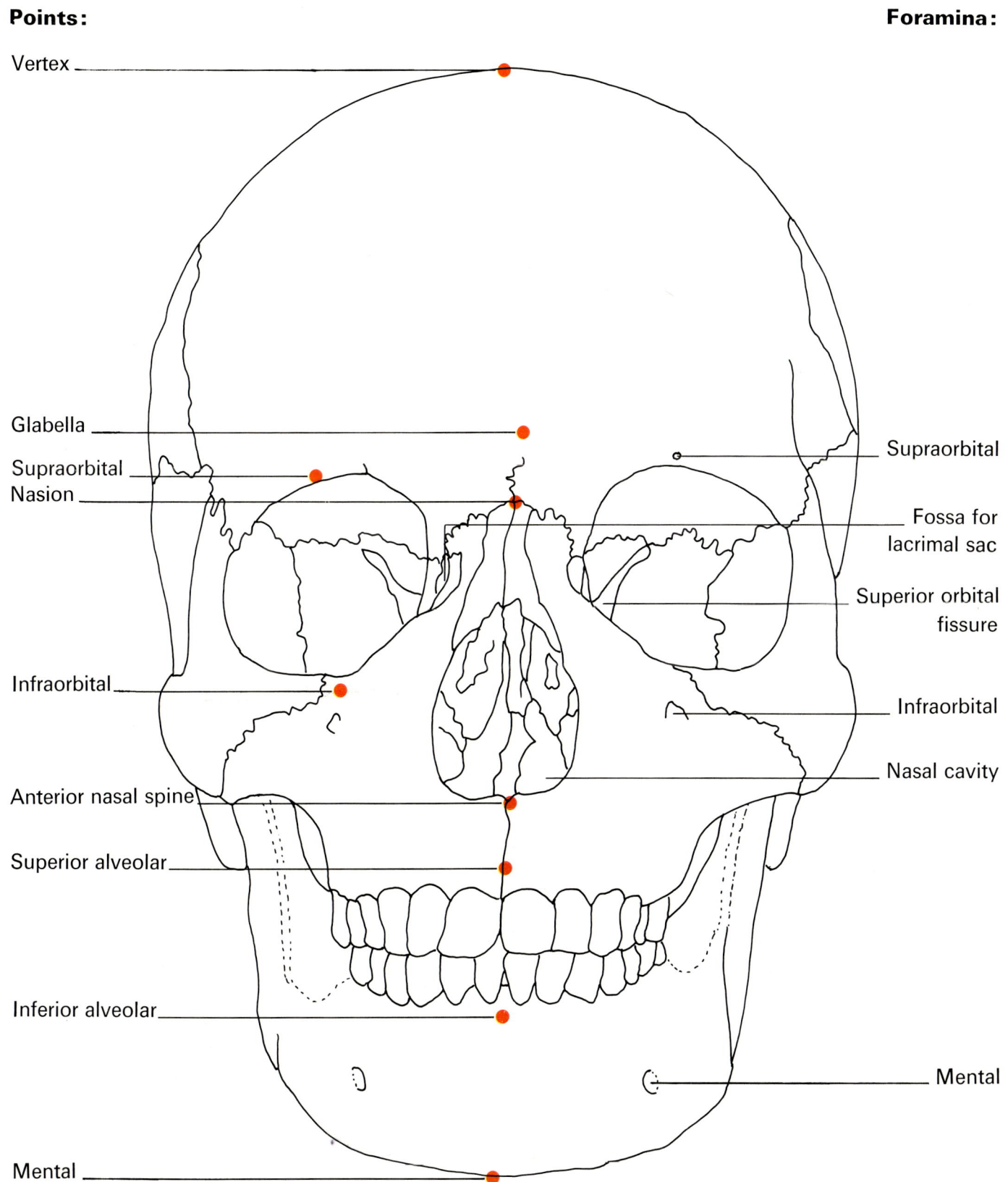

Skull from front.
Muscle attachments.

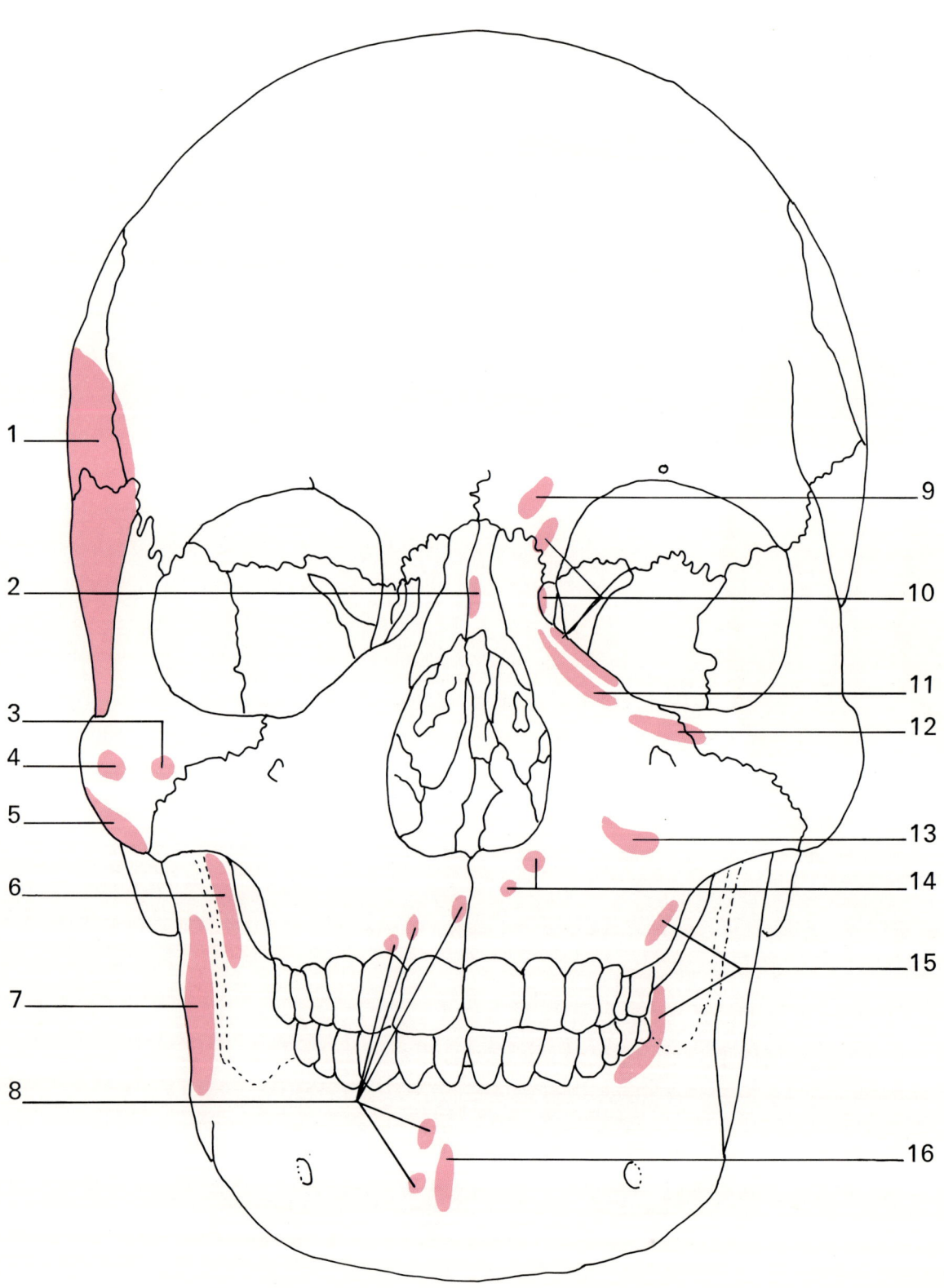

Skull from front.
Key to numbers.

Muscle attachments:

1. Temporal
2. Procerus
3. Lesser zygomatic
4. Greater zygomatic
5. Masseter
6. Temporal
7. Masseter
8. Orbicular of mouth
9. Superciliary corrugator
10. Orbicular of eye
11. Levator of upper lip and ala of nose
12. Levator of upper lip
13. Levator of angle of mouth
14. Nasal
15. Buccinator
16. Mental

SECTION 2. **Skull from side.**

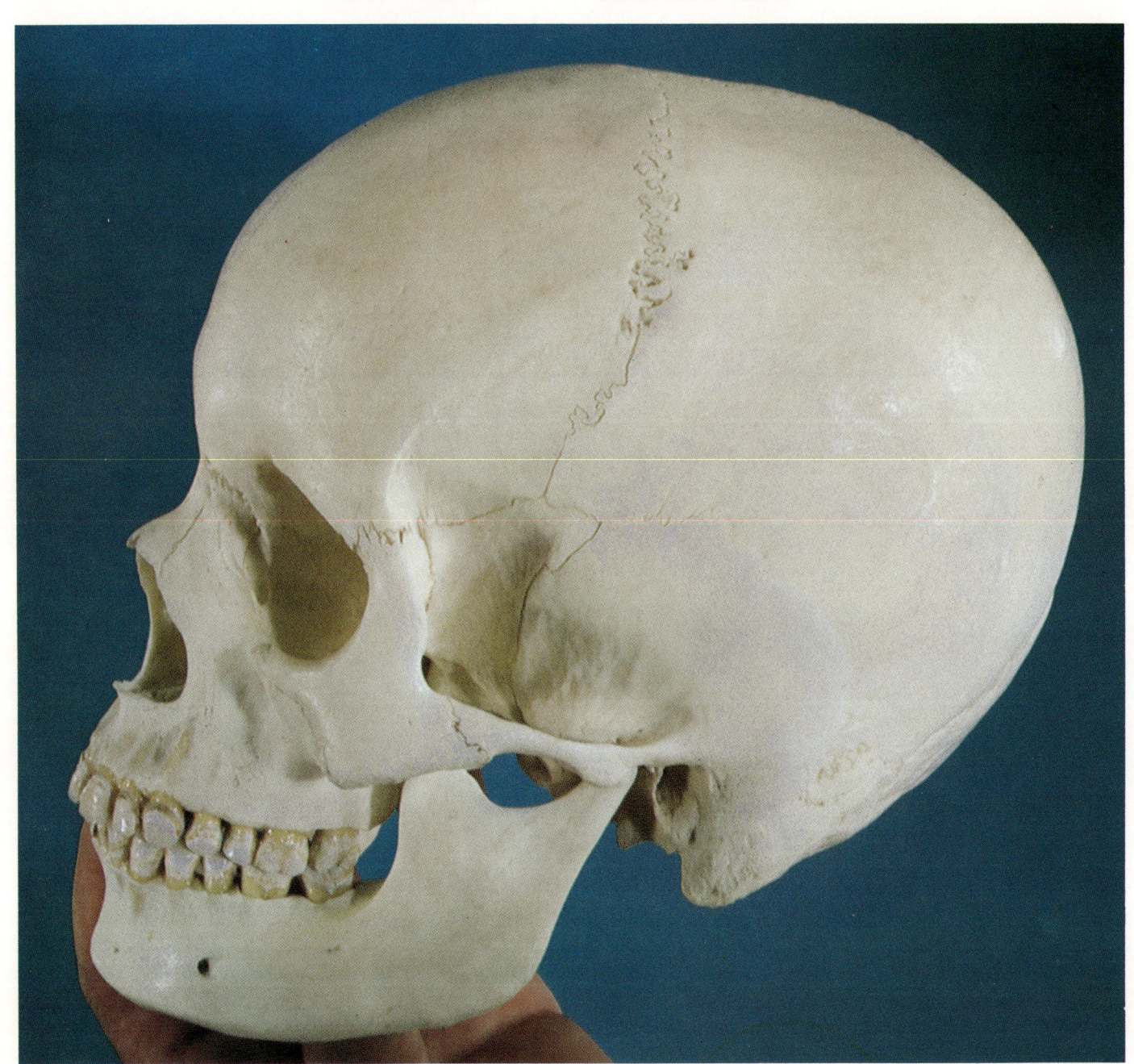

Skull from side.
A. Bones.   B. Styloid process.

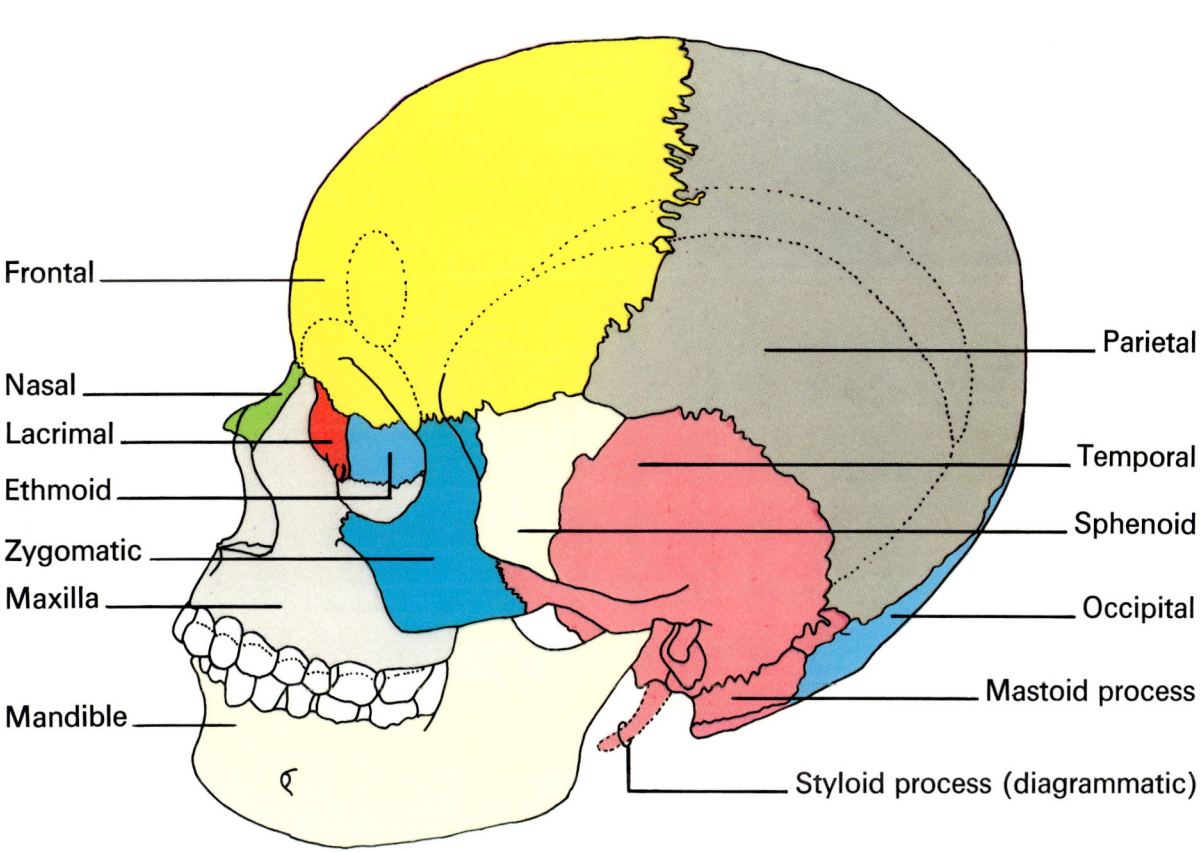

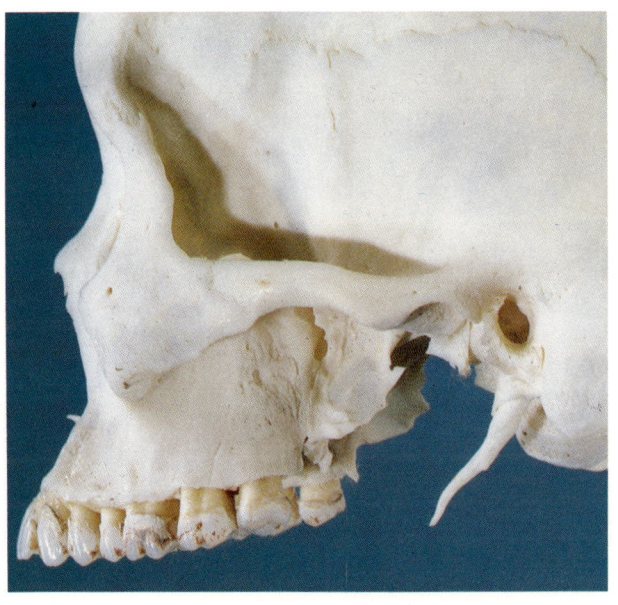
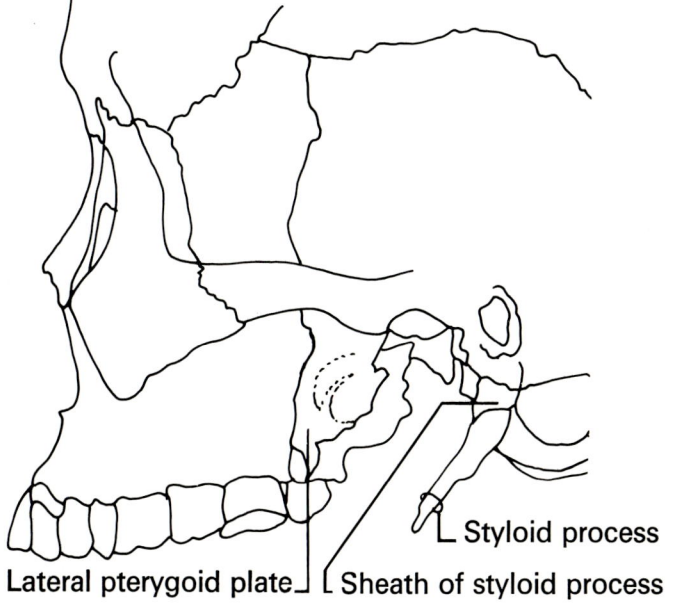

Skull from side.
Sutures.

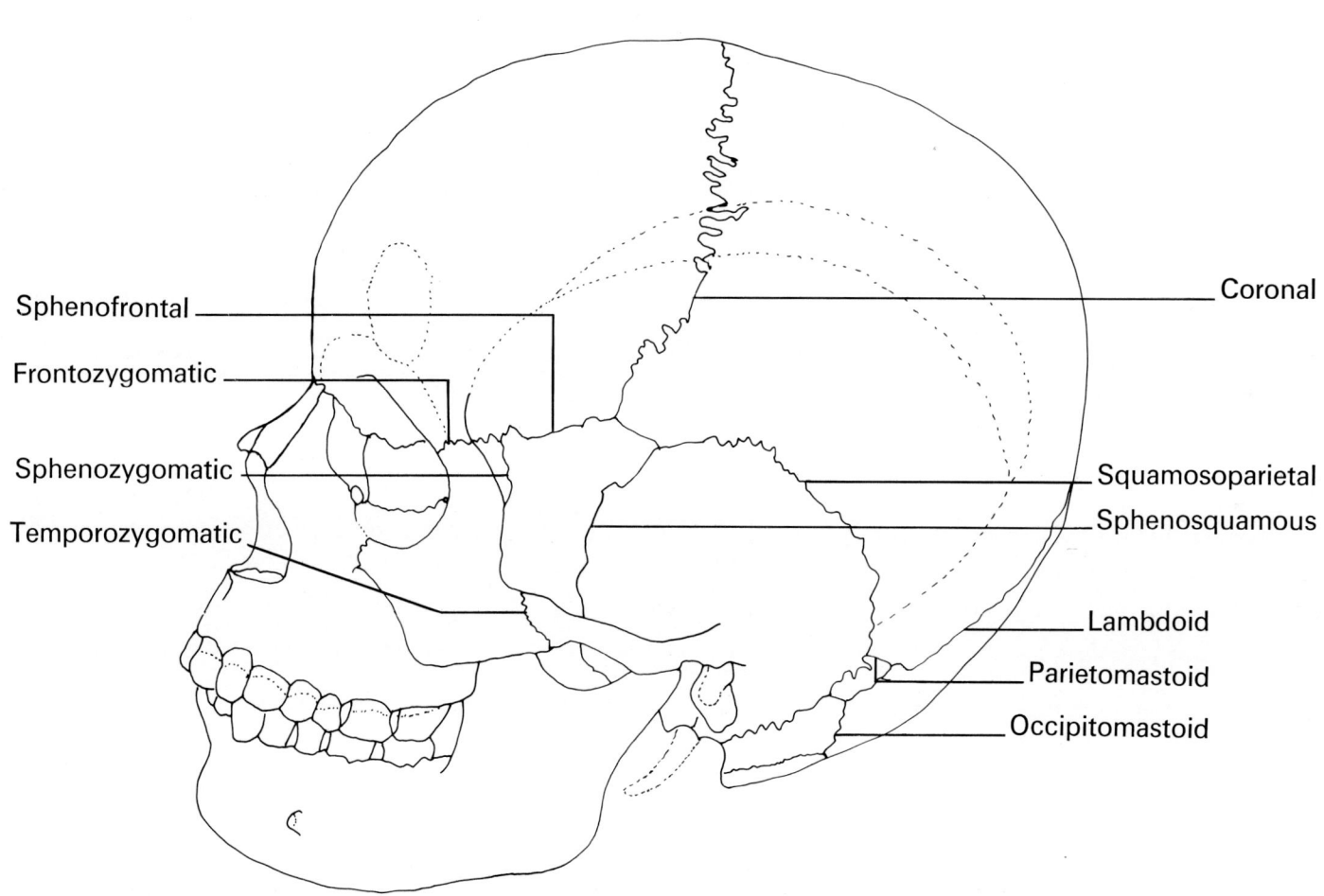

Skull from side.
Points and landmarks.

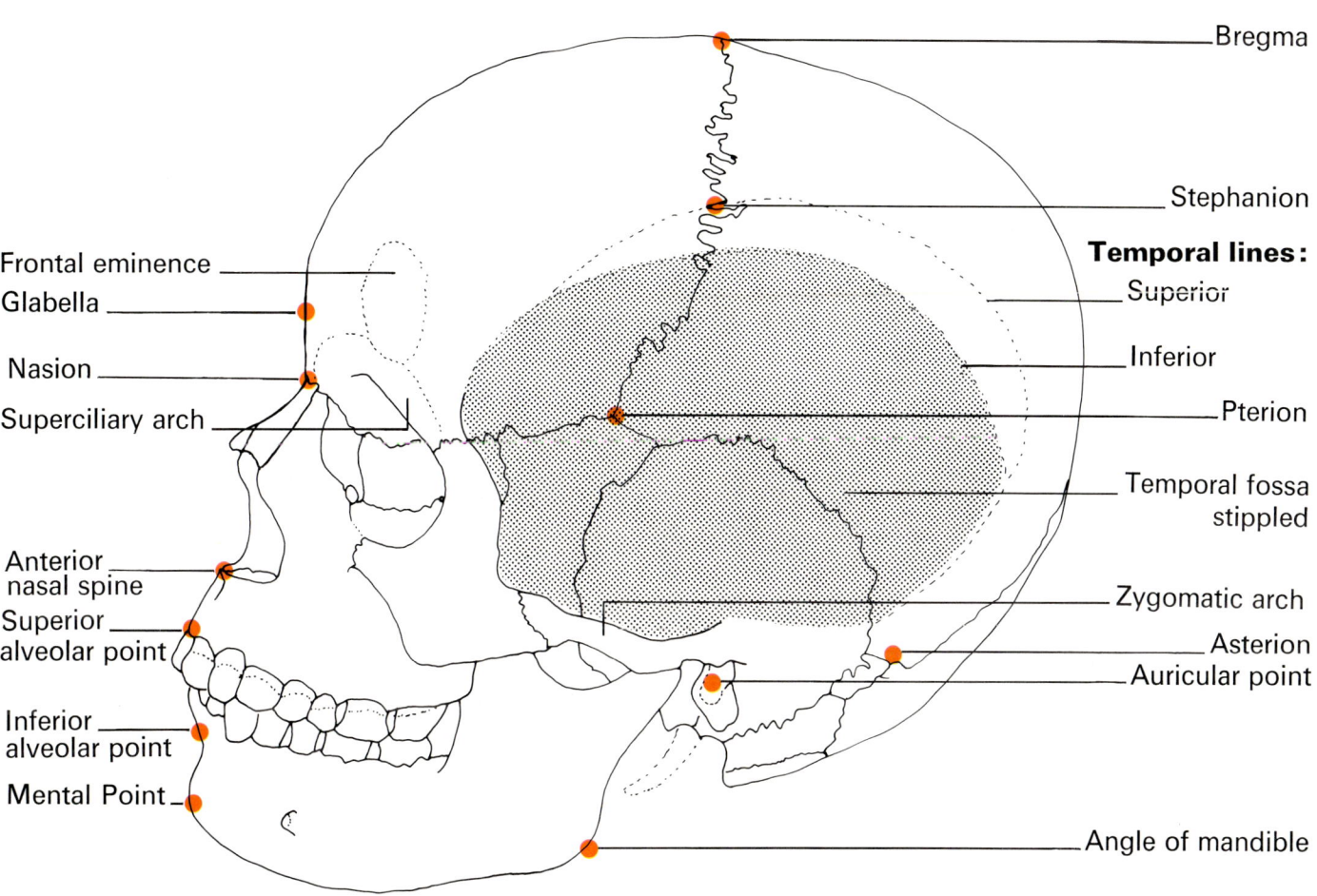

Skull from side.
Muscle attachments.

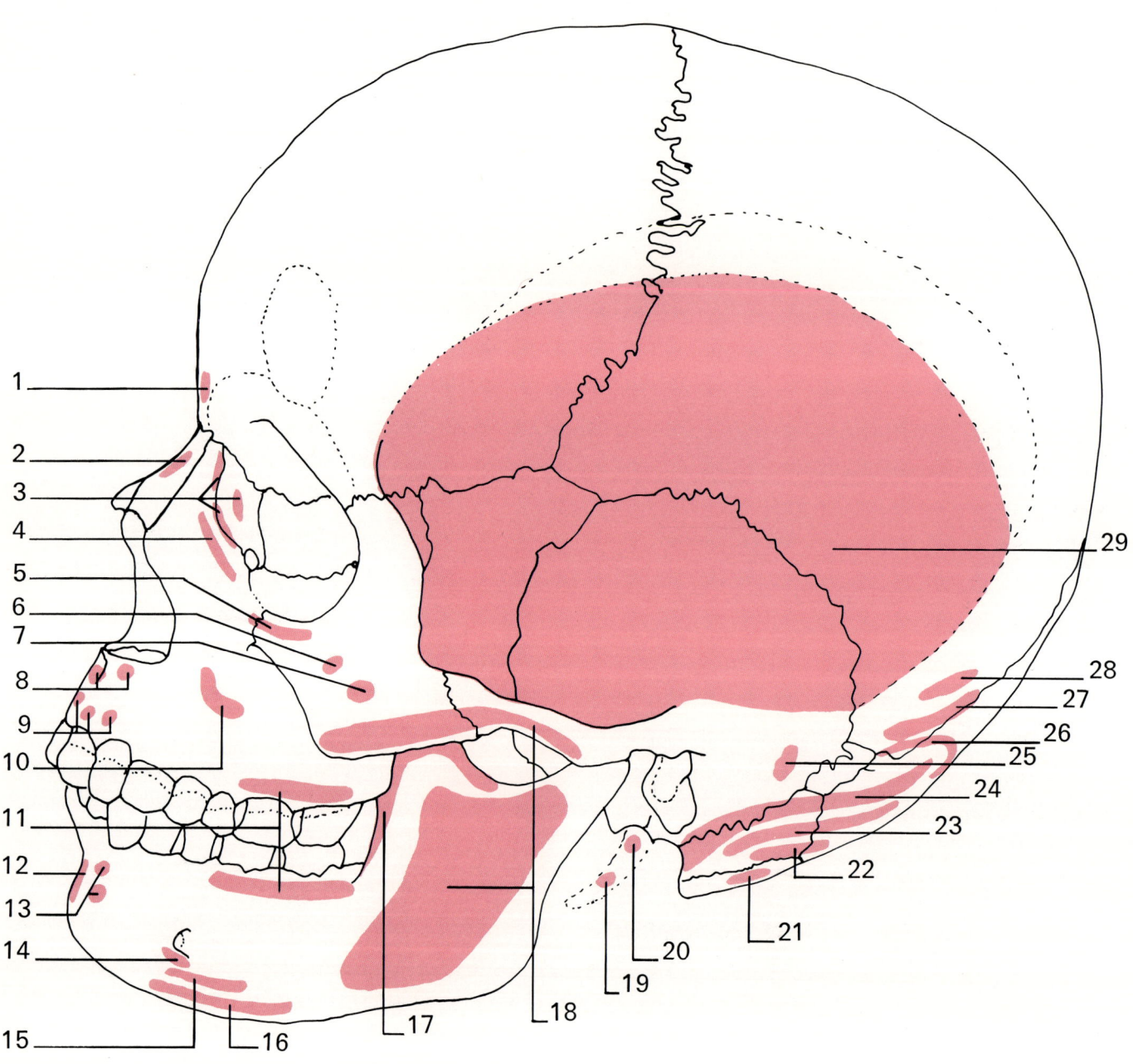

Skull from side.
Key to numbers.

Muscle attachments:
1. Superciliary corrugator
2. Procerus
3. Orbicular of eye
4. Levator of upper lip and ala of nose
5. Levator of upper lip
6. Lesser zygomatic
7. Greater zygomatic
8. Nasal
9. Orbicular of mouth
10. Levator of angle of mouth
11. Buccinator
12. Mental
13. Orbicular of mouth
14. Depressor of lower lip
15. Depressor of angle of mouth
16. Platysma
17. Temporal
18. Masseter
19. Stylohyoid
20. Styloglossus
21. Digastric
22. Longissimus of head
23. Splenius of head
24. Sternocleidomastoid
25. Posterior auricular
26. Semispinal of head
27. Trapezius
28. Occipitofrontal, occipital belly
29. Temporal

14 **SECTION 3.** **Skull from behind.**

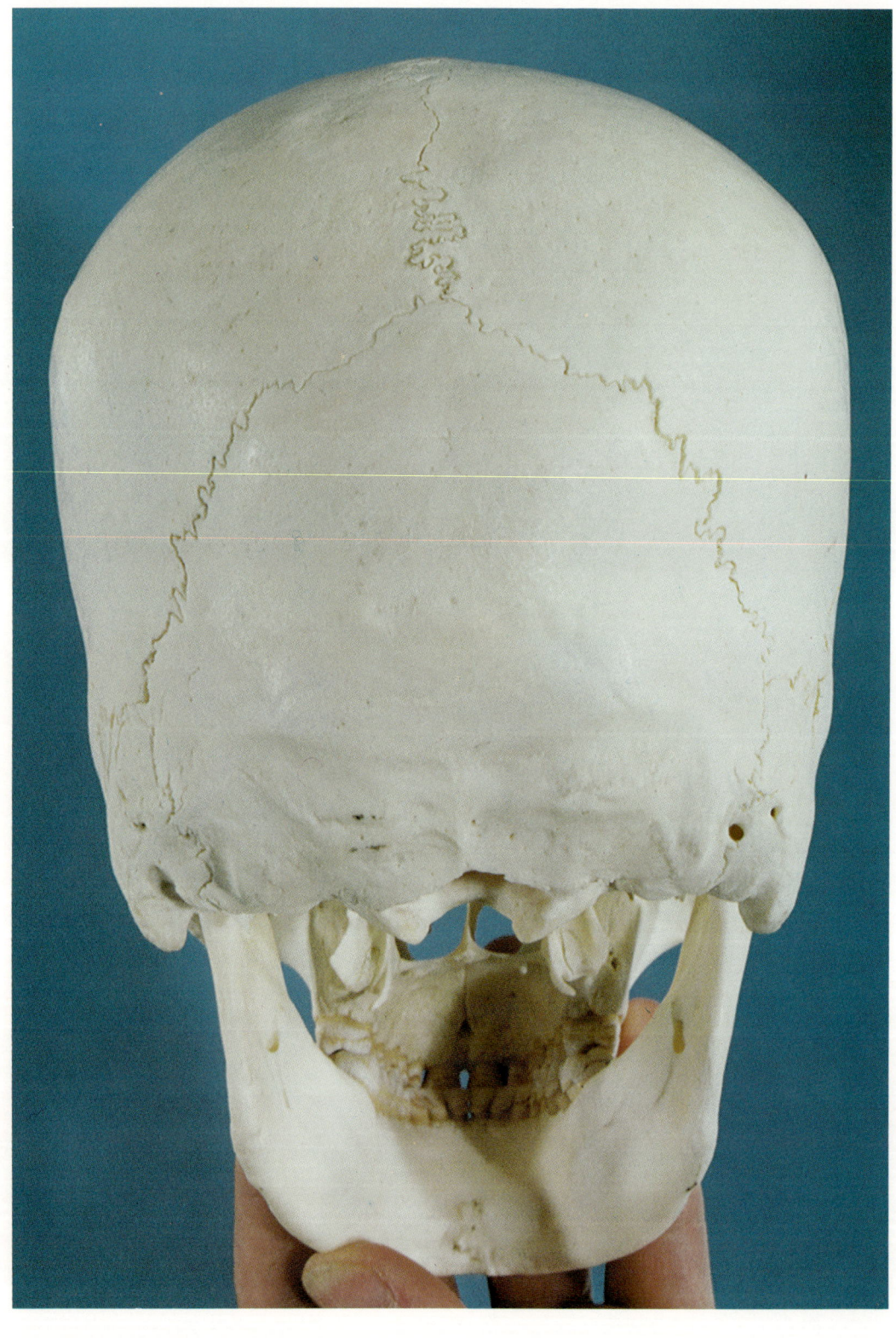

Skull from behind.
Bones.

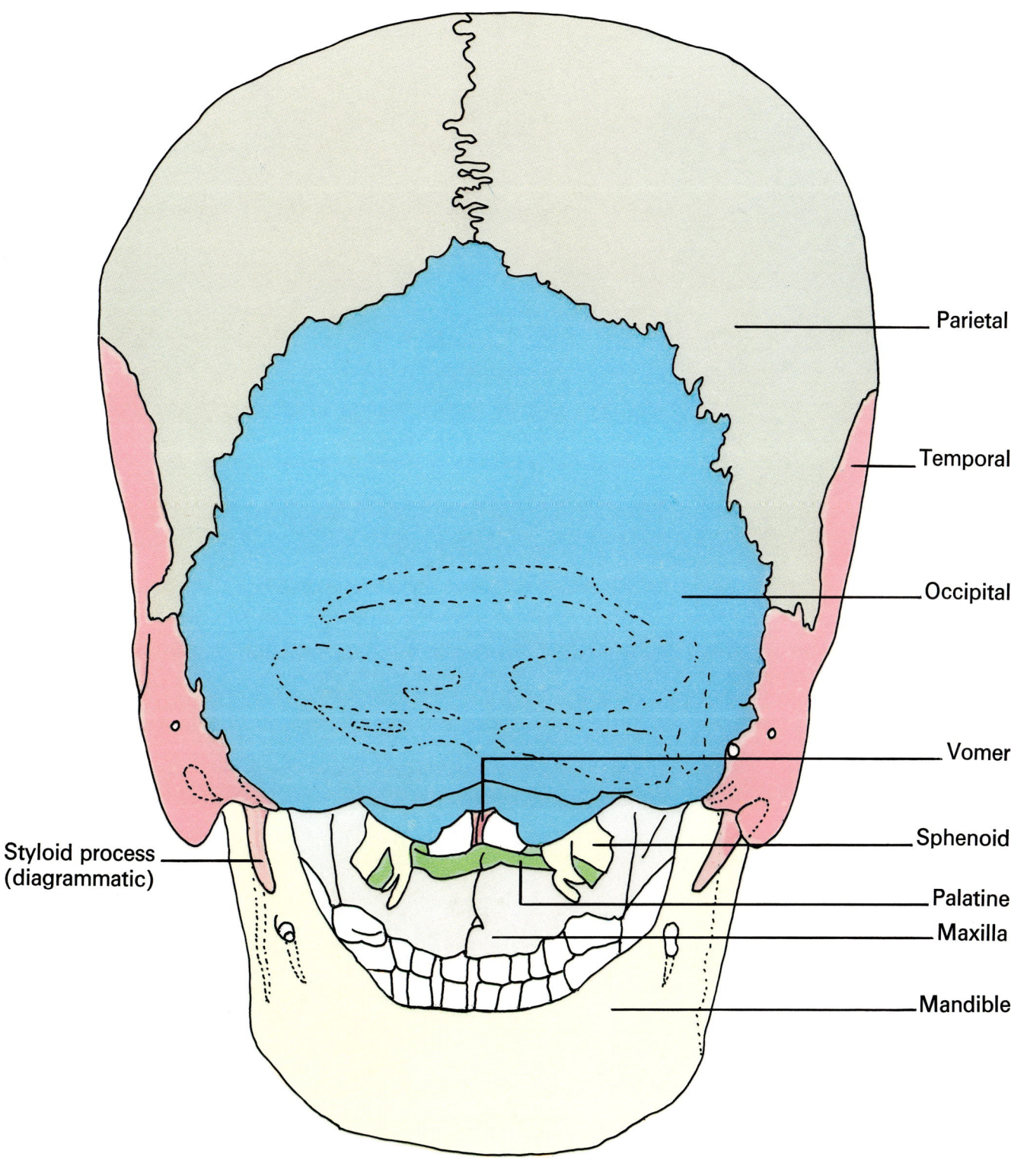

Skull from behind.
Sutures and points.

Sutures:

Points:

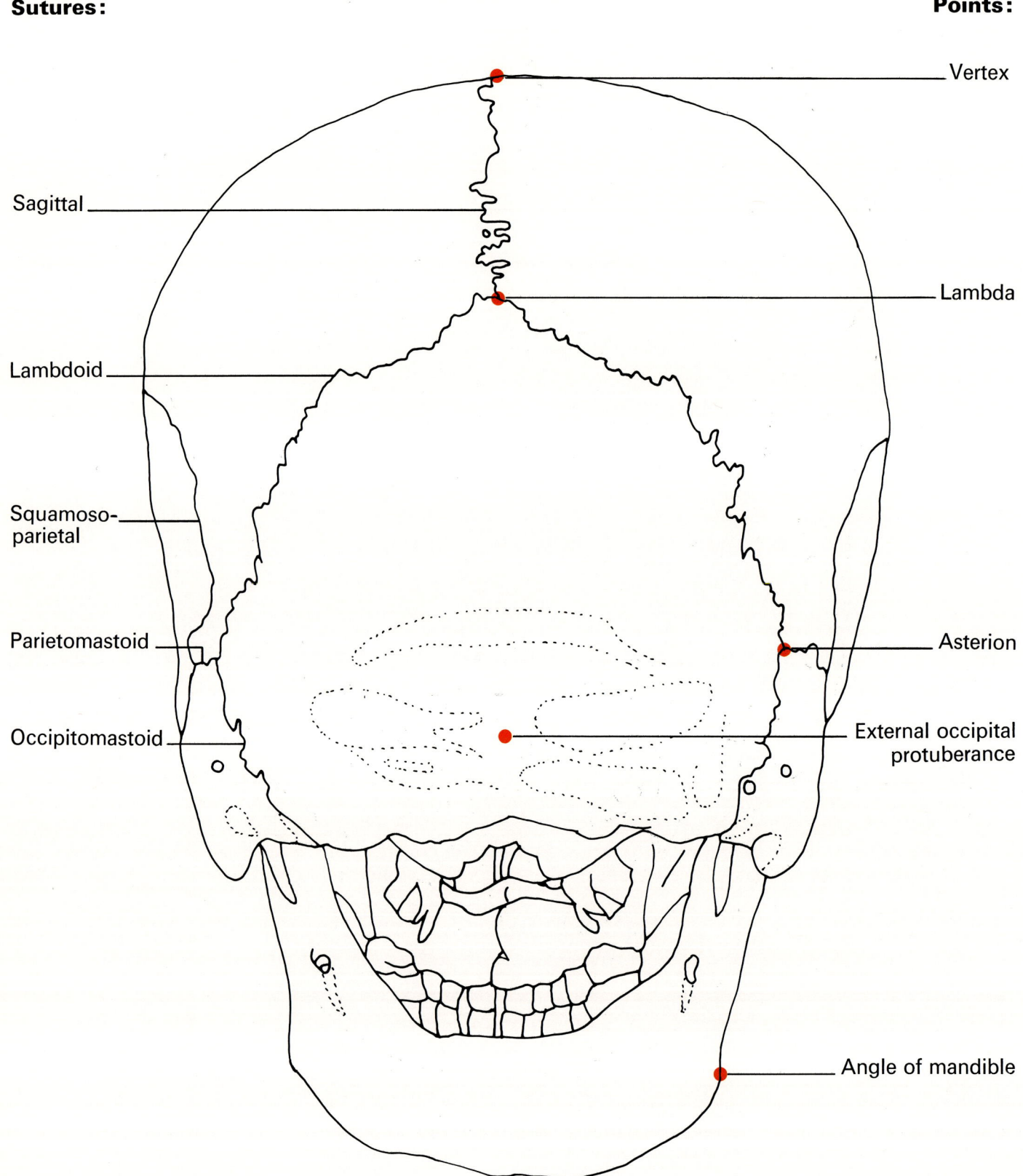

Sagittal

Lambdoid

Squamoso-
parietal

Parietomastoid

Occipitomastoid

Vertex

Lambda

Asterion

External occipital
protuberance

Angle of mandible

Skull from behind.
Foramina and landmarks.

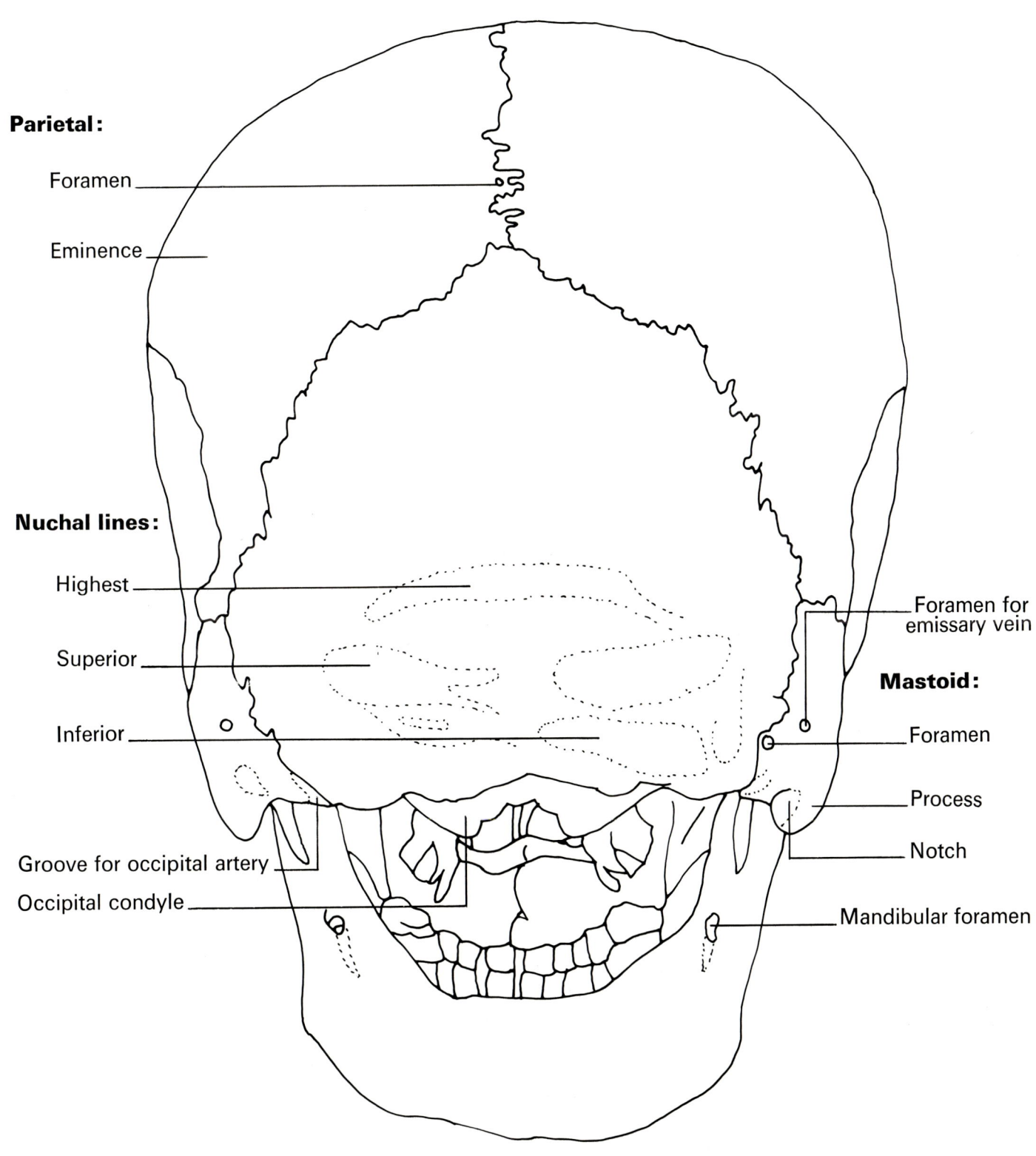

SECTION 4. Skull from below.

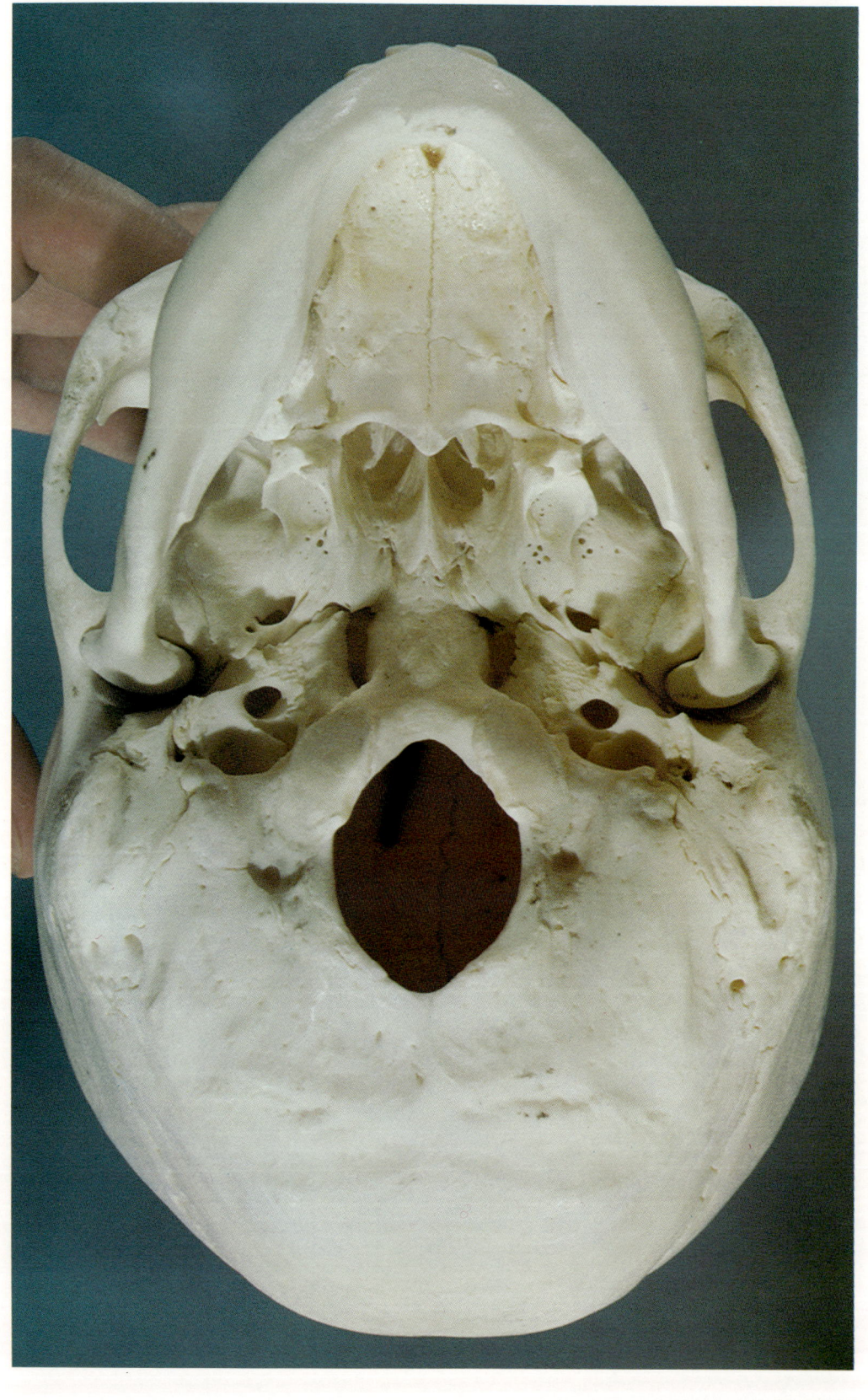

Skull from below.
Bones and landmarks.
Exterior of base of skull, mandible in place.

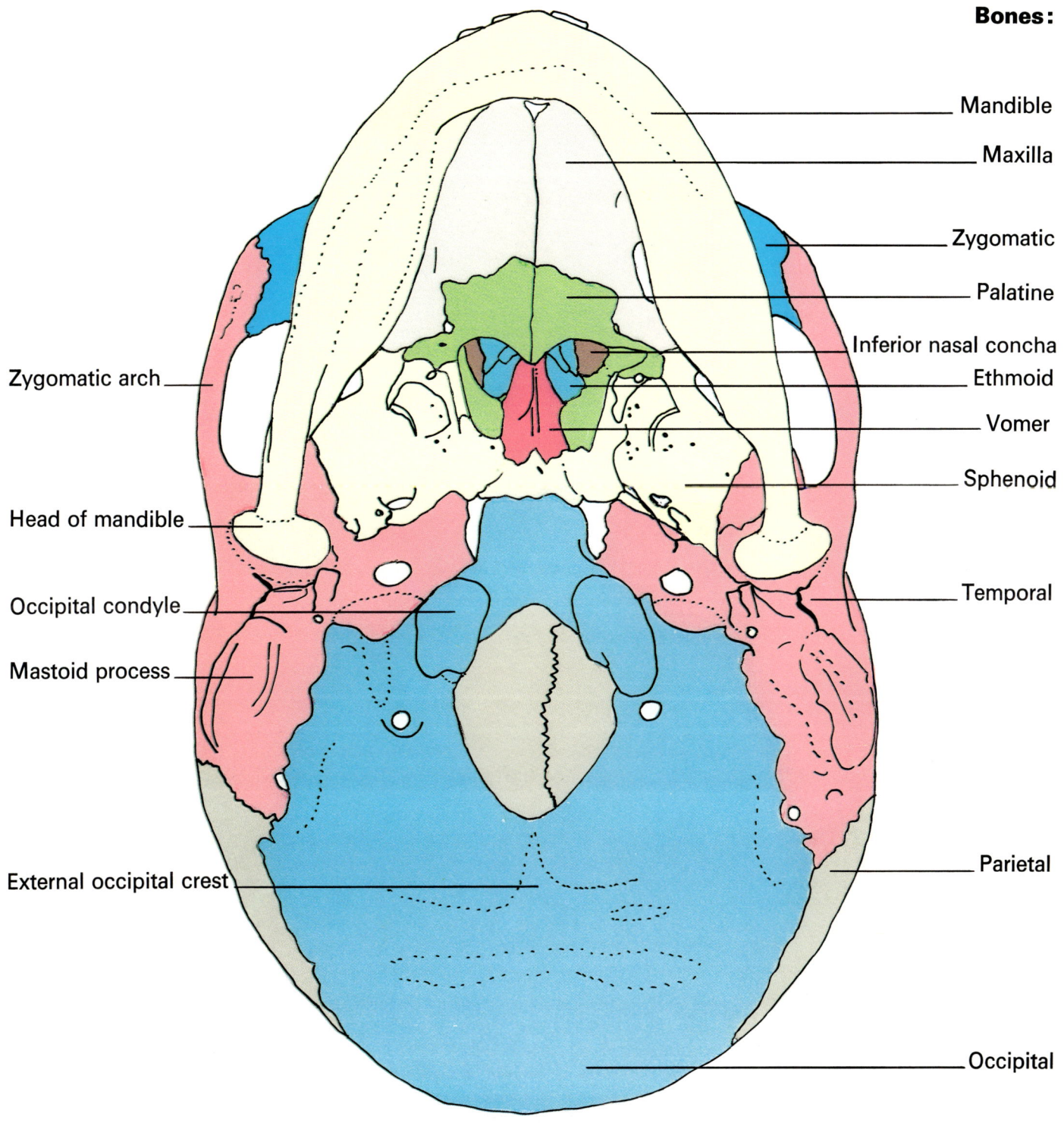

Skull from below.

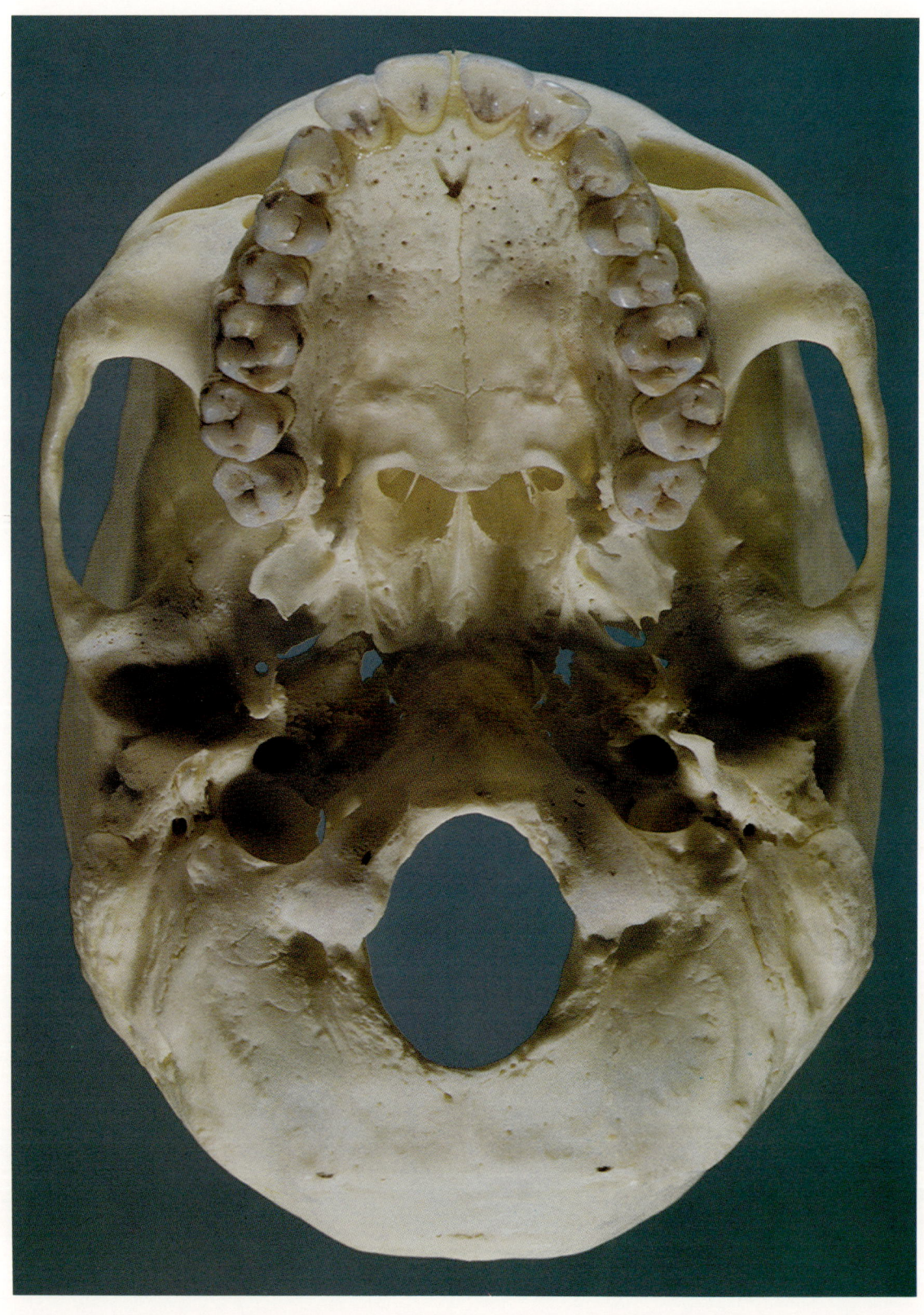

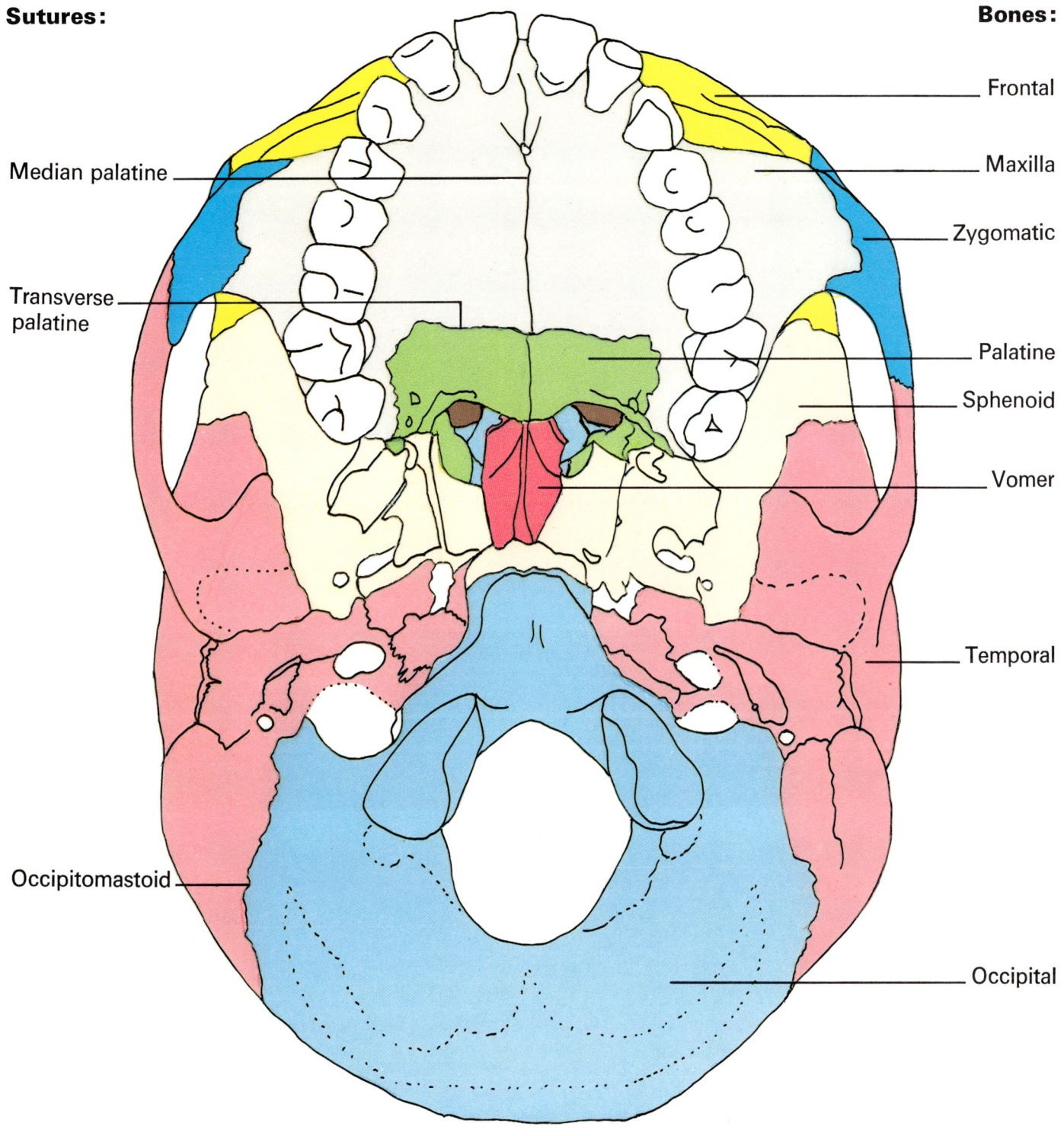

Skull from below.
Bones and sutures.
Exterior of base of skull, mandible removed.

Sutures:
- Median palatine
- Transverse palatine
- Occipitomastoid

Bones:
- Frontal
- Maxilla
- Zygomatic
- Palatine
- Sphenoid
- Vomer
- Temporal
- Occipital

Skull from below.
Foramina and canals.

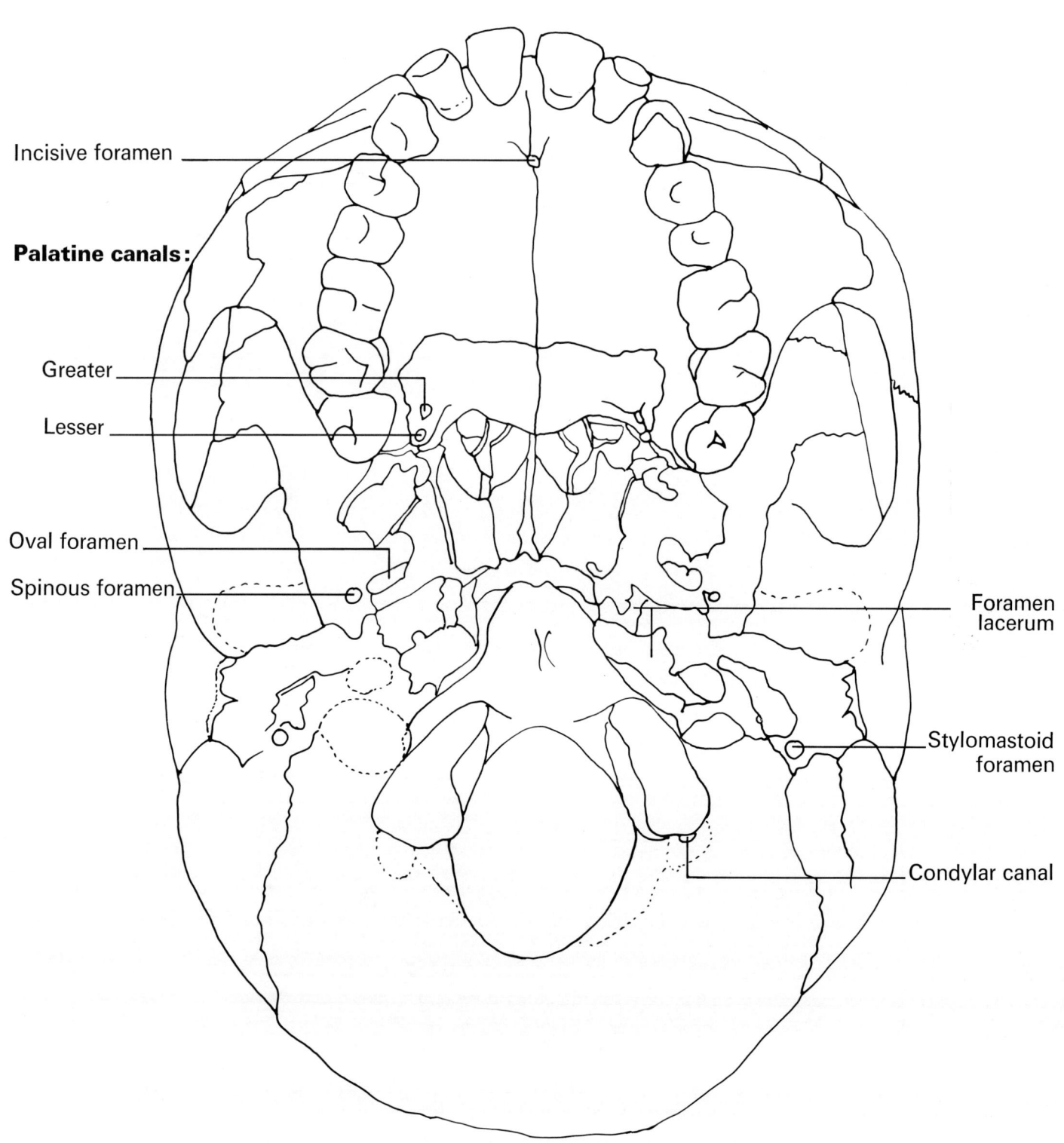

Skull from below.
Points and landmarks.

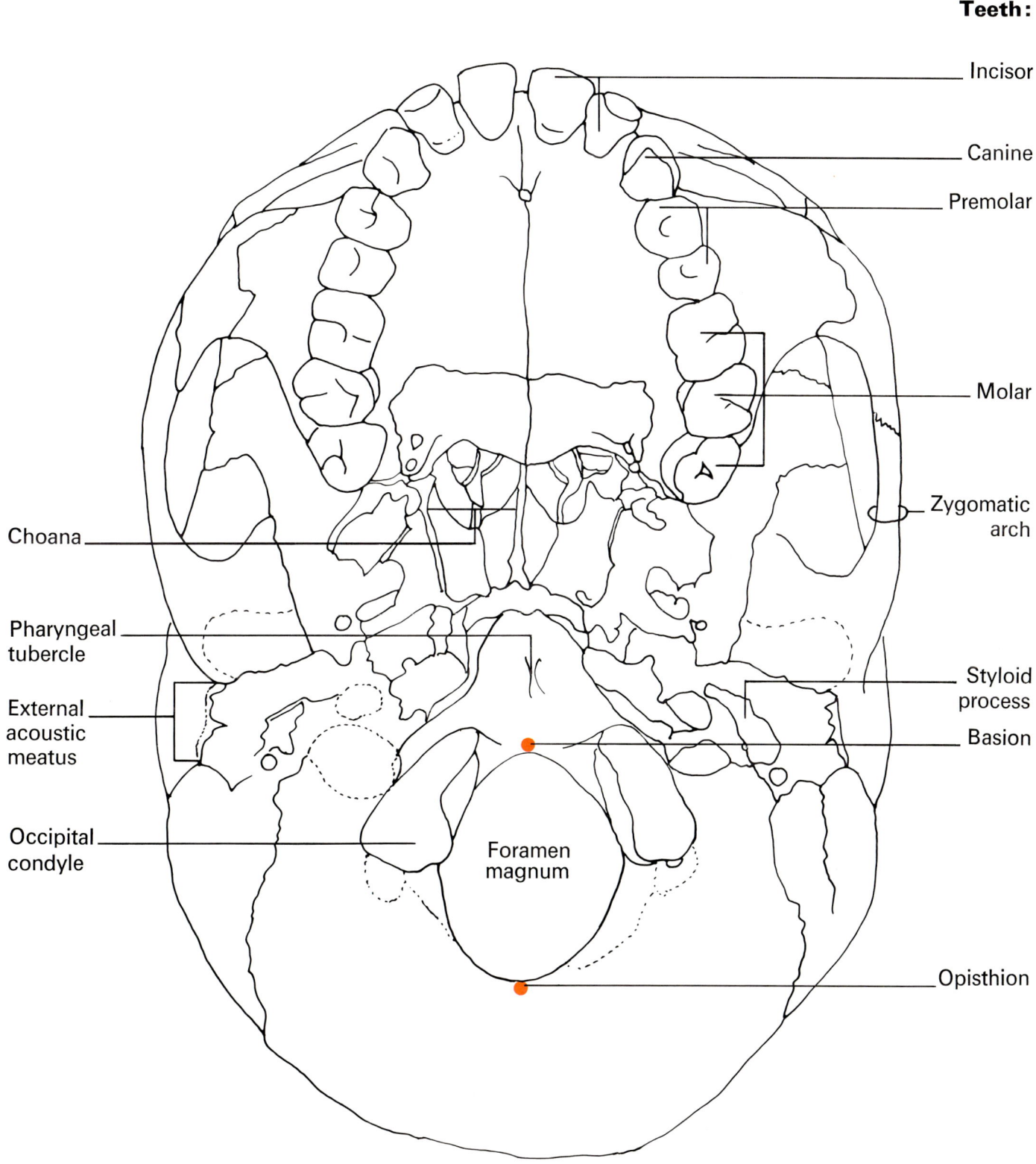

Skull from below.
Muscle attachments.

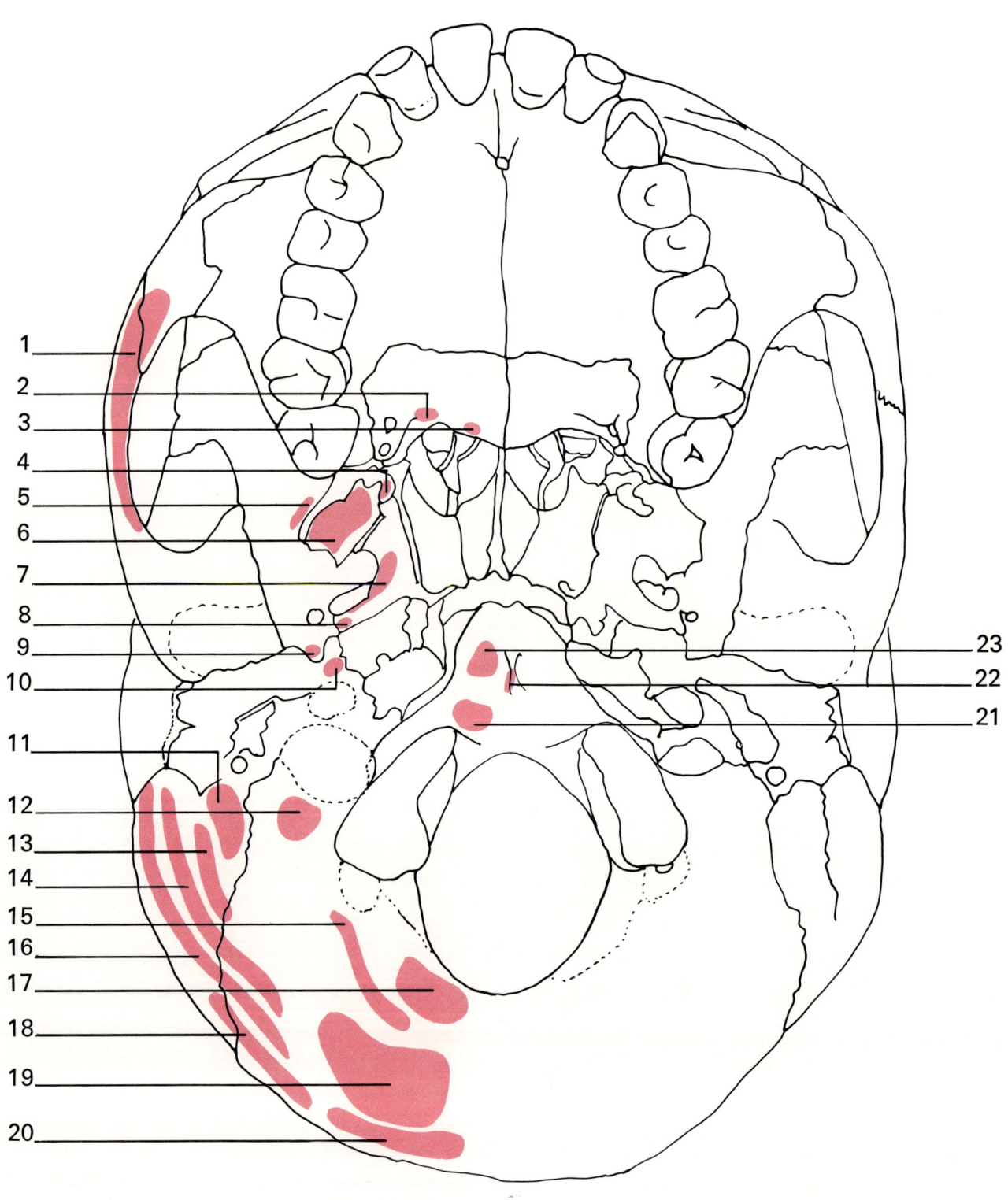

Skull from below.
Key to numbers.

Muscle attachments:
1. Masseter
2. Tensor of velum palatini
3. Uvular
4. Superior constrictor of pharynx
5. Lateral pterygoid, lower head
6. Medial pterygoid
7. Tensor of velum palatini
8. Tensor of tympanic membrane
9. Tensor of velum palatini
10. Levator of velum palatini
11. Digastric
12. Rectus capitis lateralis
13. Long (of head)
14. Splenius of head
15. Rectus capitis posterior major
16. Sternocleidomastoid
17. Rectus capitis posterior minor
18. Occipitofrontal, occipital belly
19. Semispinal of head
20. Trapezius
21. Long (of head)
22. Pharyngeal raphe
23. Rectus capitis anterior

Skull from below.

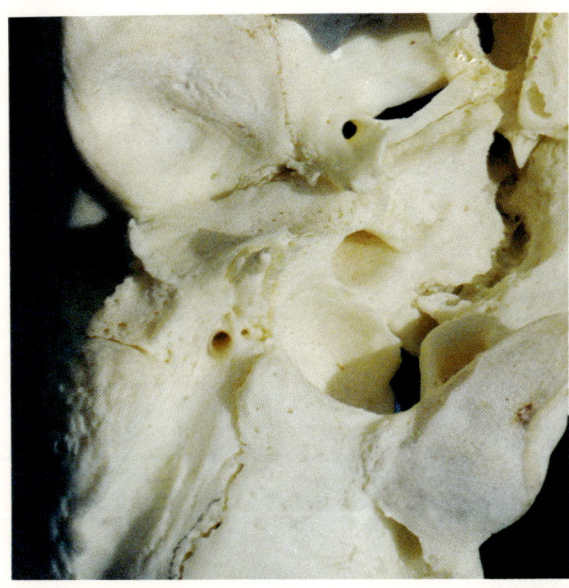

Closeup view, foramina and adjacent areas.

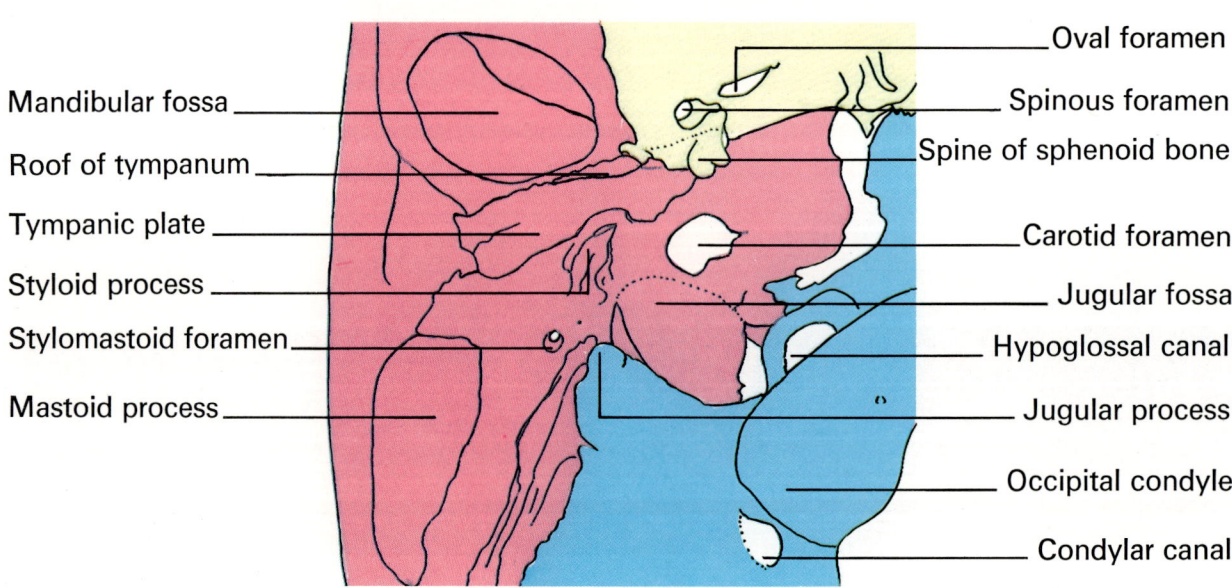

Skull from below.

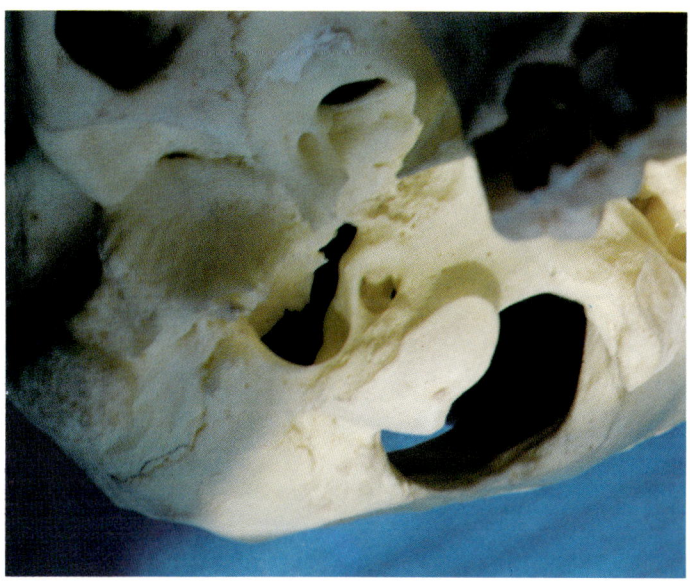

Skull slightly oblique, jugular fossa and foramen (three divisions; anterior, smallest; middle; and posterior, largest).

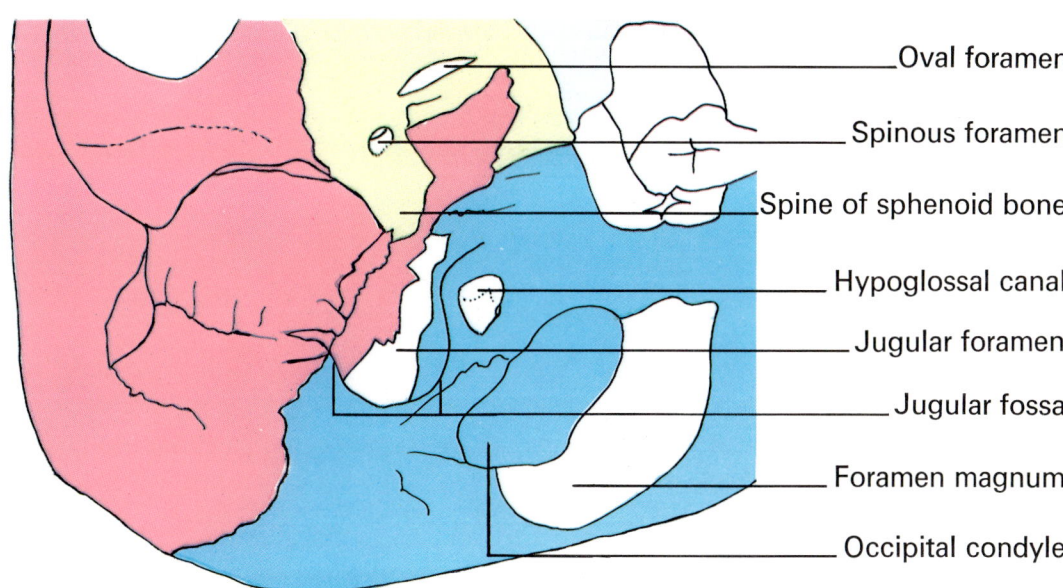

Skull from below.

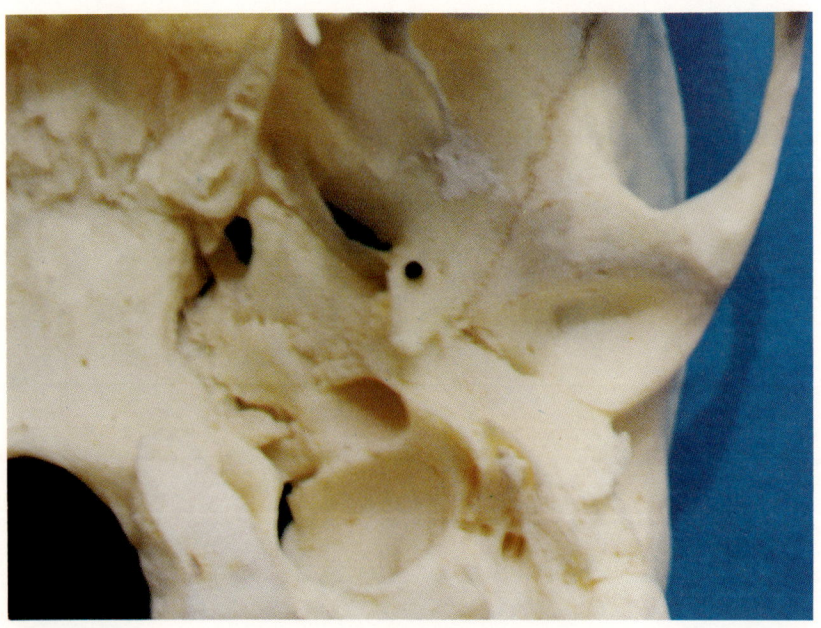

Closeup view, foramina and adjacent areas.

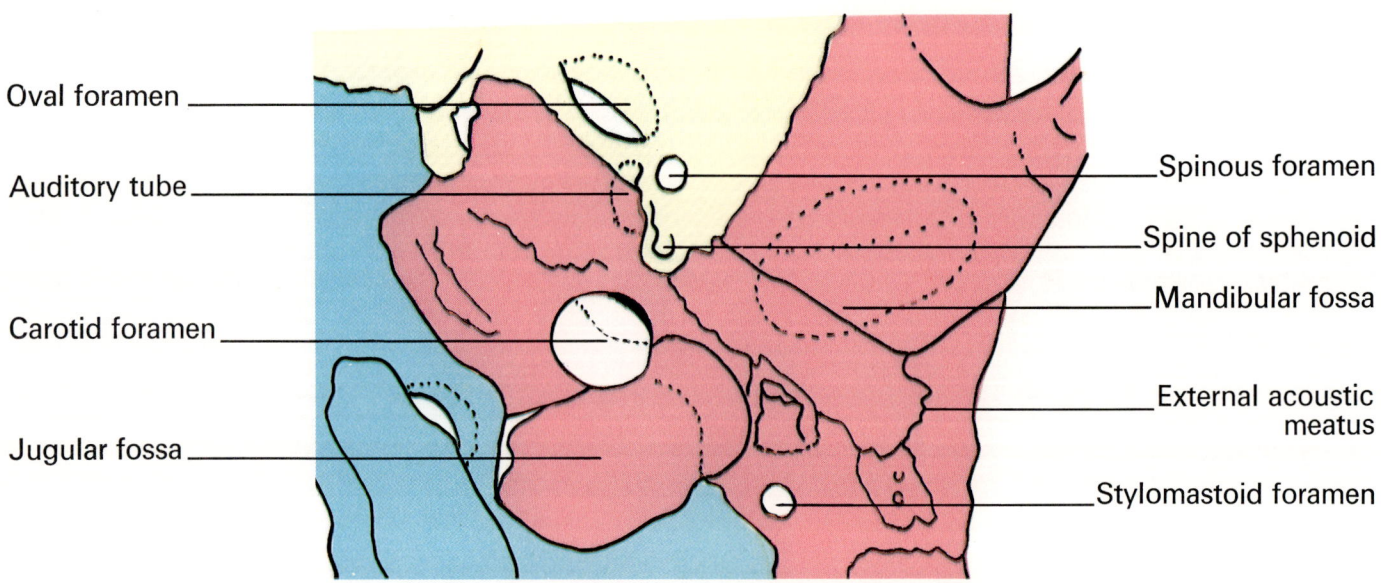

Skull from below.

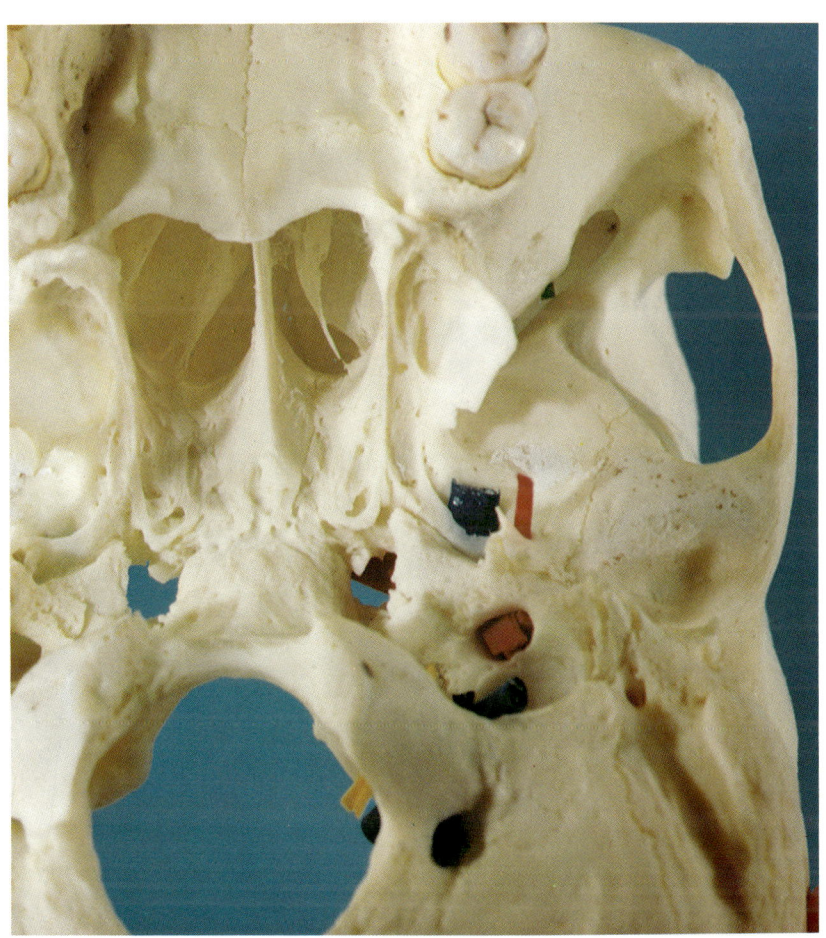

Colored rods in foramina and canals to indicate direction.

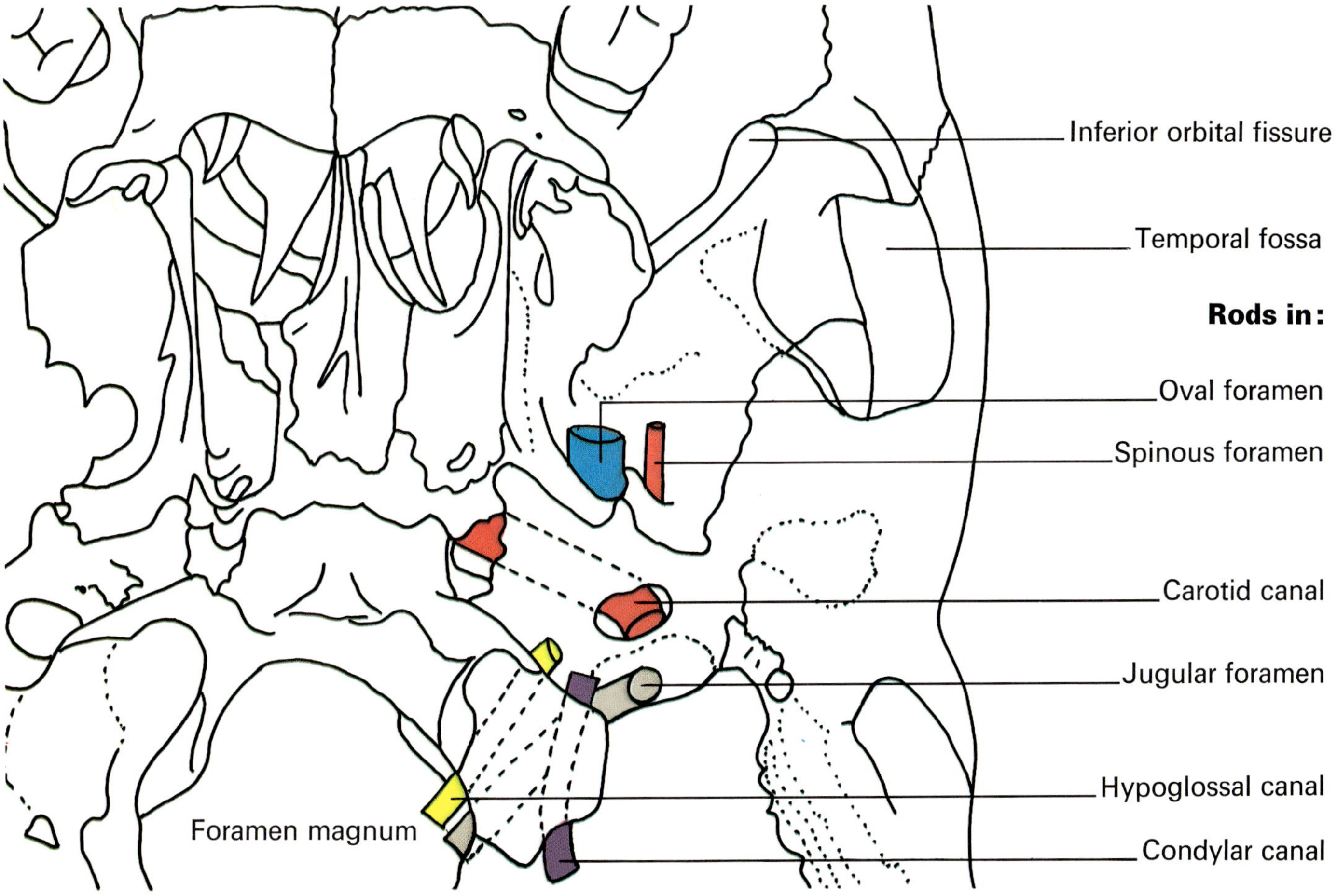

- Inferior orbital fissure
- Temporal fossa

Rods in:

- Oval foramen
- Spinous foramen
- Carotid canal
- Jugular foramen
- Hypoglossal canal
- Condylar canal
- Foramen magnum

SECTION 5.   Skull lines.

Lines:

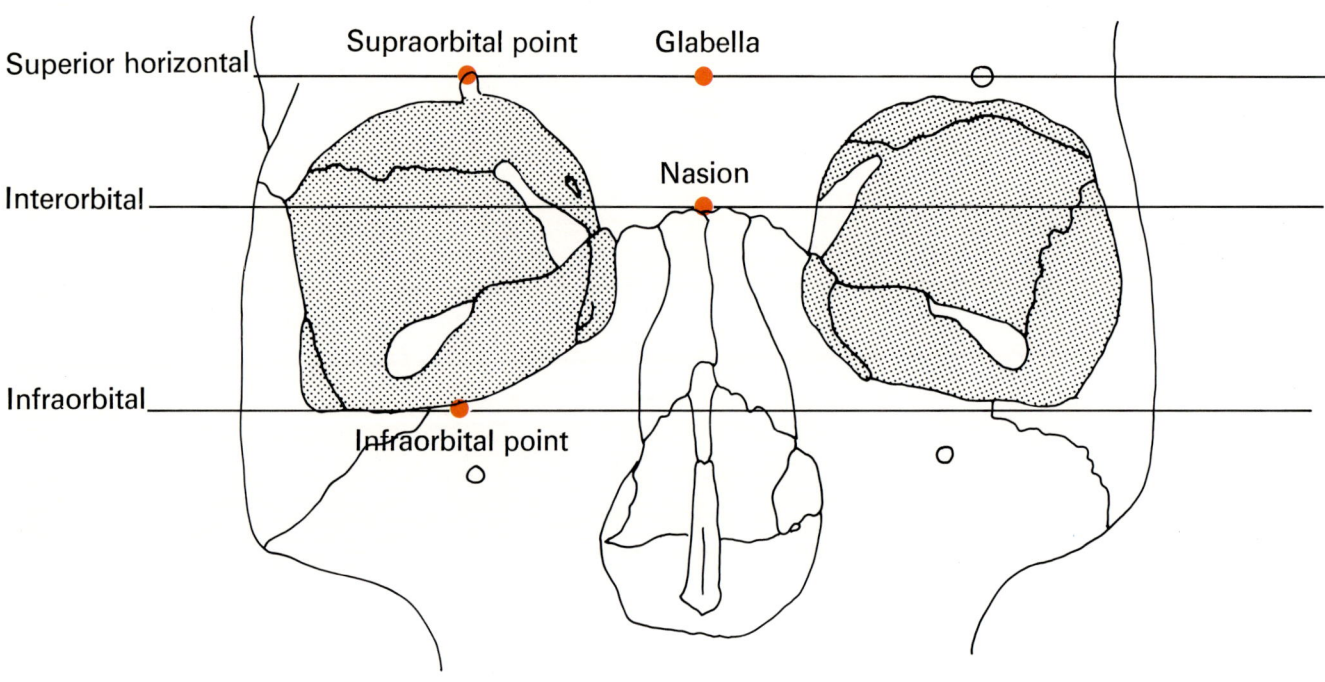

- Superior horizontal
- Interorbital
- Infraorbital

Supraorbital point, Glabella, Nasion, Infraorbital point

Lines:

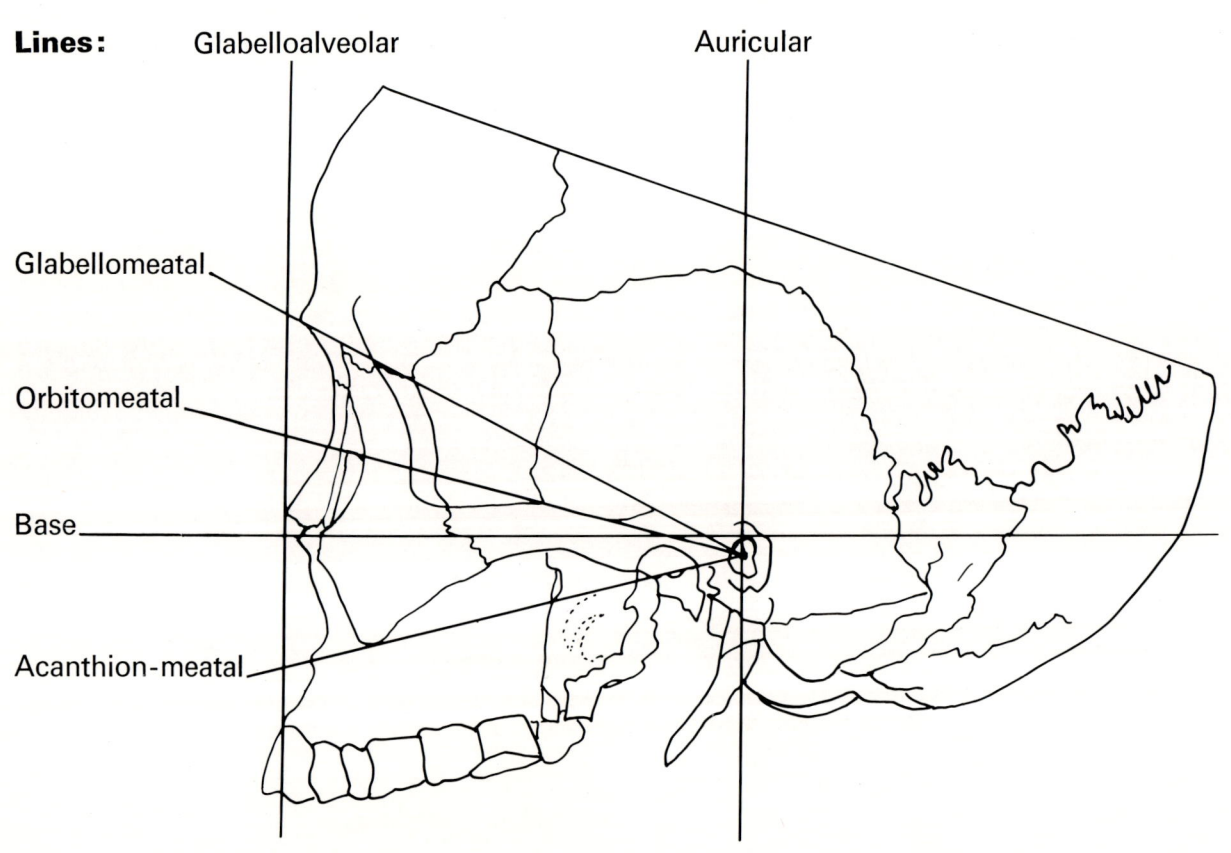

- Glabelloalveolar
- Auricular
- Glabellomeatal
- Orbitomeatal
- Base
- Acanthion-meatal

Skull lines.

Transverse lines:

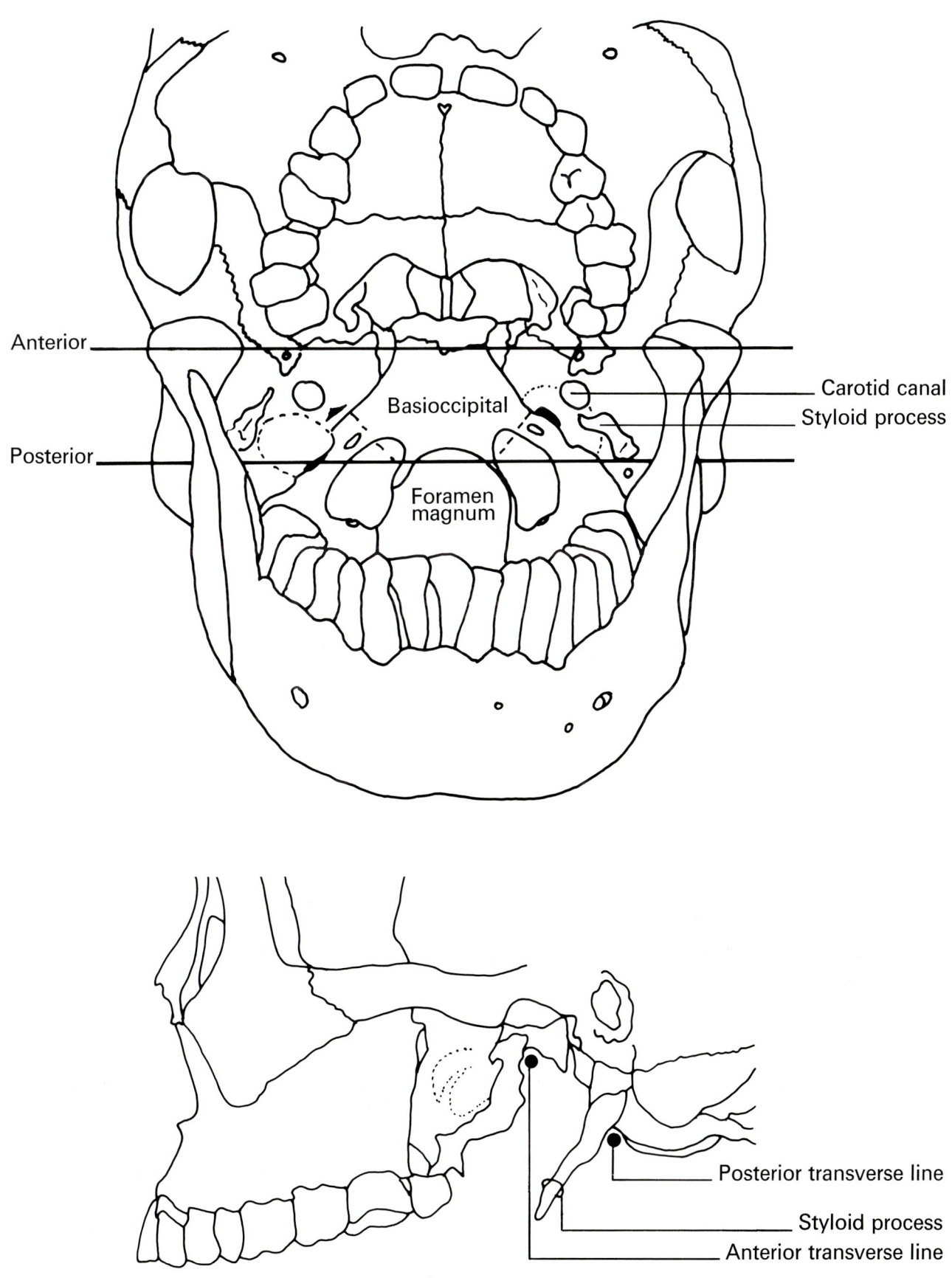

Side view of skull.

SECTION 6.   Skull from above.

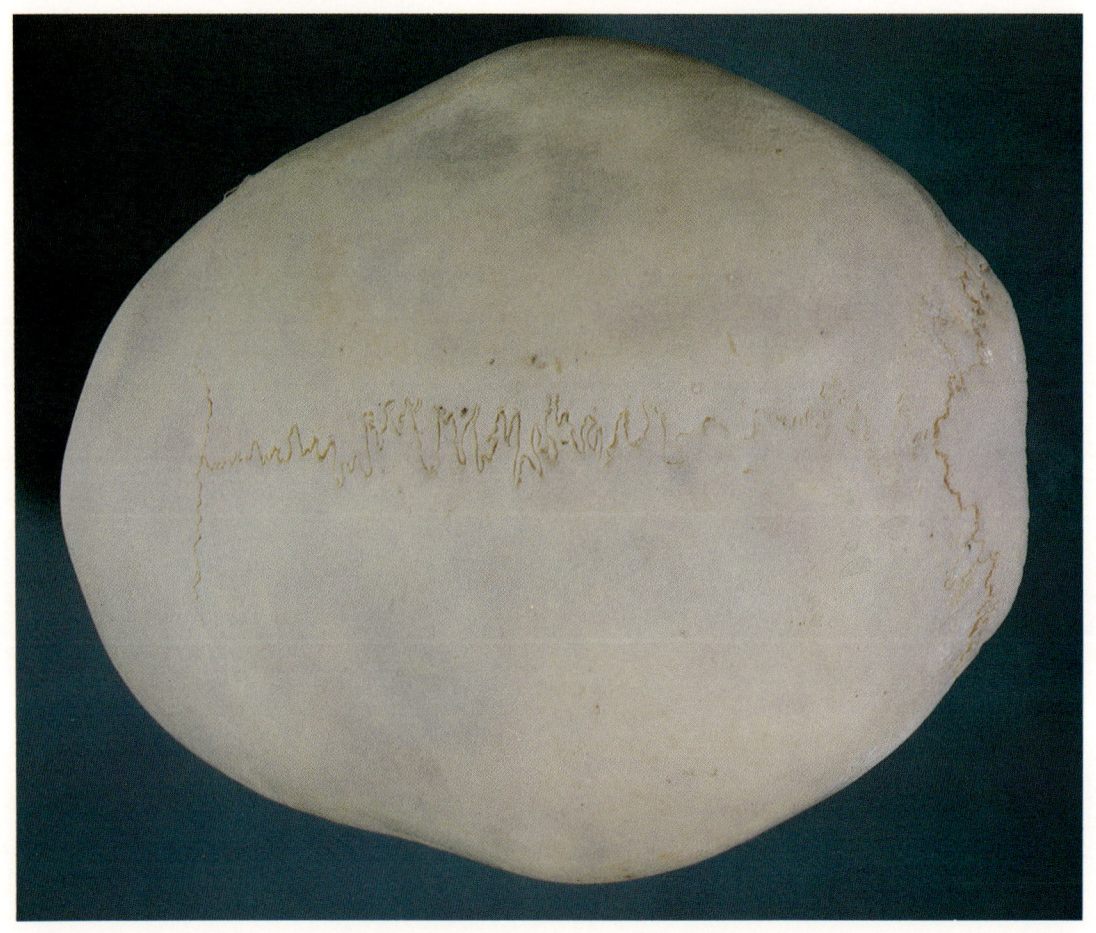

Cranial vault, bones and sutures.

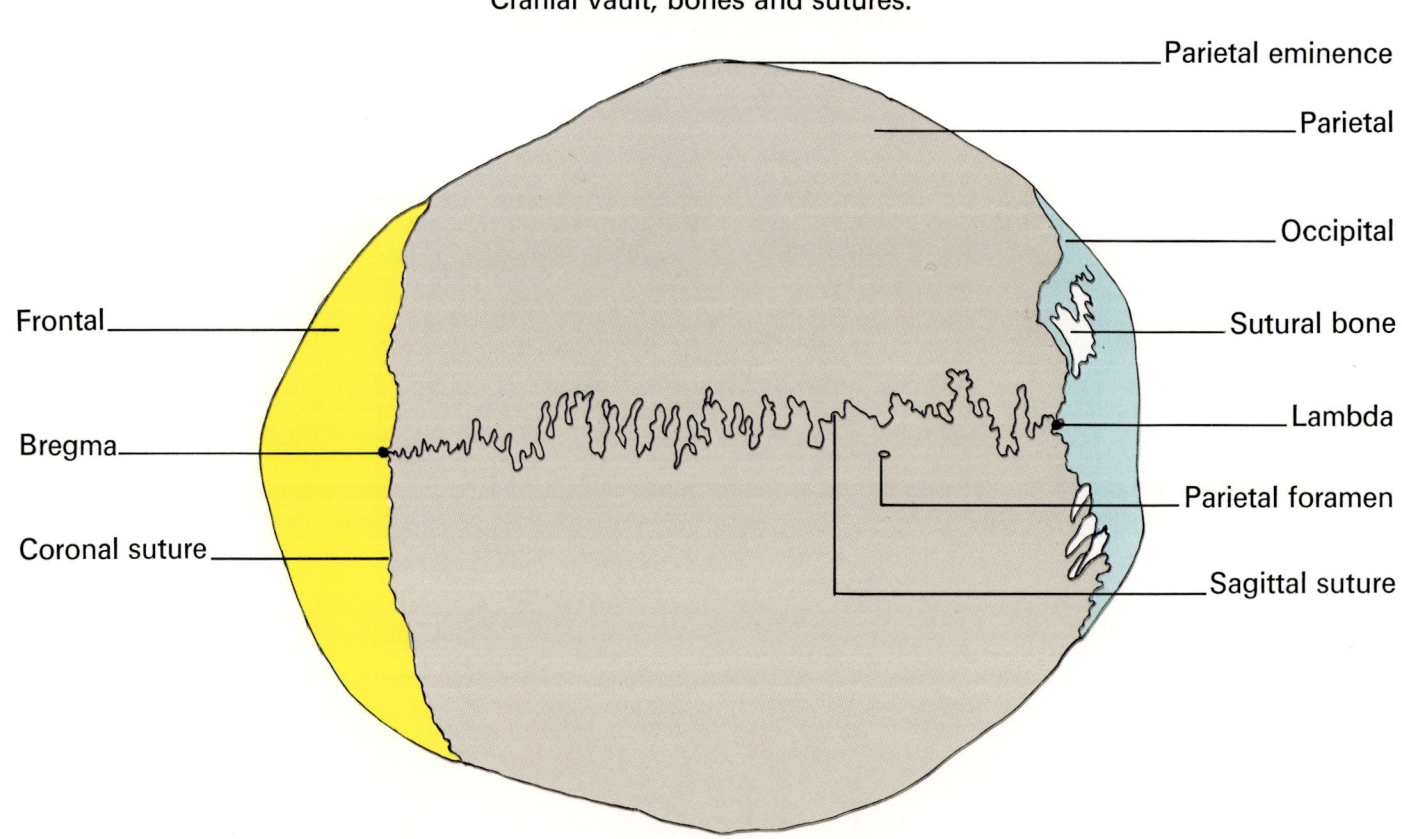

Skull from above.

Cranial vault, internal surface. Bones and vascular grooves.

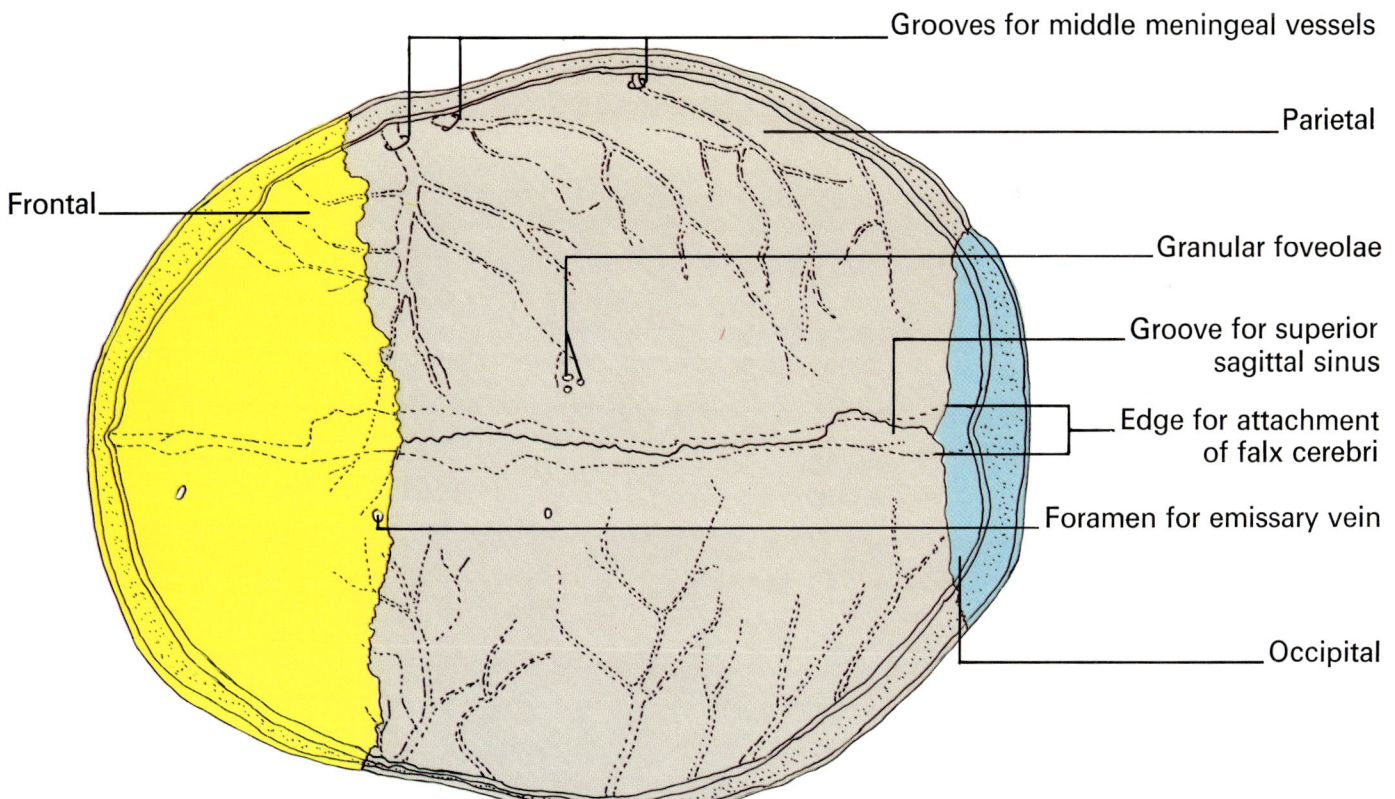

SECTION 7. Base of skull, internal surface.

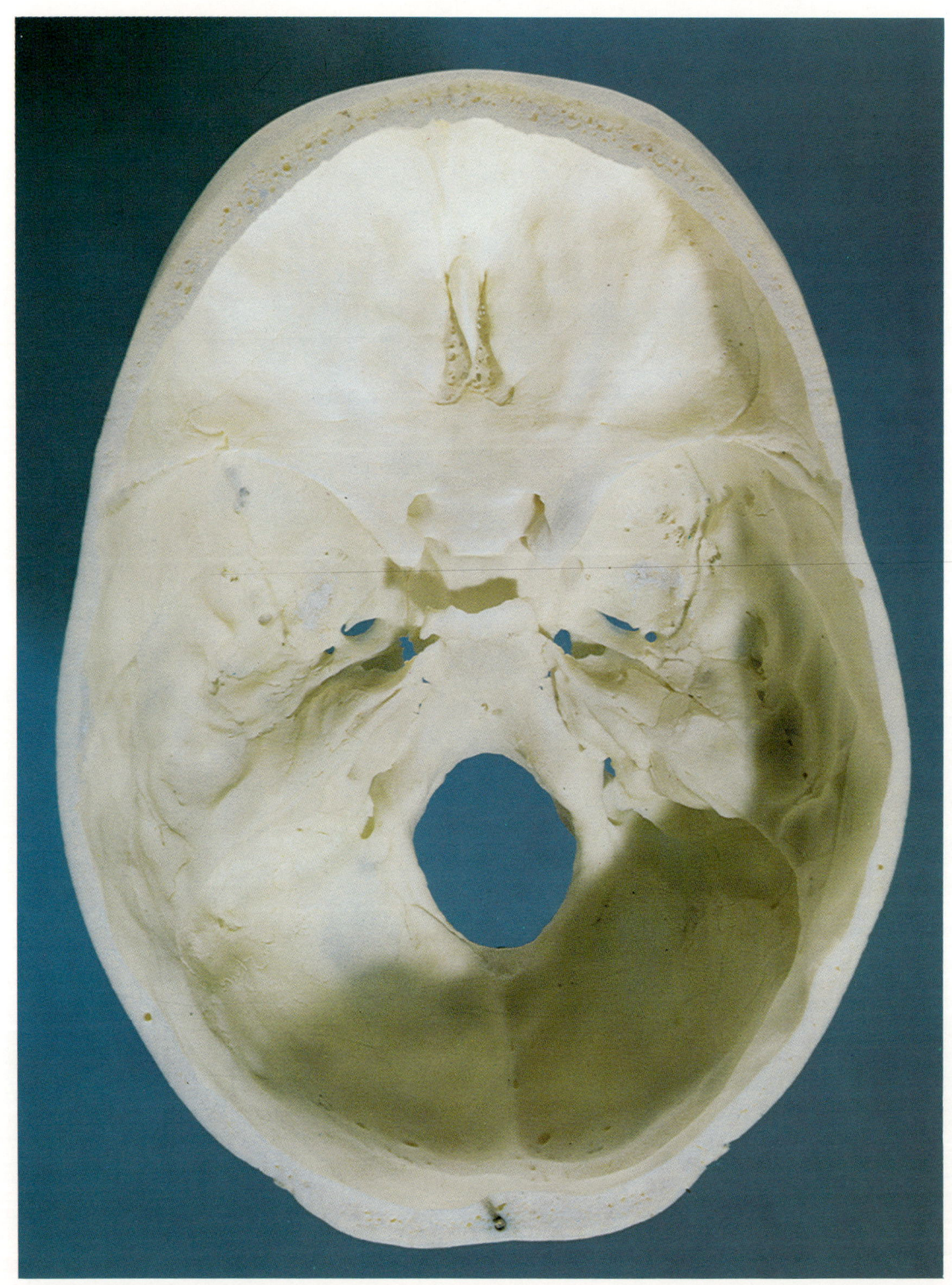

Base of skull, internal surface.
Bones and vascular grooves.

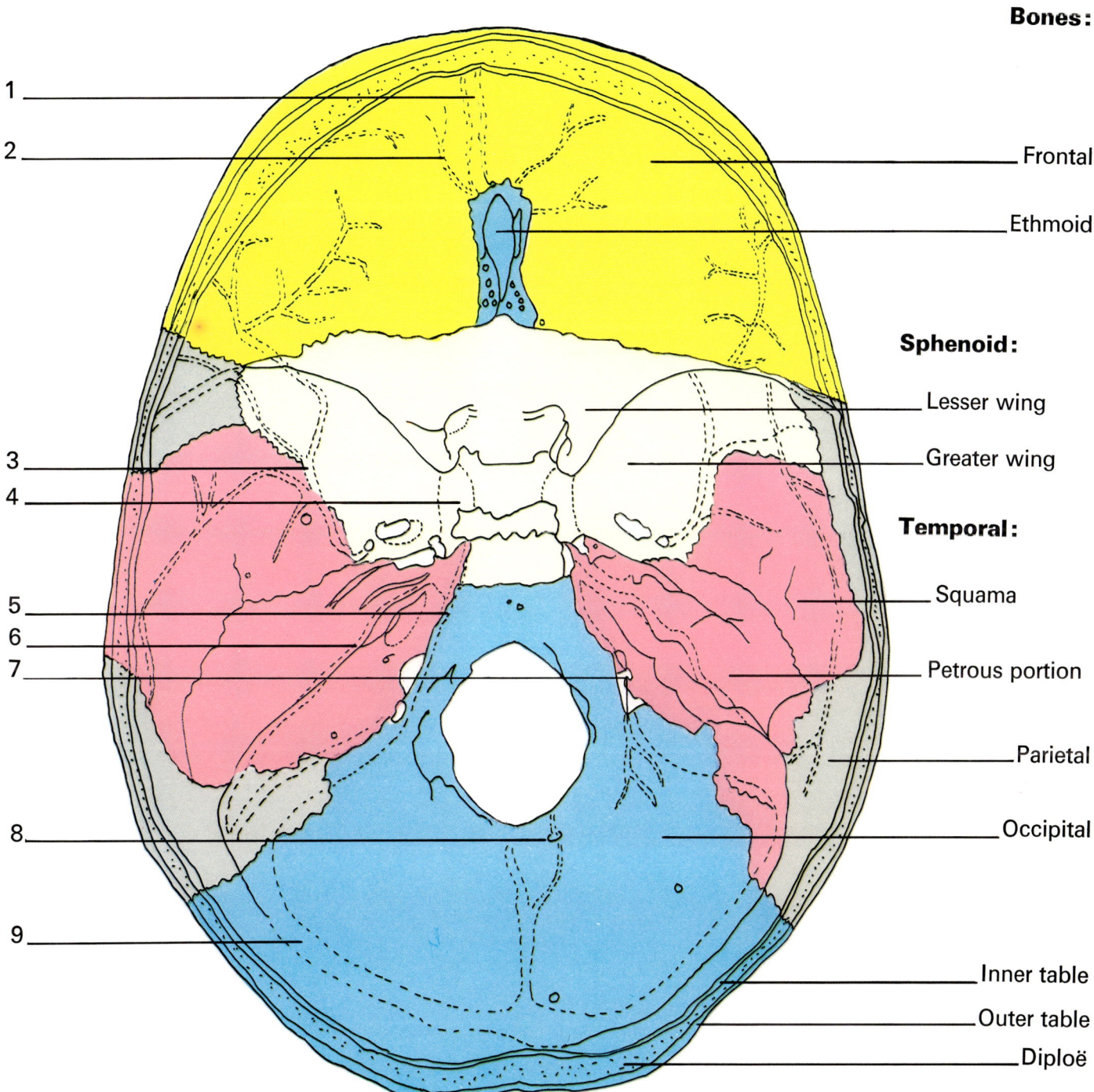

Key to numbers
Grooves for:

1. Superior sagittal sinus
2. Anterior meningeal vessels
3. Middle meningeal vessels
4. Internal carotid artery
5. Inferior petrosal sinus
6. Superior petrosal sinus
7. Posterior meningeal vessels
8. Occipital sinus
9. Transverse sinus

Base of skull, internal surface.

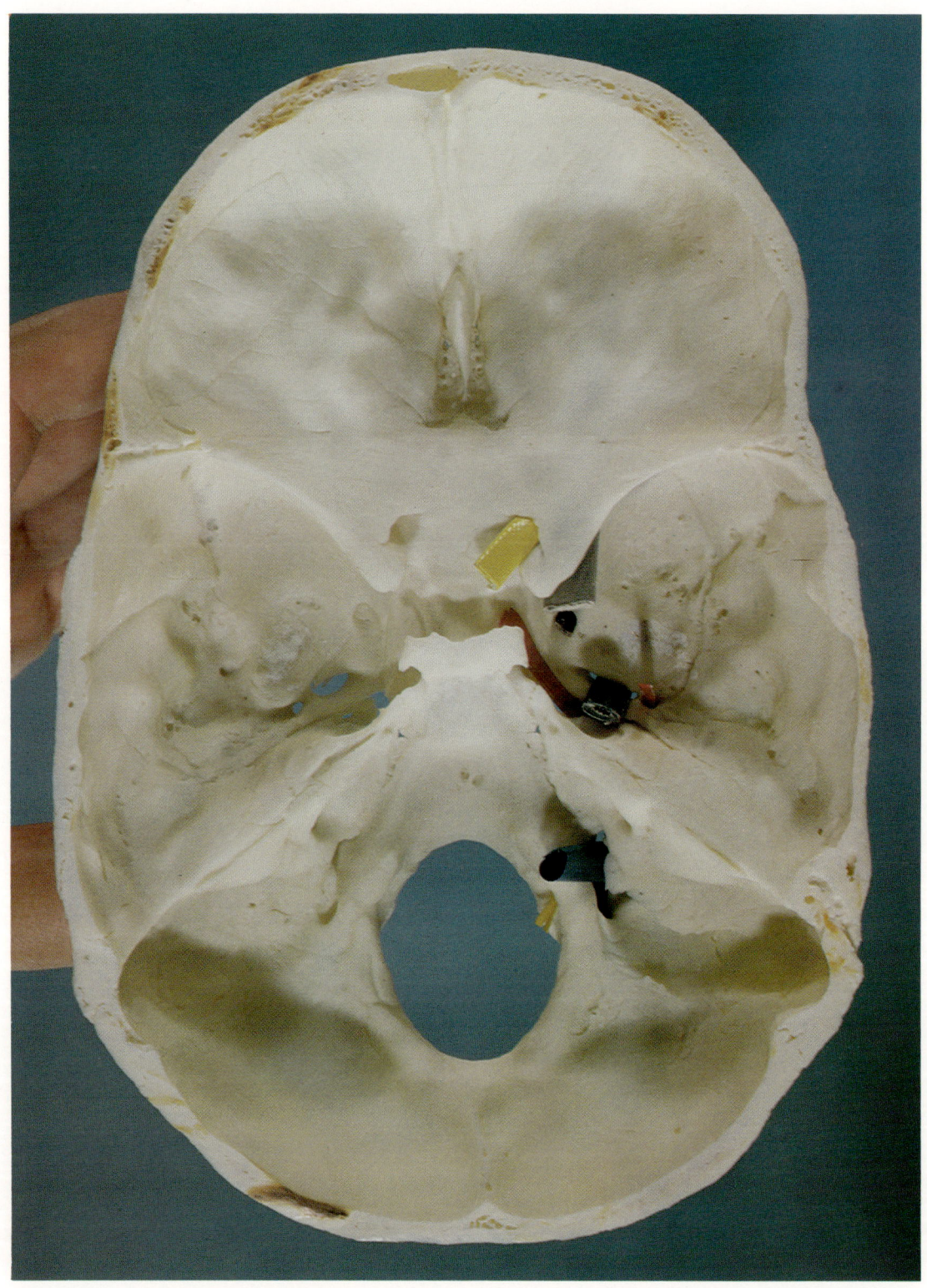

37

Base of skull, internal surface.
Cranial fossae, foramina and canals.
Colored rods in foramina and canals.

Cranial fossae:

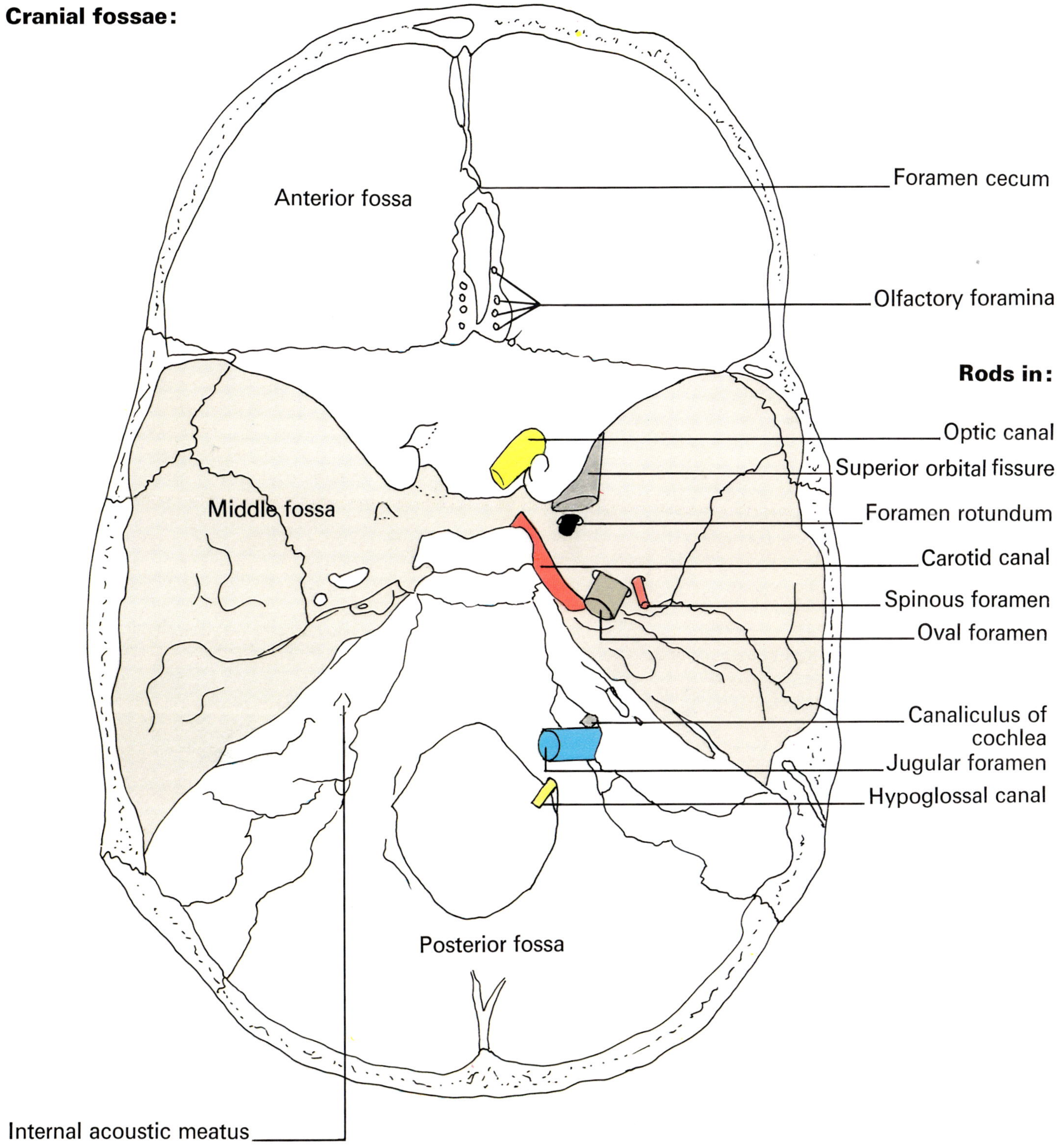

Anterior fossa

Middle fossa

Posterior fossa

Internal acoustic meatus

Foramen cecum

Olfactory foramina

Rods in:

Optic canal
Superior orbital fissure
Foramen rotundum
Carotid canal
Spinous foramen
Oval foramen

Canaliculus of cochlea
Jugular foramen
Hypoglossal canal

SECTION 8.   Hypophyseal fossa.

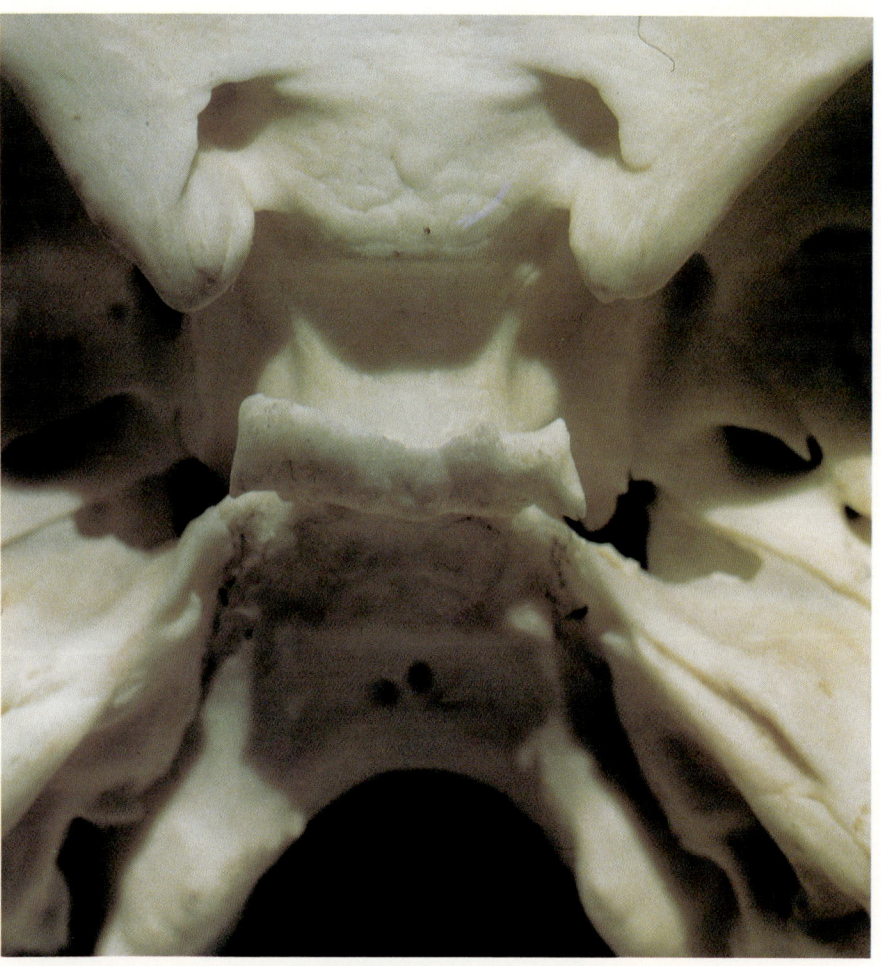

View from above and slightly behind. Landmarks.
Magnification 2.5X

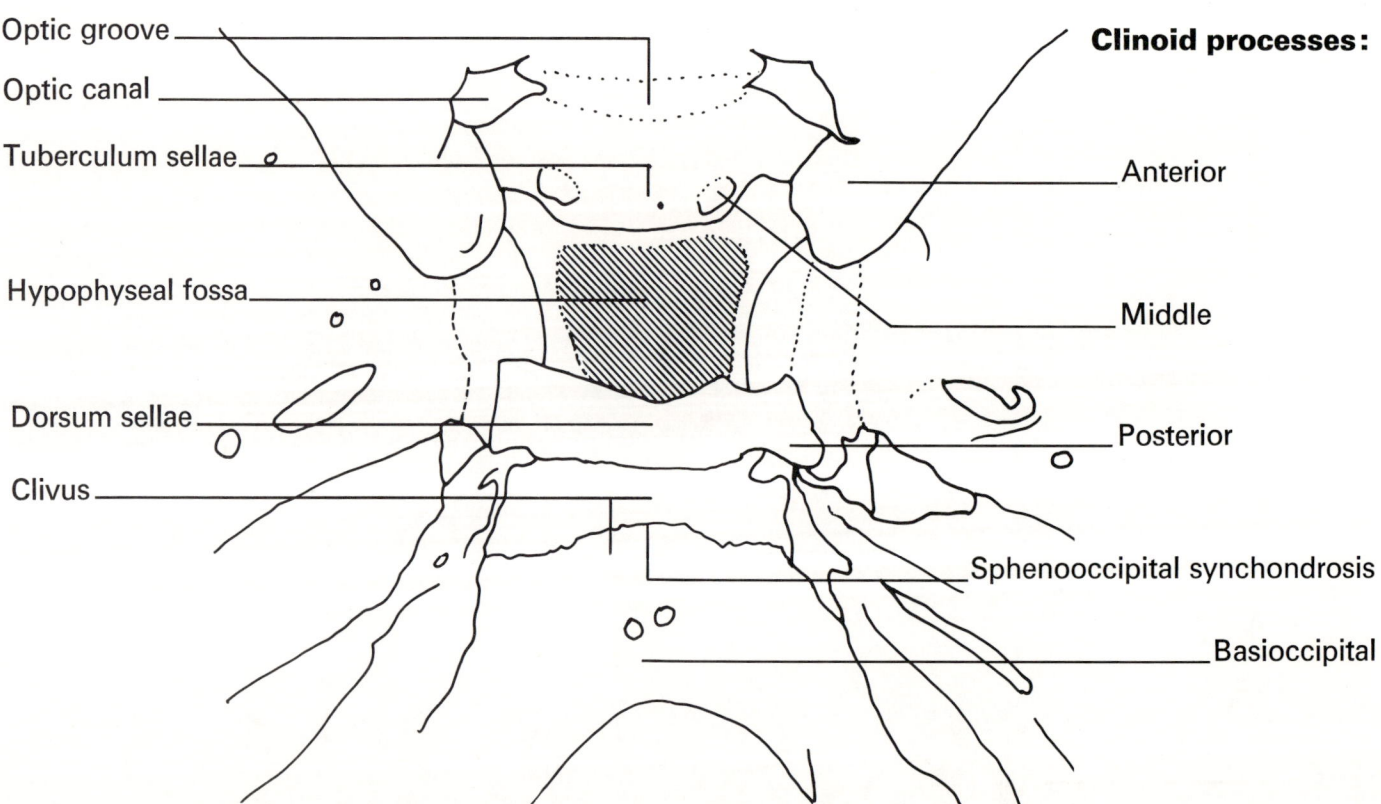

- Optic groove
- Optic canal
- Tuberculum sellae
- Hypophyseal fossa
- Dorsum sellae
- Clivus

Clinoid processes:
- Anterior
- Middle
- Posterior
- Sphenooccipital synchondrosis
- Basioccipital

Hypophyseal fossa.

From side: skull with vault and portions of squama resected. Bones and landmarks.

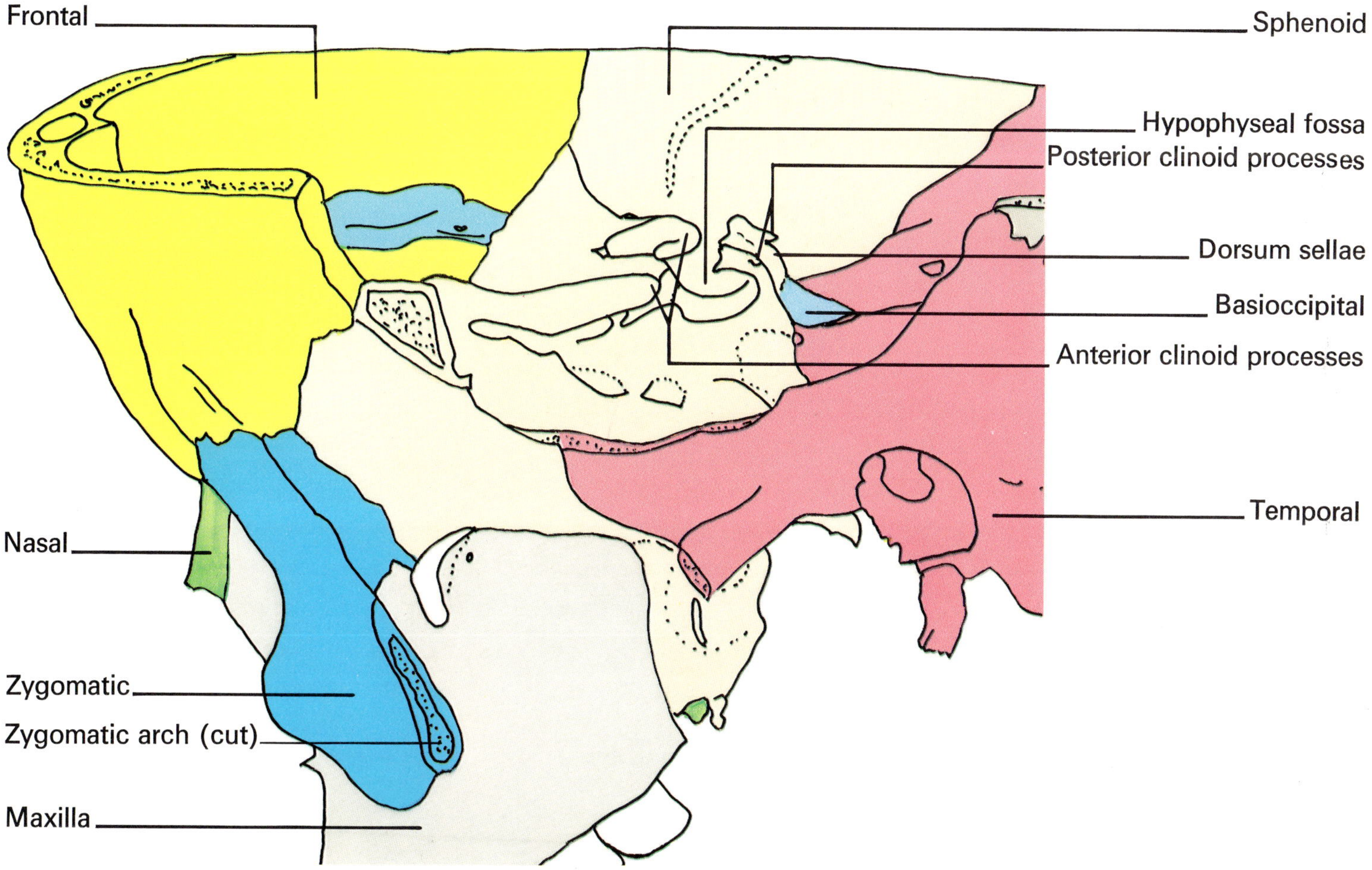

Hypophyseal fossa.

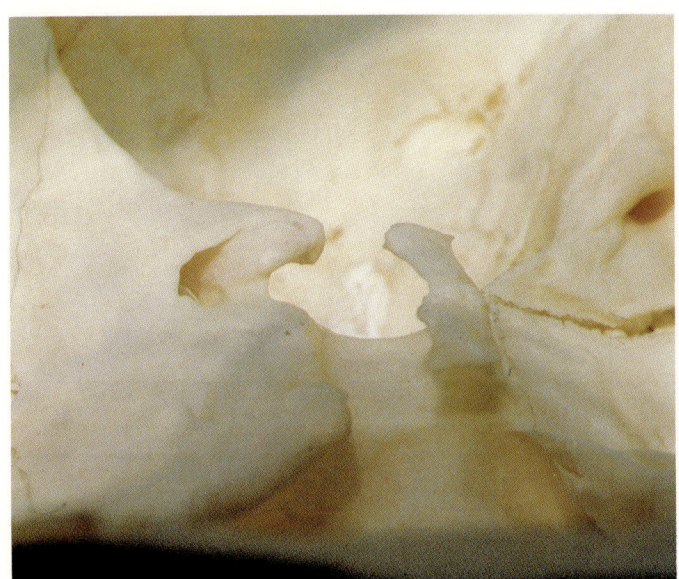

A. From side, closeup view.

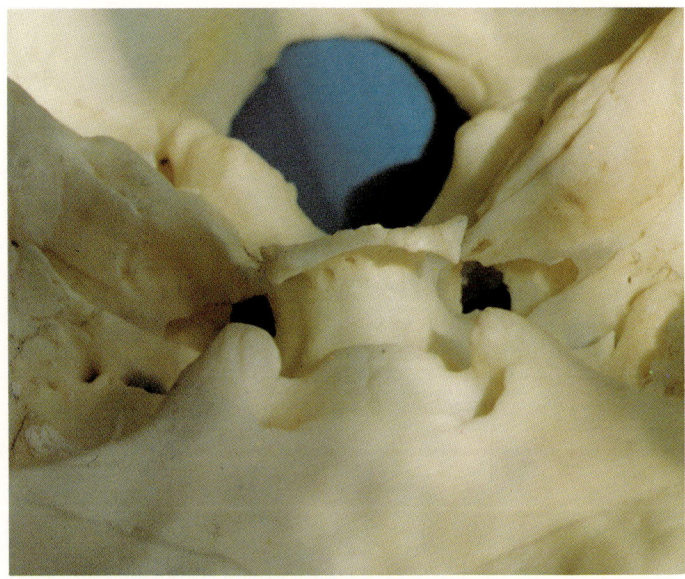

B. From front and above.

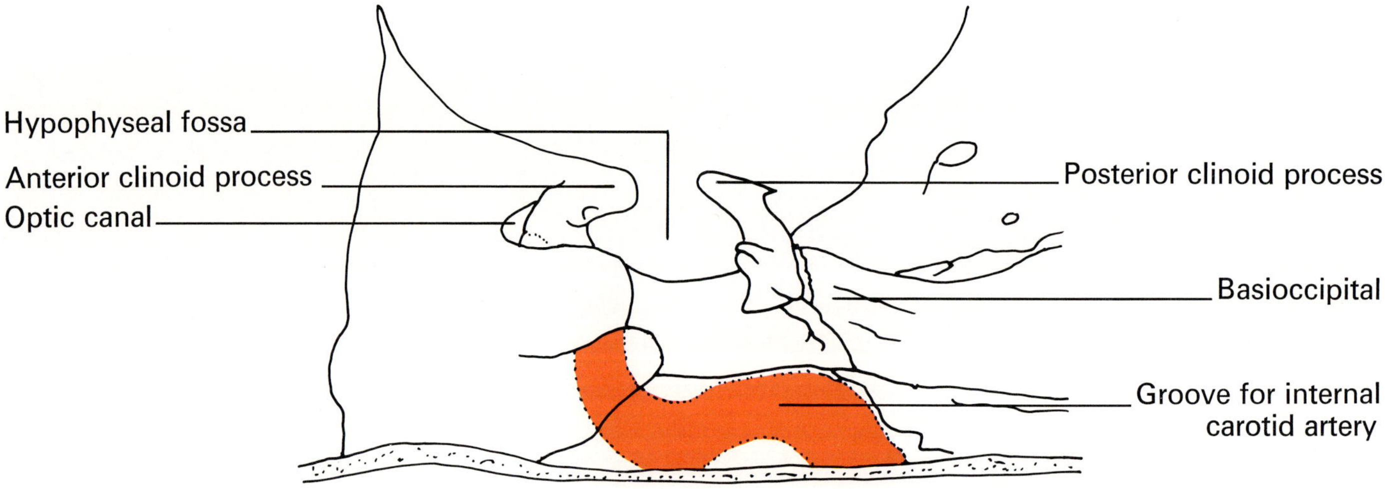

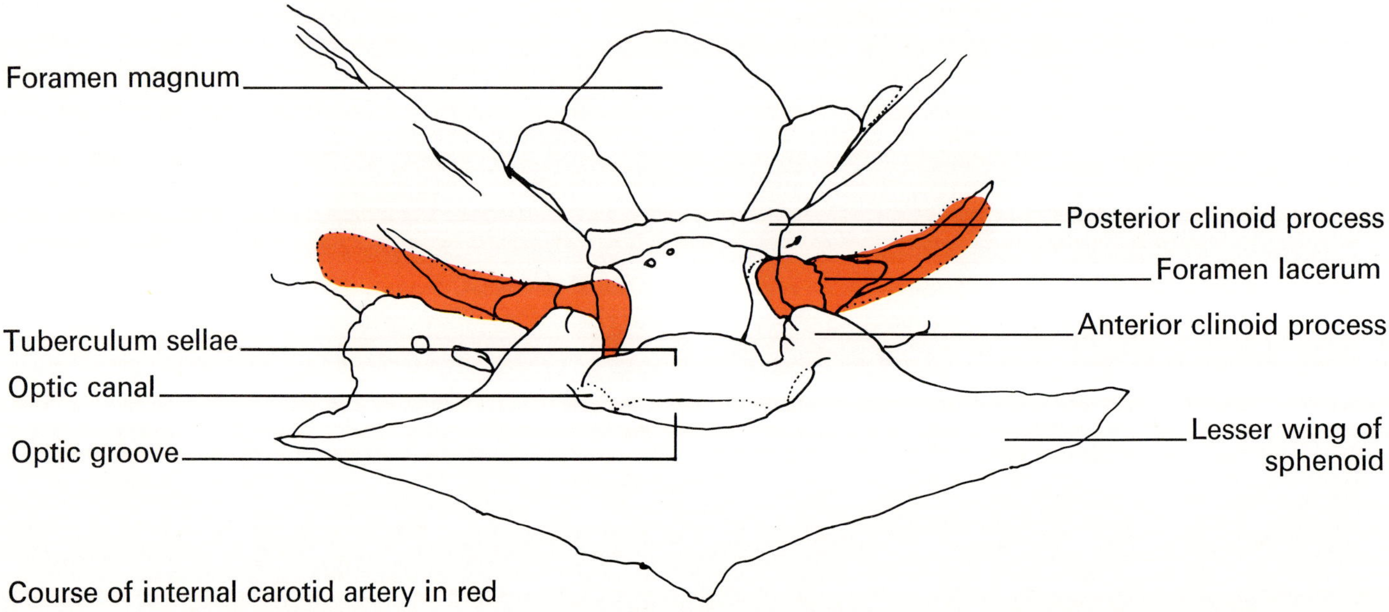

Hypophyseal fossa
Anterior clinoid process
Optic canal

Posterior clinoid process

Basioccipital

Groove for internal carotid artery

Foramen magnum

Posterior clinoid process
Foramen lacerum
Anterior clinoid process

Tuberculum sellae
Optic canal
Optic groove

Lesser wing of sphenoid

Course of internal carotid artery in red

Hypophyseal fossa.

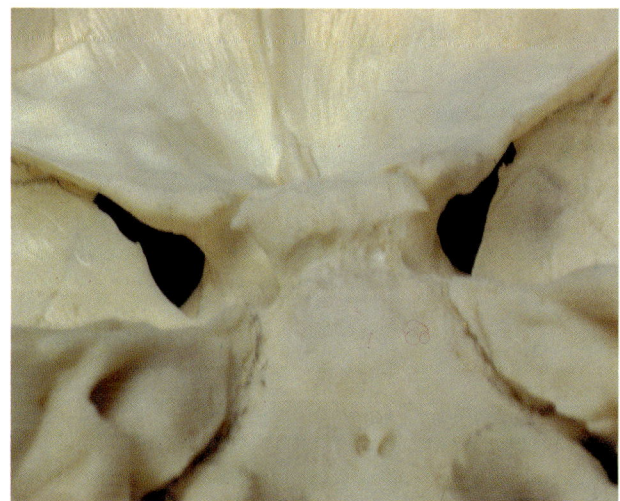

A. From behind.

B. Lateral oblique, view of sphenoid bone.

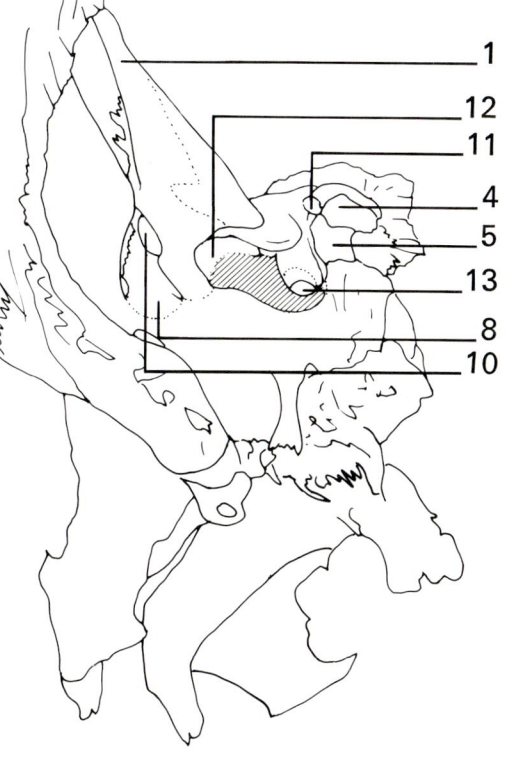

Key to numbers:

1. Lesser wing of sphenoid
2. Crista galli
3. Anterior clinoid process
4. Dorsum sellae
5. Clivus
6. Basiocciptal
7. Notch for abducens nerve
8. Superior orbital fissure
9. Middle cranial fossa
10. Optic canal
11. Posterior clinoid process
12. Optic groove
13. Caroticoclinoid foramen

SECTION 9. Orbital cavity.

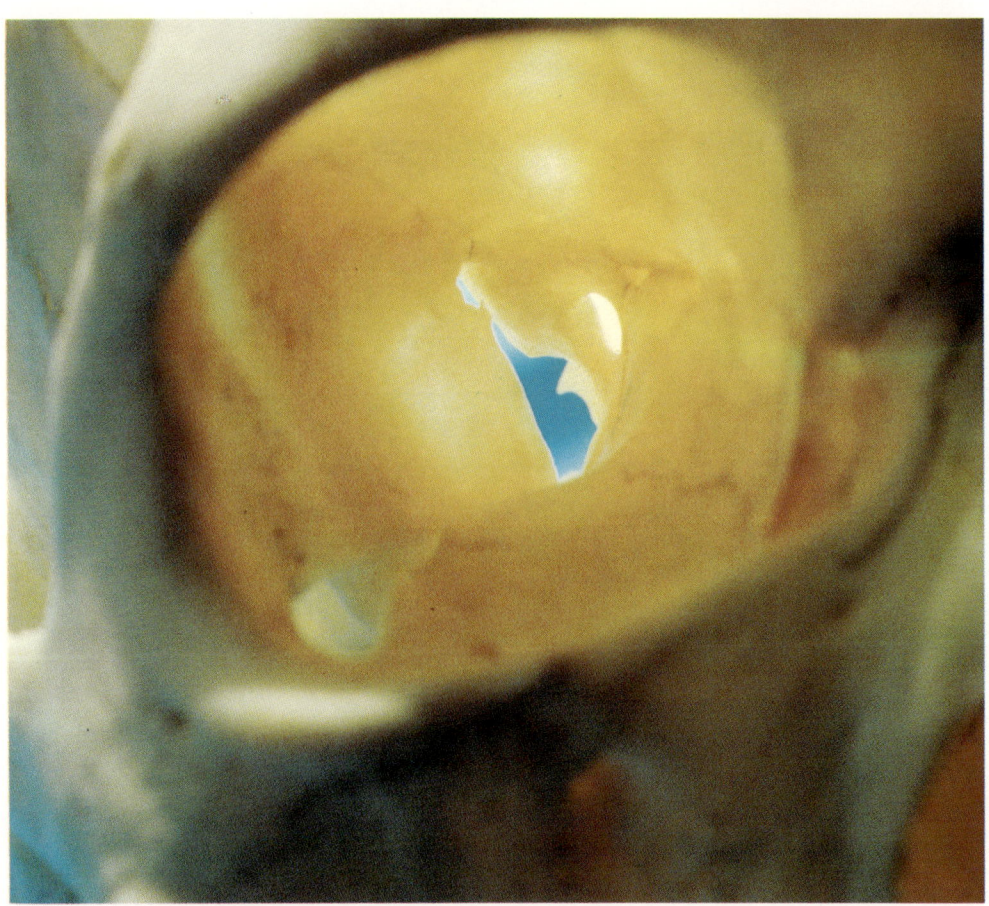

From front. Bones and foramina.
Magnification 2.5 X.

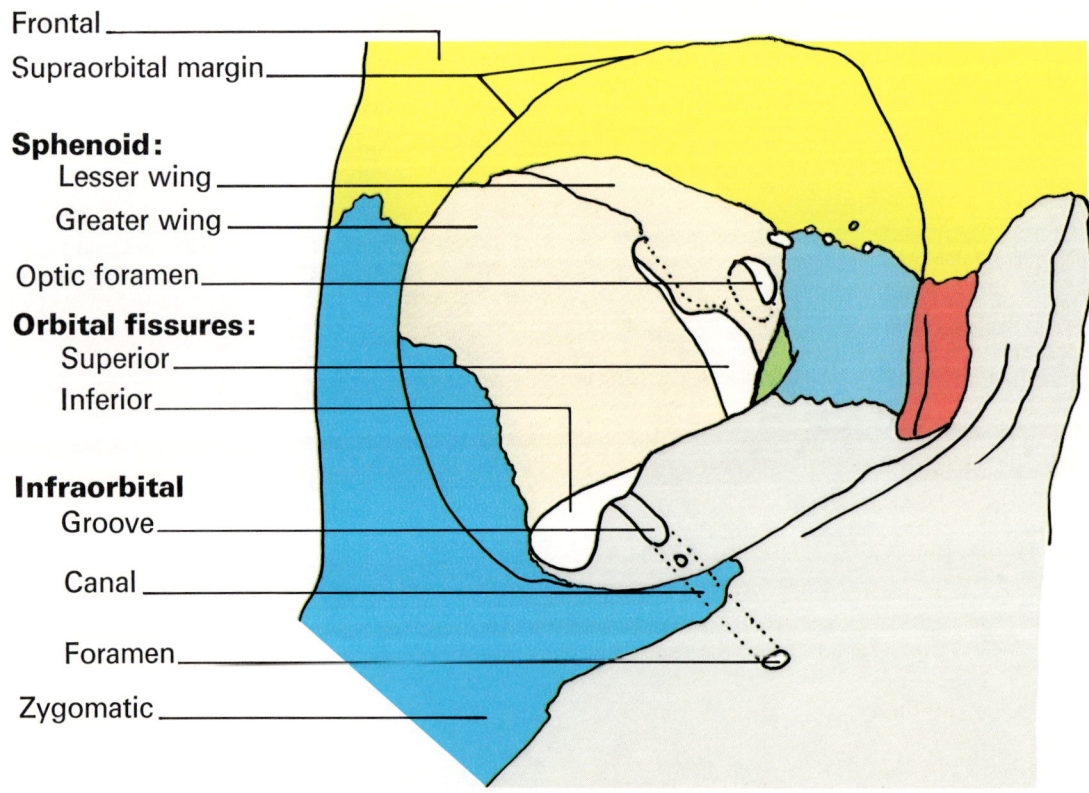

Frontal
Supraorbital margin

Sphenoid:
　Lesser wing
　Greater wing
　Optic foramen
Orbital fissures:
　Superior
　Inferior

Infraorbital
　Groove
　Canal
　Foramen
Zygomatic

Orbital cavity.

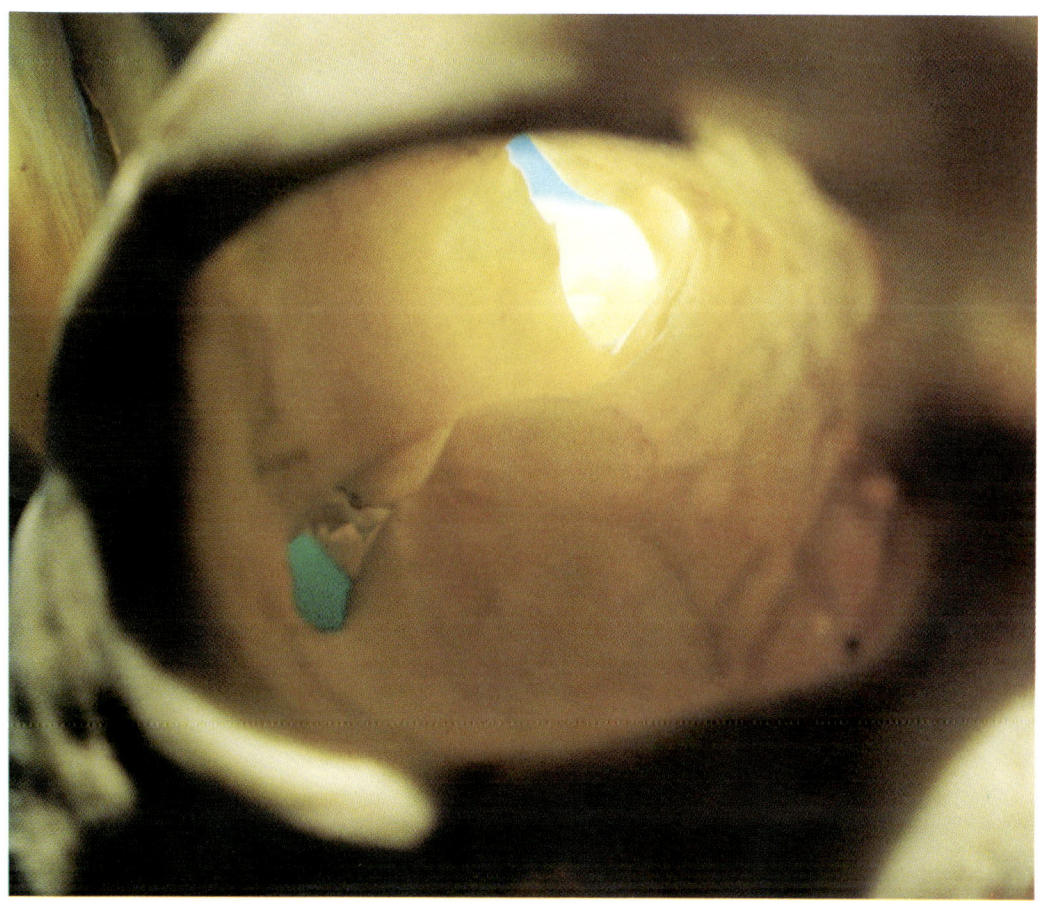

From front and slightly above, closeup view of orbital floor. Margins, fissures and foramina.

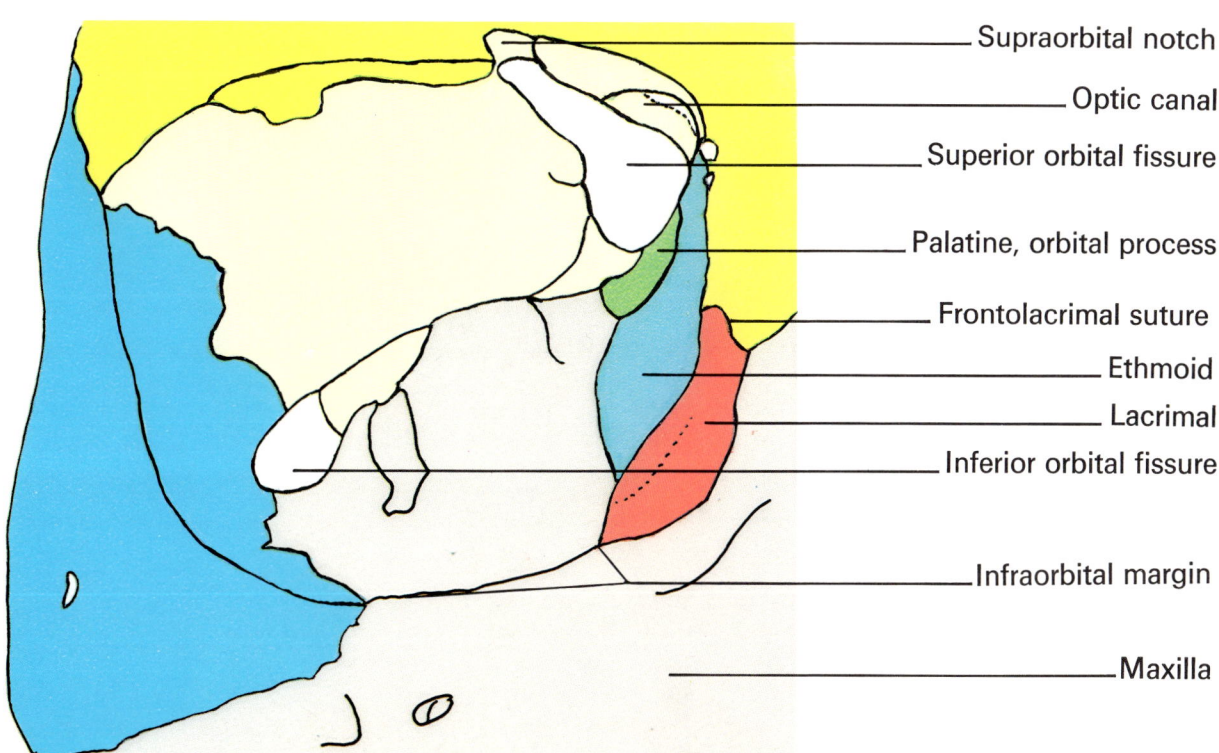

- Supraorbital notch
- Optic canal
- Superior orbital fissure
- Palatine, orbital process
- Frontolacrimal suture
- Ethmoid
- Lacrimal
- Inferior orbital fissure
- Infraorbital margin
- Maxilla

Orbital cavity.

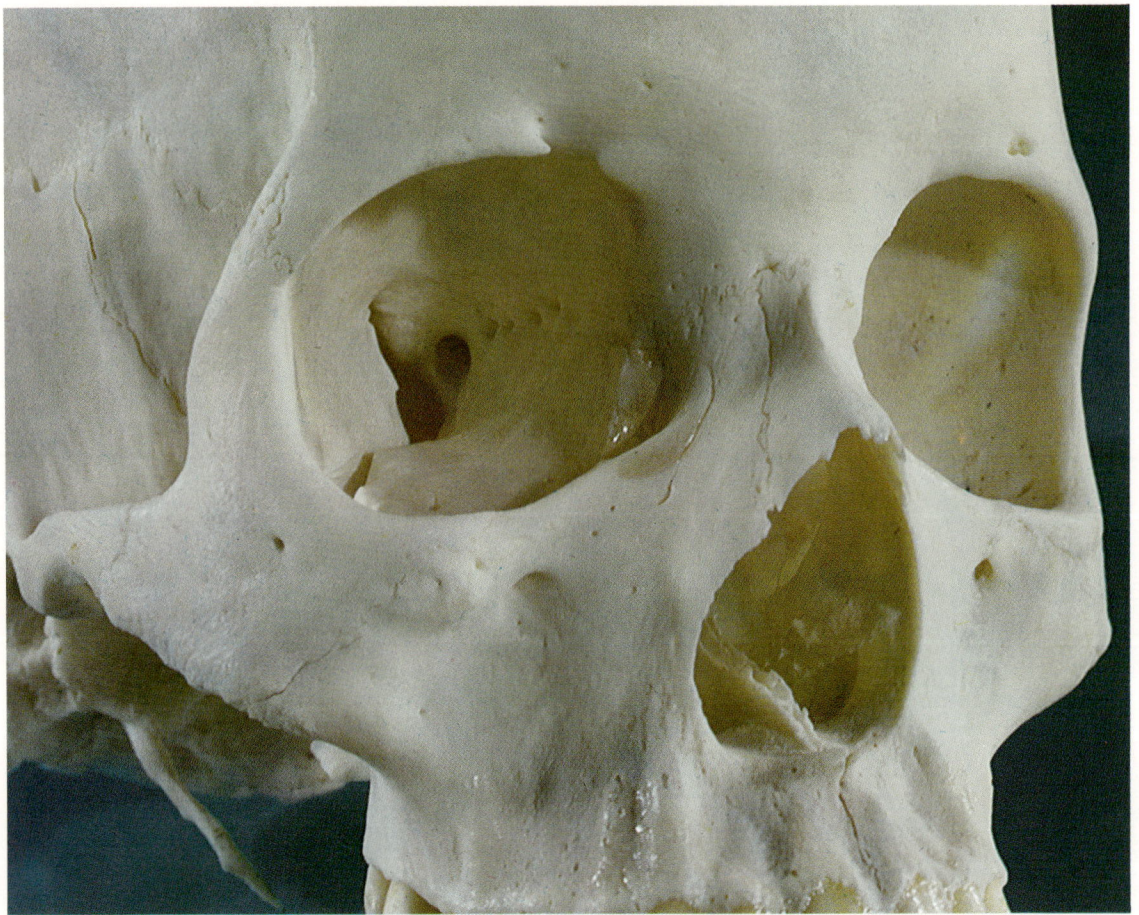

From front and slightly lateral, view into optic canal and adjacent foramina.

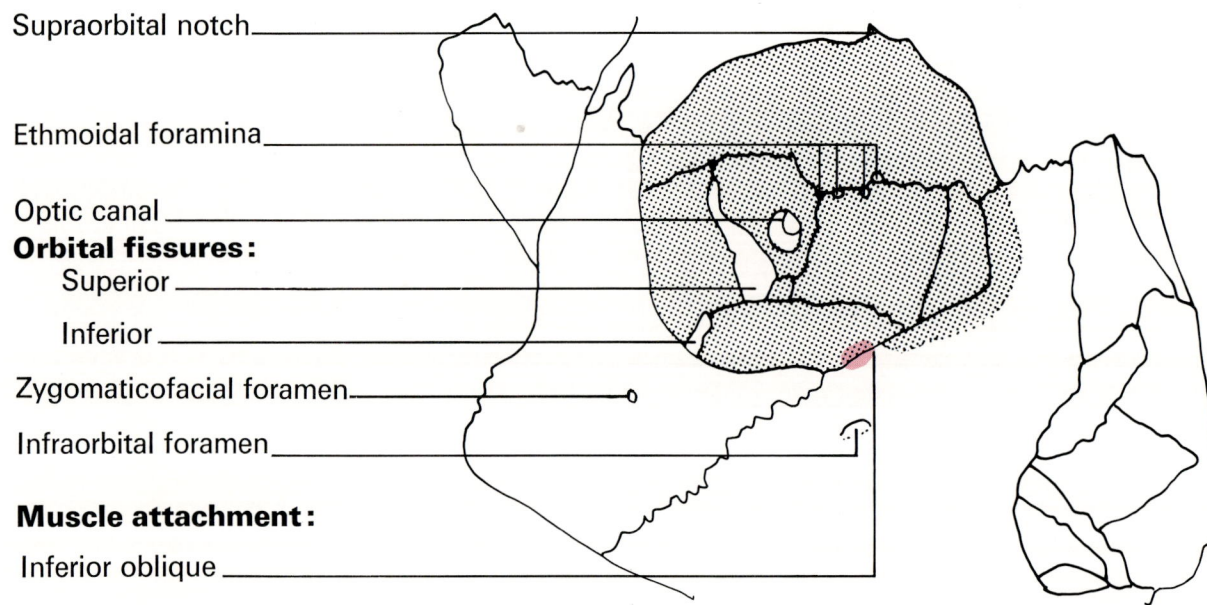

Supraorbital notch

Ethmoidal foramina

Optic canal
Orbital fissures:
    Superior
    Inferior

Zygomaticofacial foramen

Infraorbital foramen

Muscle attachment:
    Inferior oblique

Orbital cavity.

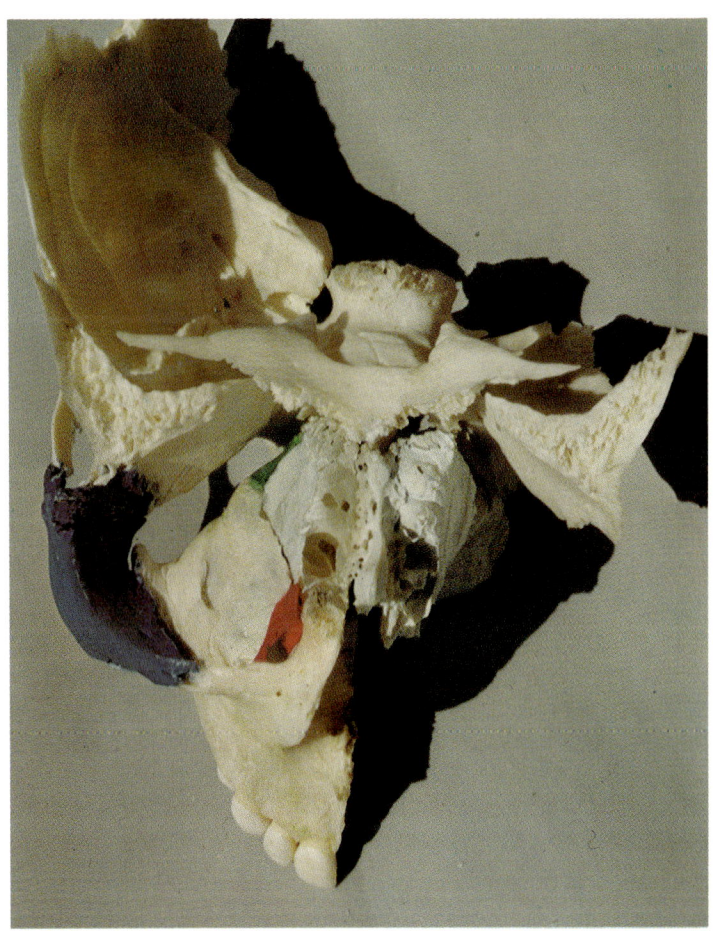

Orbital floor from above. Composite of maxilla and adjacent bones.

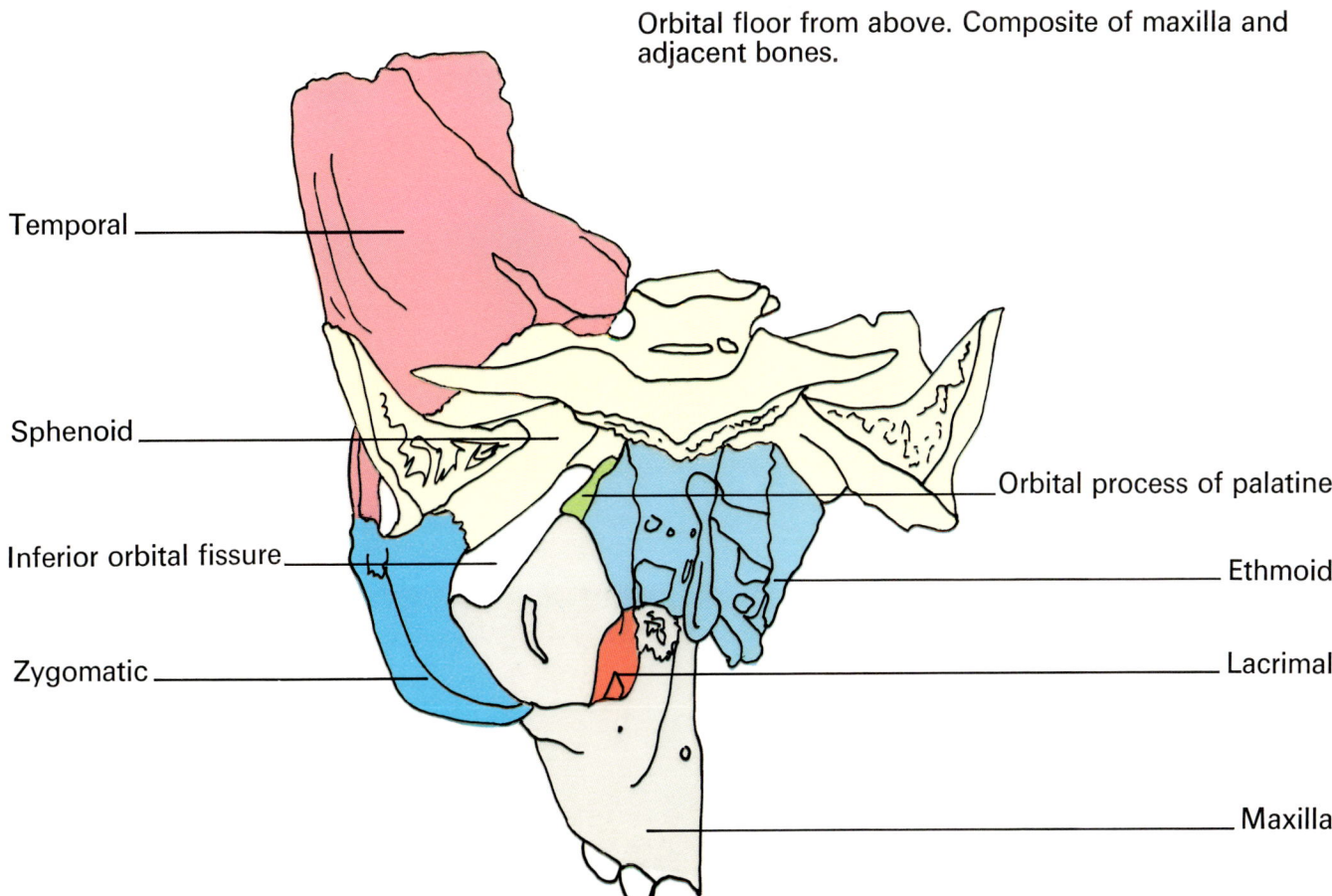

Orbital cavity.

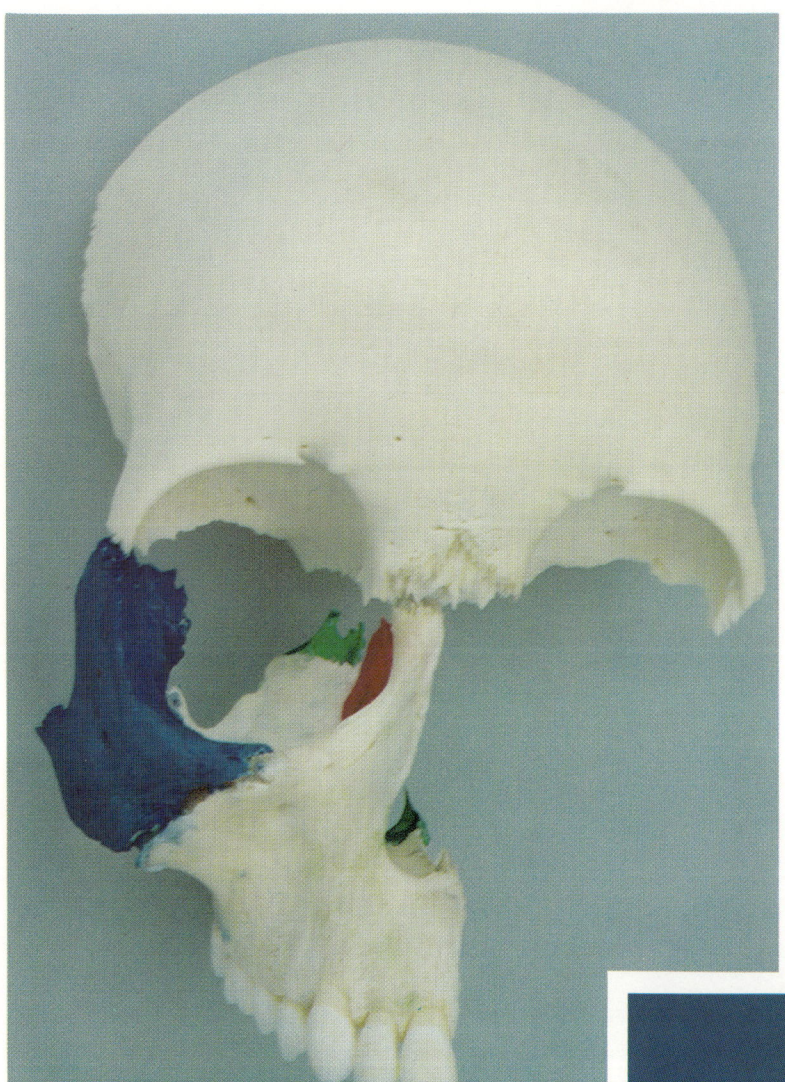

A. Orbital rim from front, composite of frontal and adjacent bones.

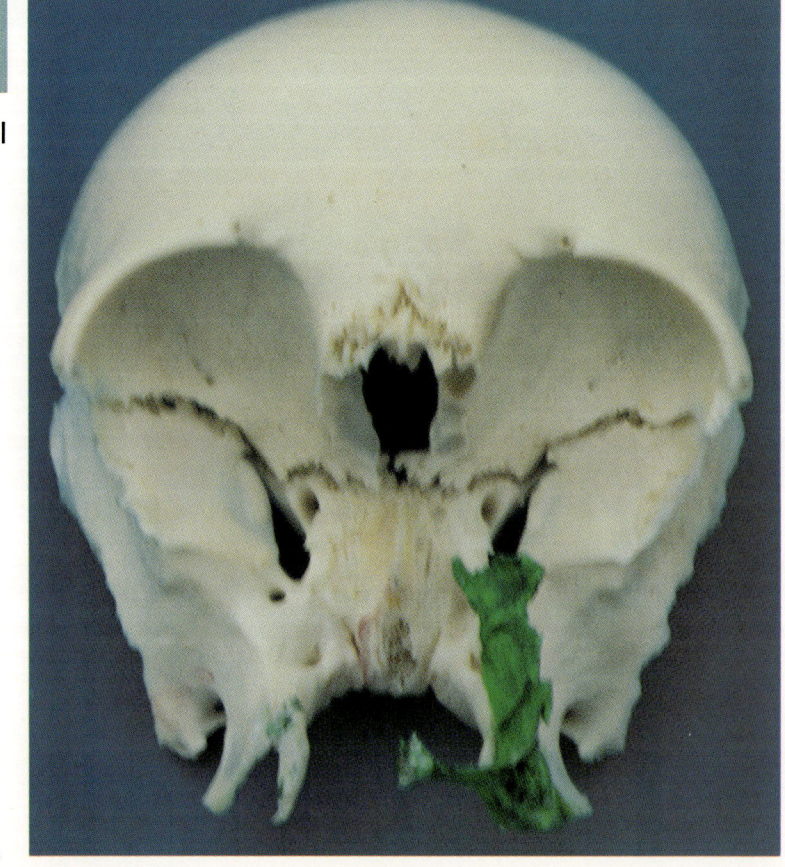

B. Orbital roof from front and below. Composite of frontal, sphenoid and palatine.

Orbital cavity.
Bones and landmarks.

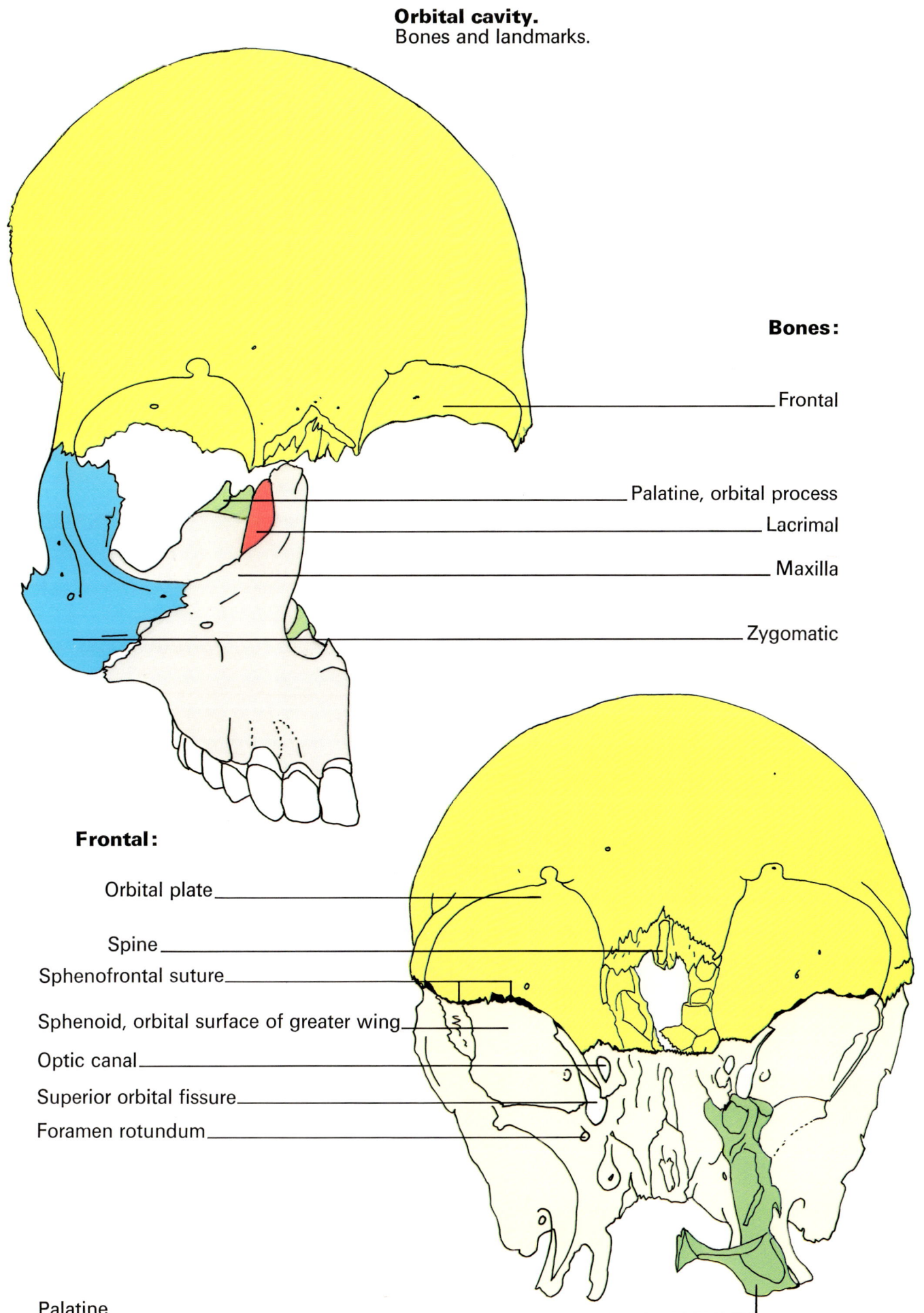

Bones:

Frontal

Palatine, orbital process

Lacrimal

Maxilla

Zygomatic

Frontal:

Orbital plate

Spine

Sphenofrontal suture

Sphenoid, orbital surface of greater wing

Optic canal

Superior orbital fissure

Foramen rotundum

Palatine

Orbital cavity.

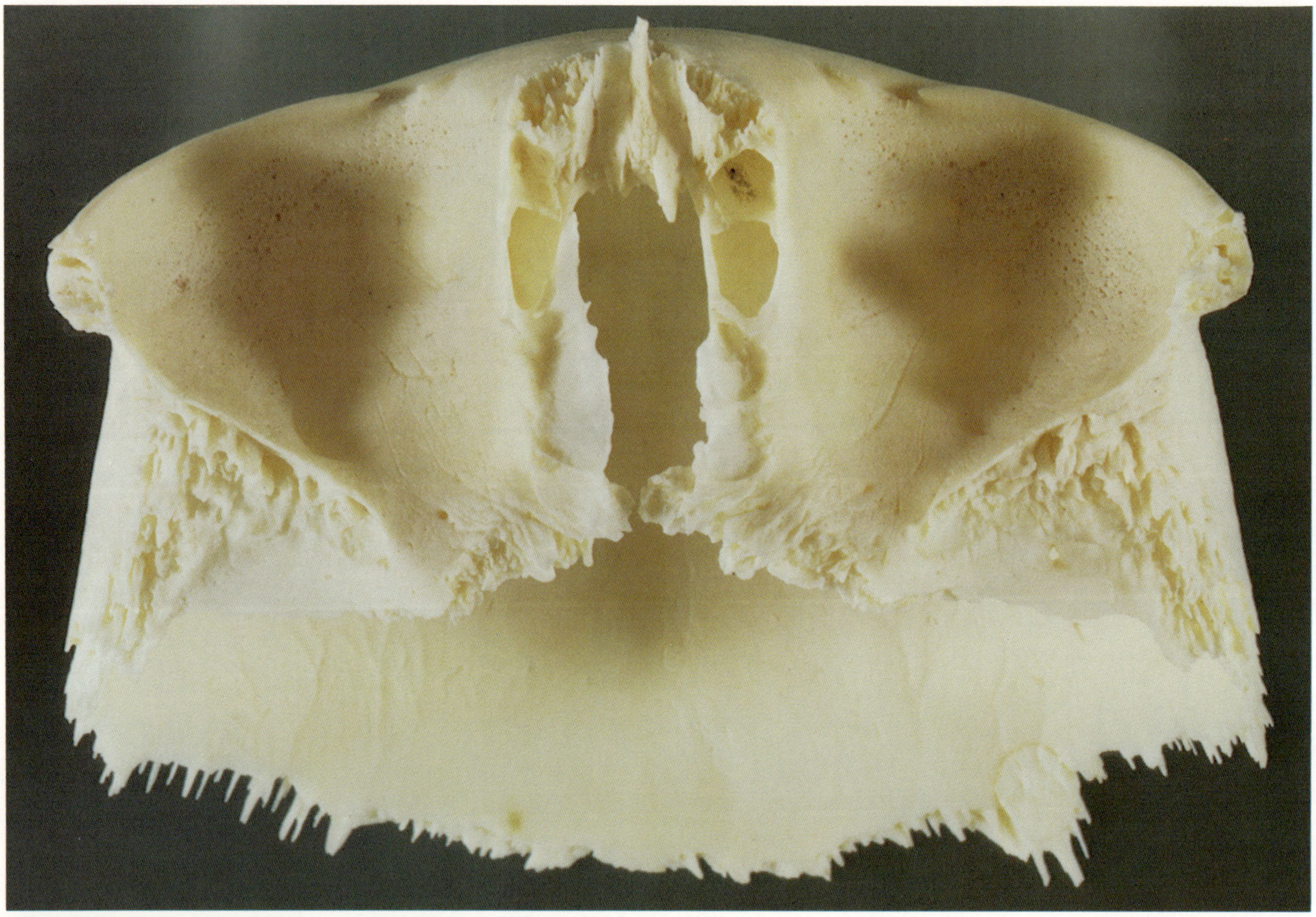

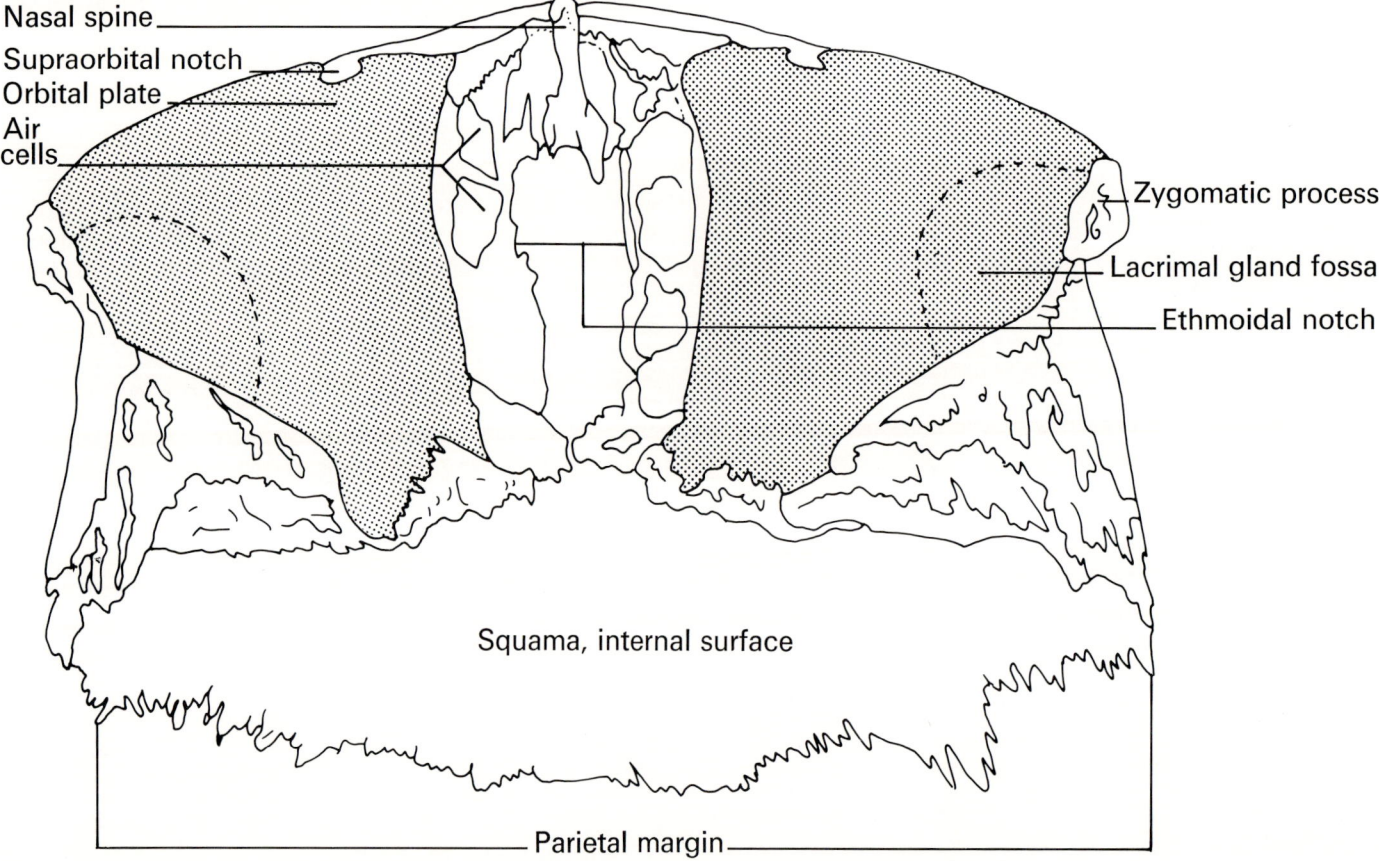

Frontal bone from below. Orbital plates stippled.

- Nasal spine
- Supraorbital notch
- Orbital plate
- Air cells
- Zygomatic process
- Lacrimal gland fossa
- Ethmoidal notch
- Squama, internal surface
- Parietal margin

Orbital cavity.

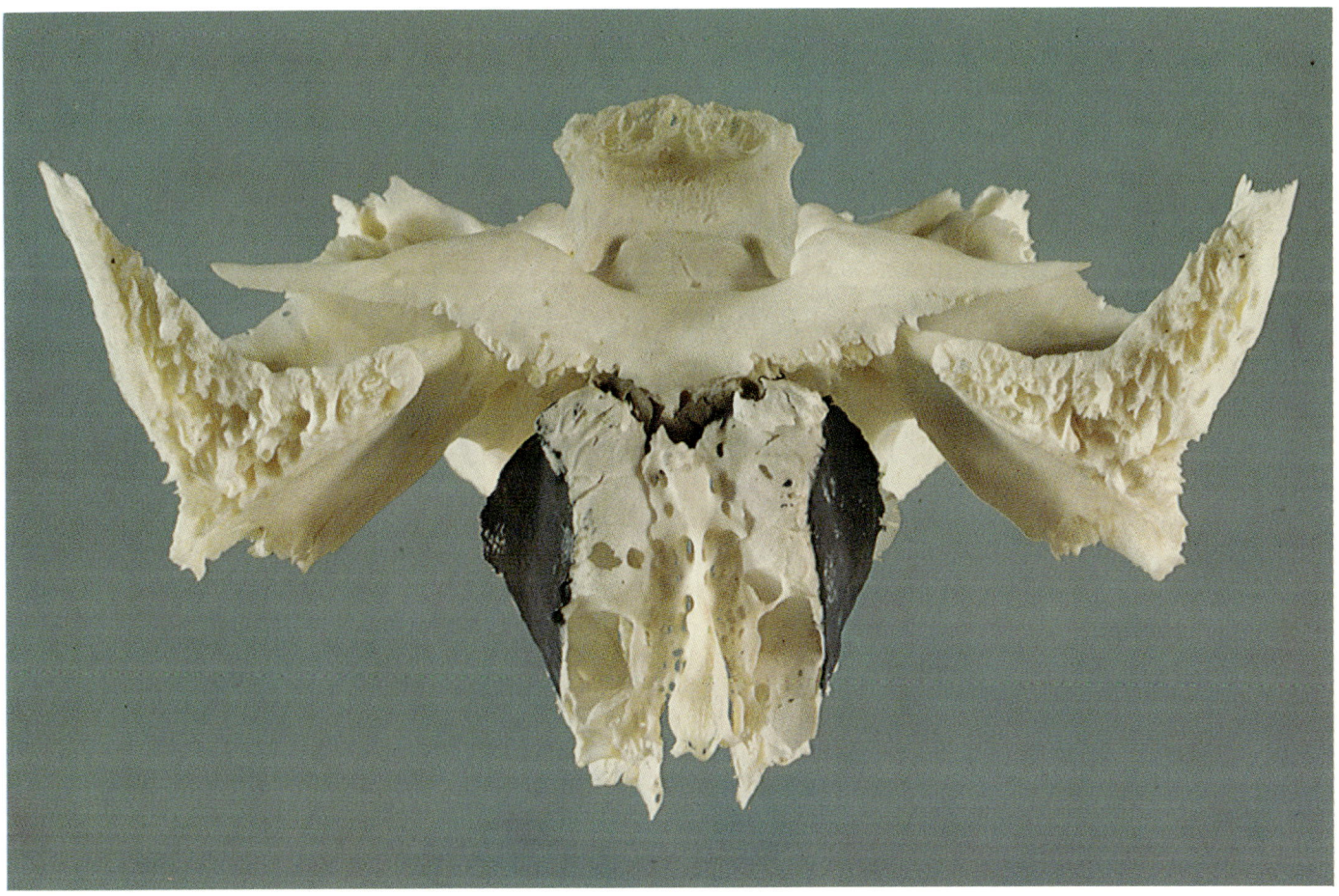

Orbital walls from above, stippled. Composite of sphenoid and ethmoid.

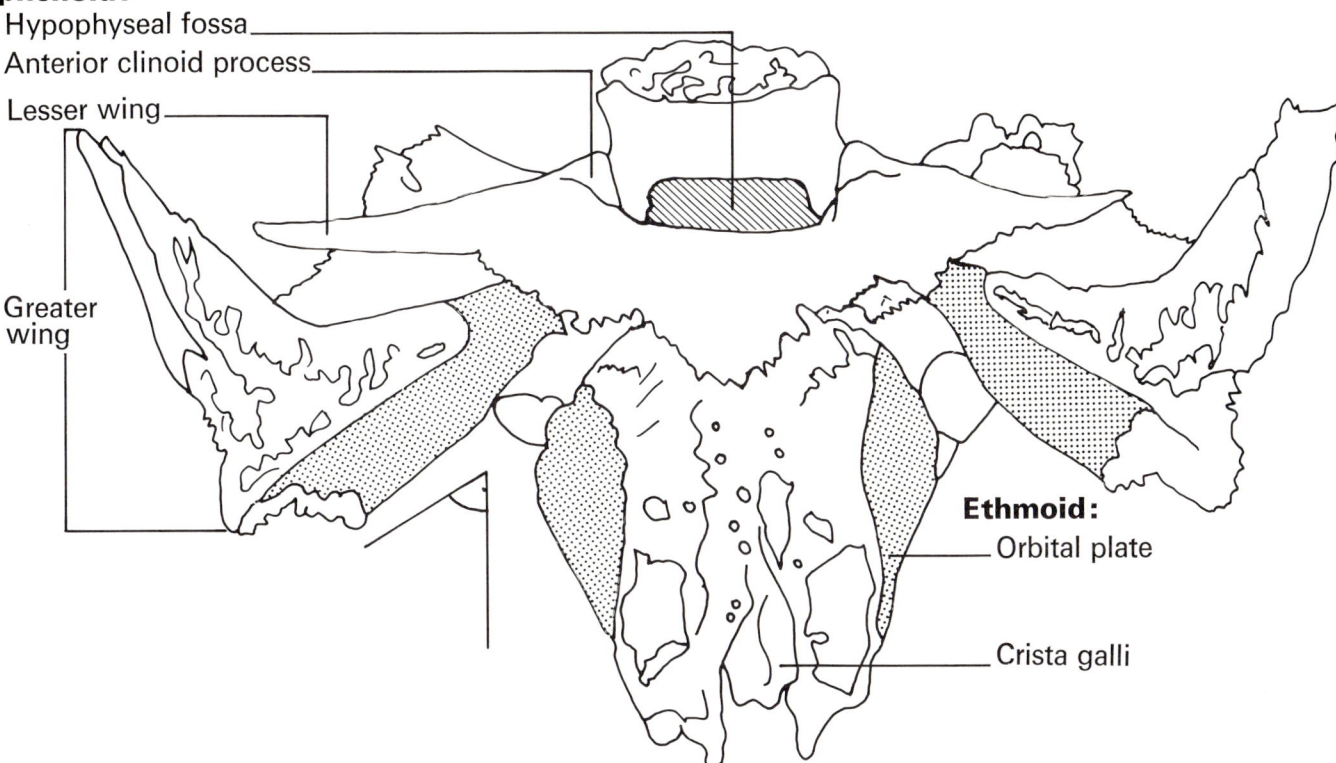

Sphenoid:
- Hypophyseal fossa
- Anterior clinoid process
- Lesser wing
- Greater wing

Ethmoid:
- Orbital plate
- Crista galli

Orbital cavity.

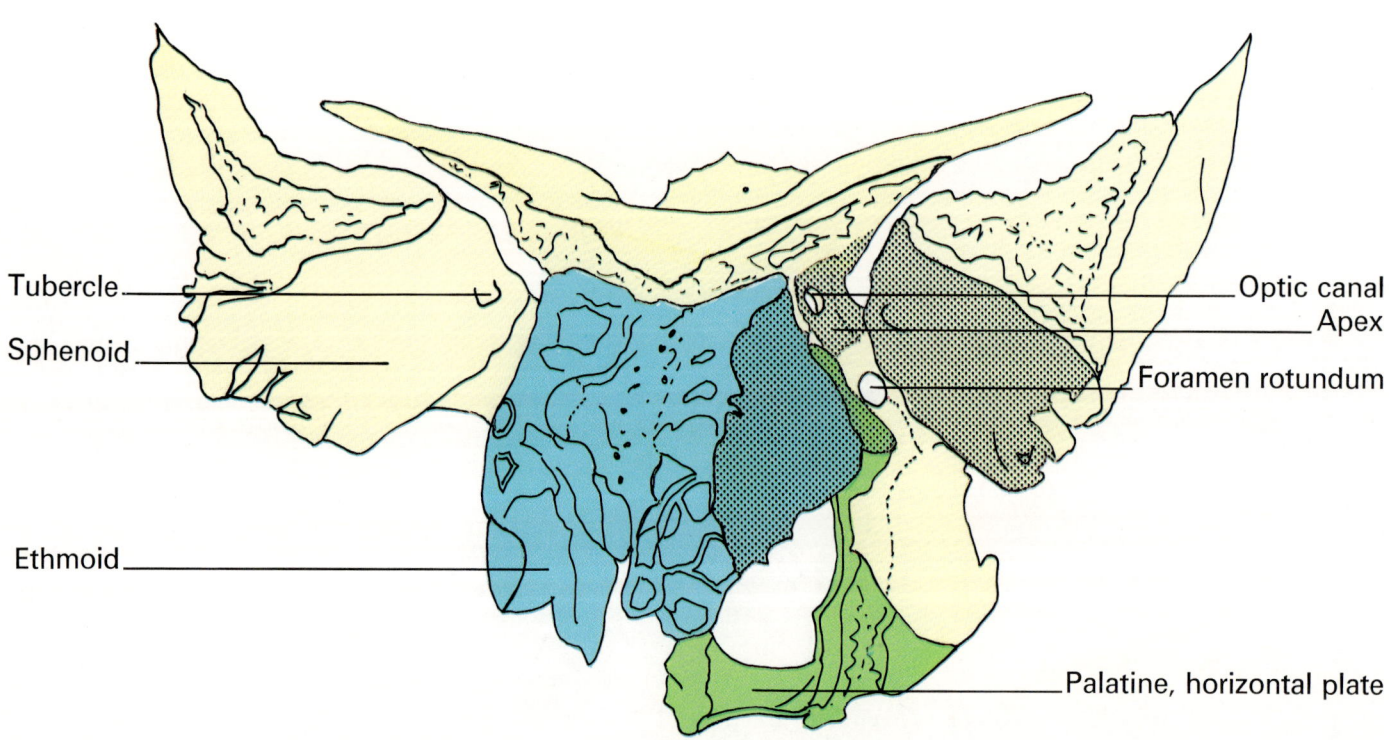

Orbital walls from front and to the right. Composite of sphenoid, ethmoid and palatine. View of apex of orbit and optic canal.

Orbital cavity.

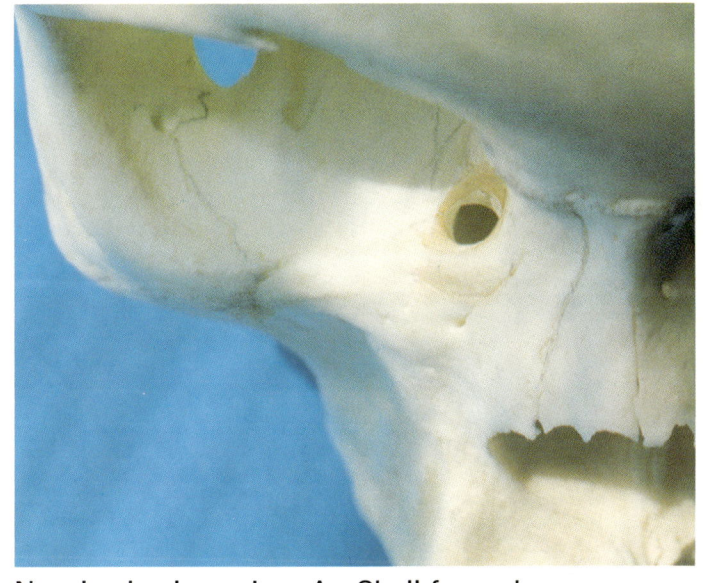
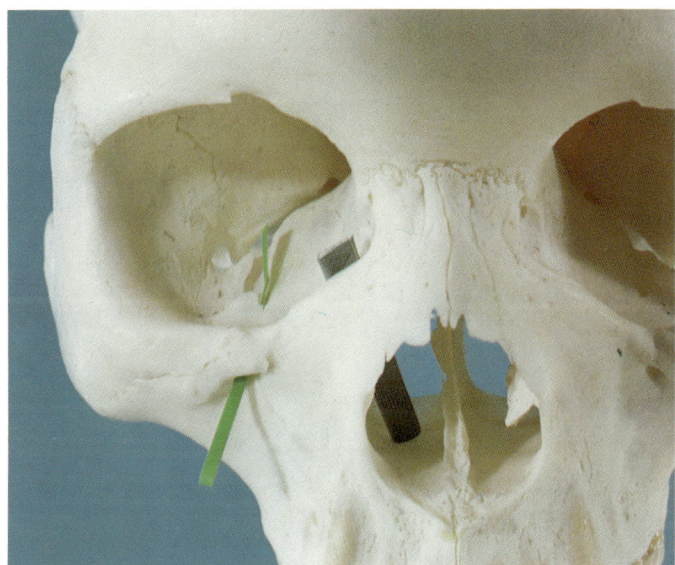

Nasolacrimal canal: A. Skull from above.   B. From front.

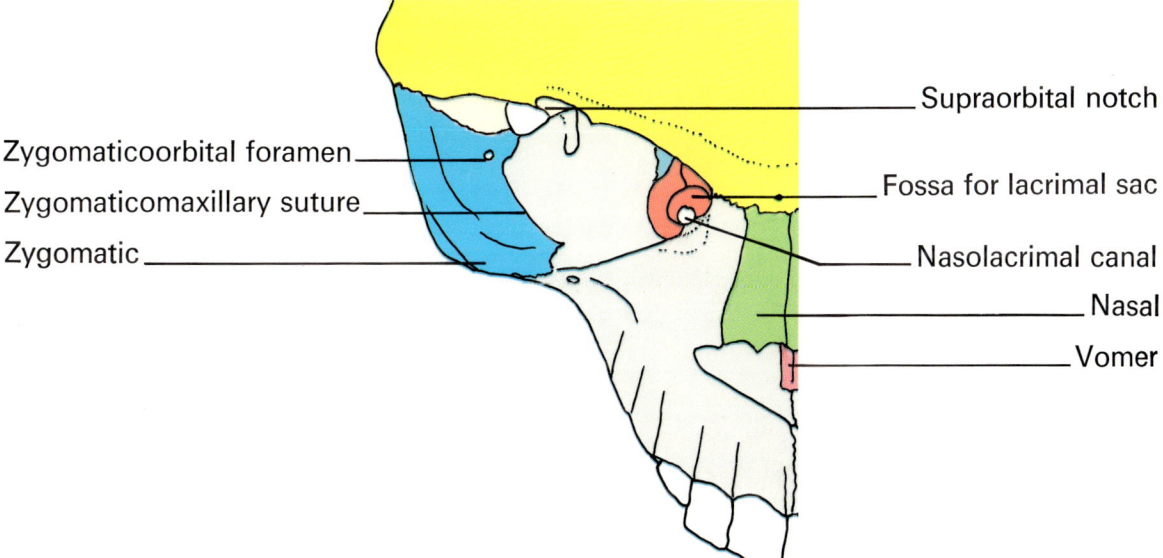

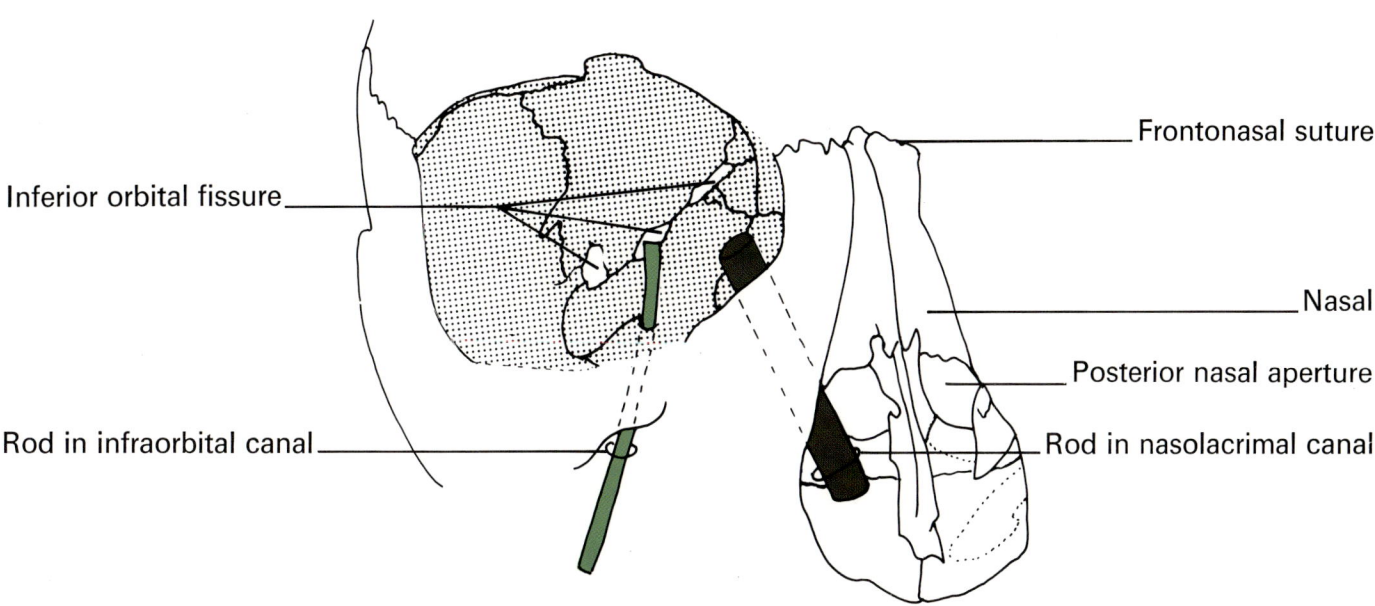

Orbital cavity.

Floor and walls of orbital cavity. Composite of right maxilla and adjacent bones.
A. From side and above.

B. From nasal surface, laterally.

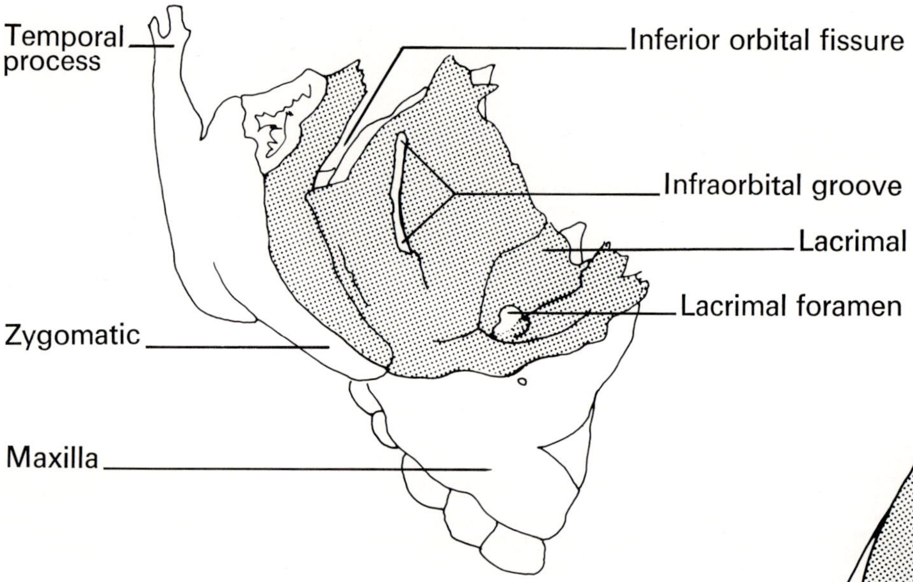

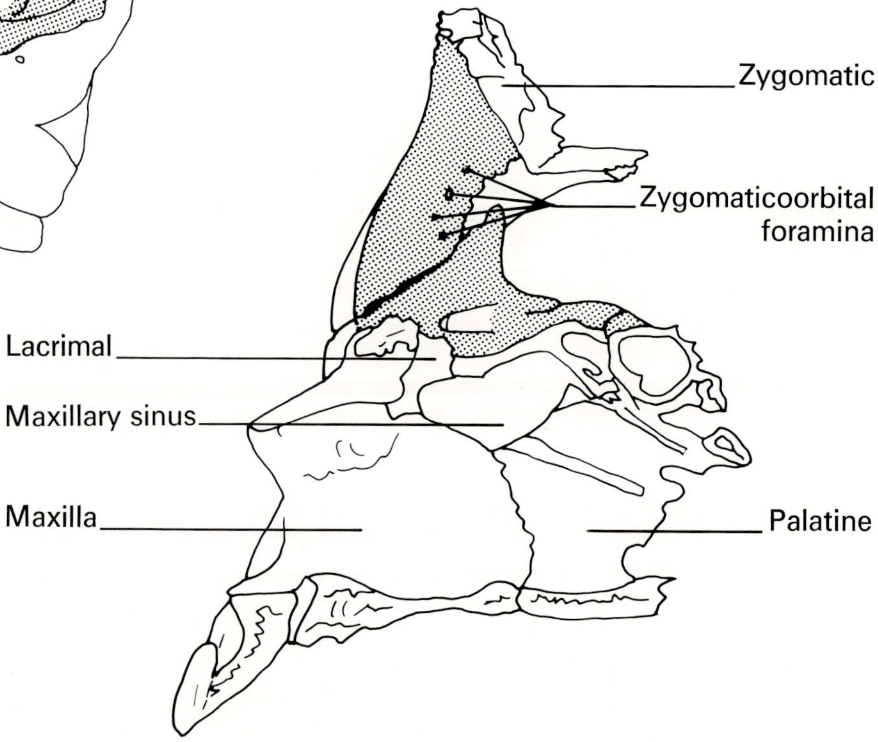

Orbital cavity.

A. From side, medially.

B. From behind and slightly above.

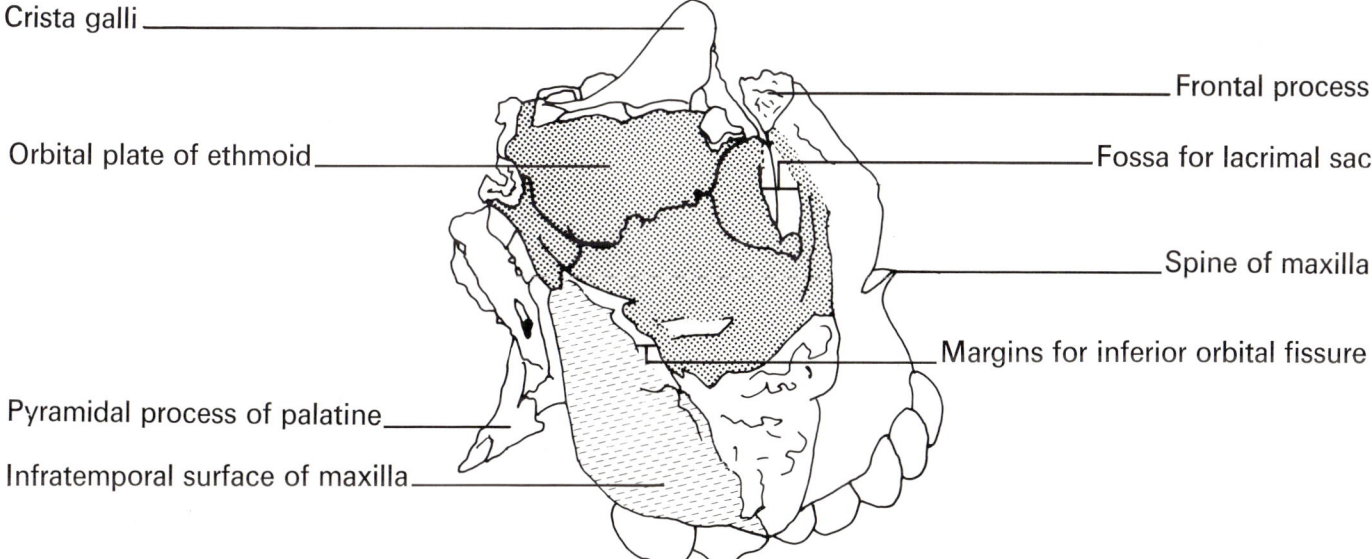

- Crista galli
- Orbital plate of ethmoid
- Pyramidal process of palatine
- Infratemporal surface of maxilla
- Frontal process
- Fossa for lacrimal sac
- Spine of maxilla
- Margins for inferior orbital fissure

Palatine:

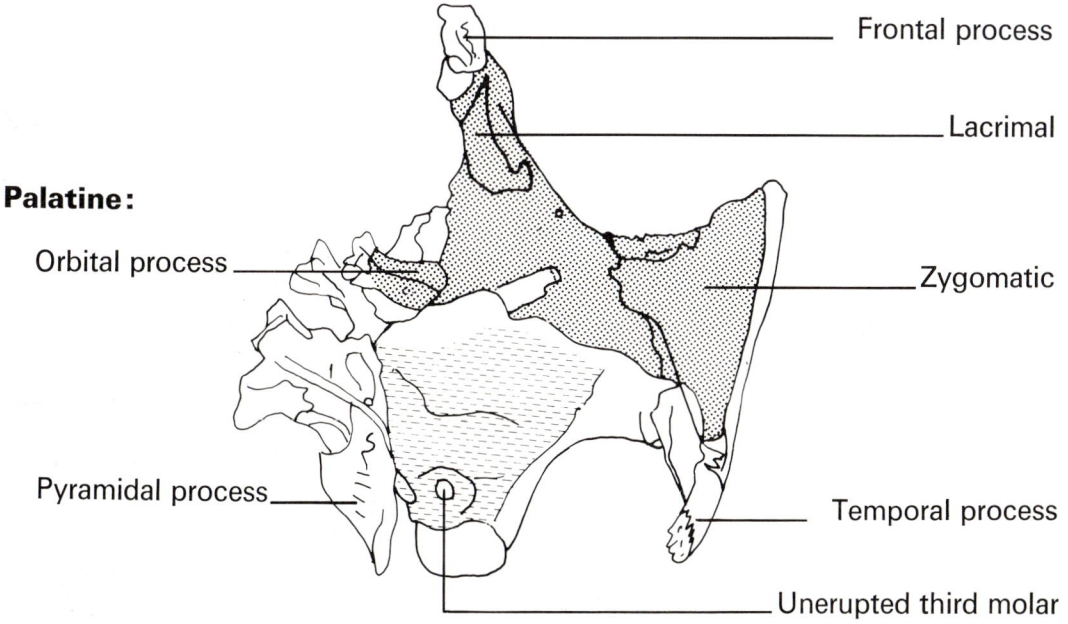

- Orbital process
- Pyramidal process
- Frontal process
- Lacrimal
- Zygomatic
- Temporal process
- Unerupted third molar

Orbital cavity.

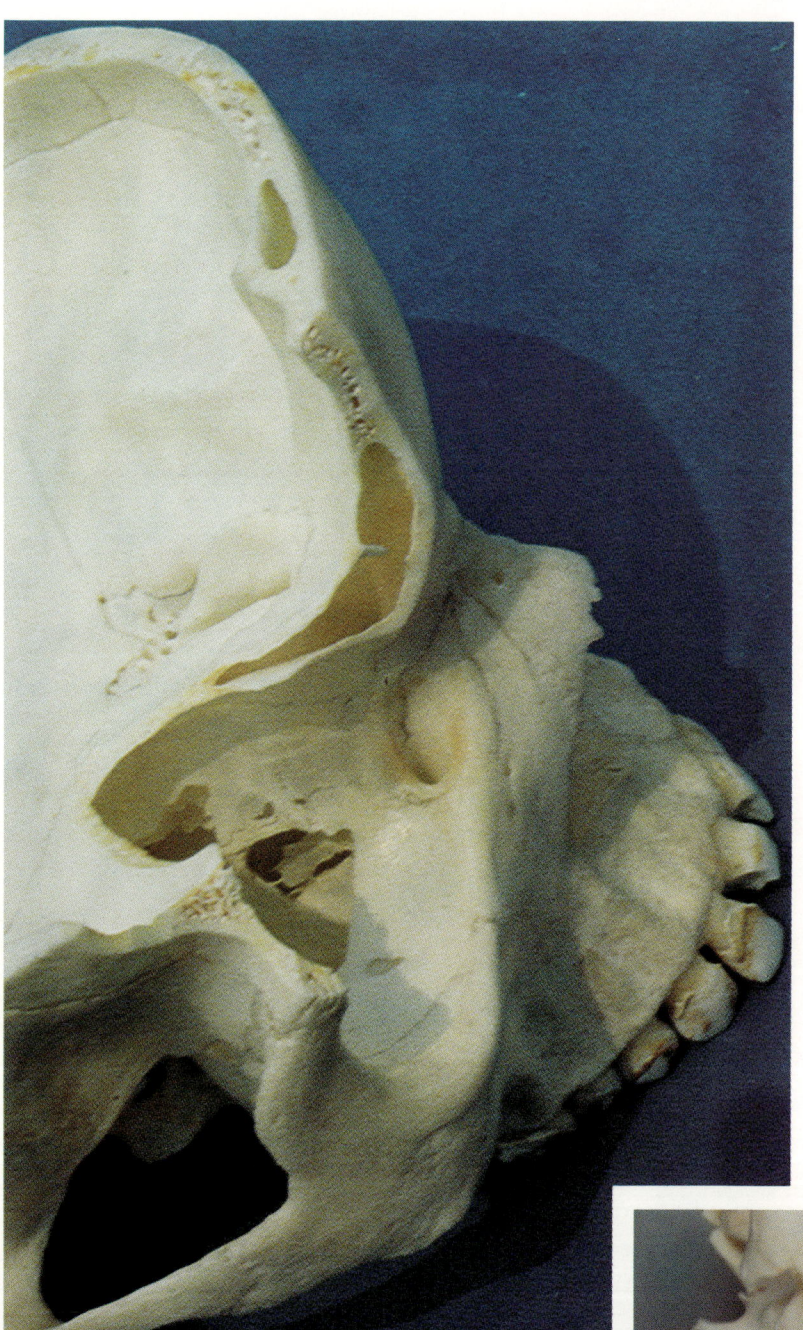

A. Skull with frontal bone resected, view of right orbital cavity from side and above.

B. Inferior orbital fissure and temporal fossa from above, composite of bones forming boundaries.

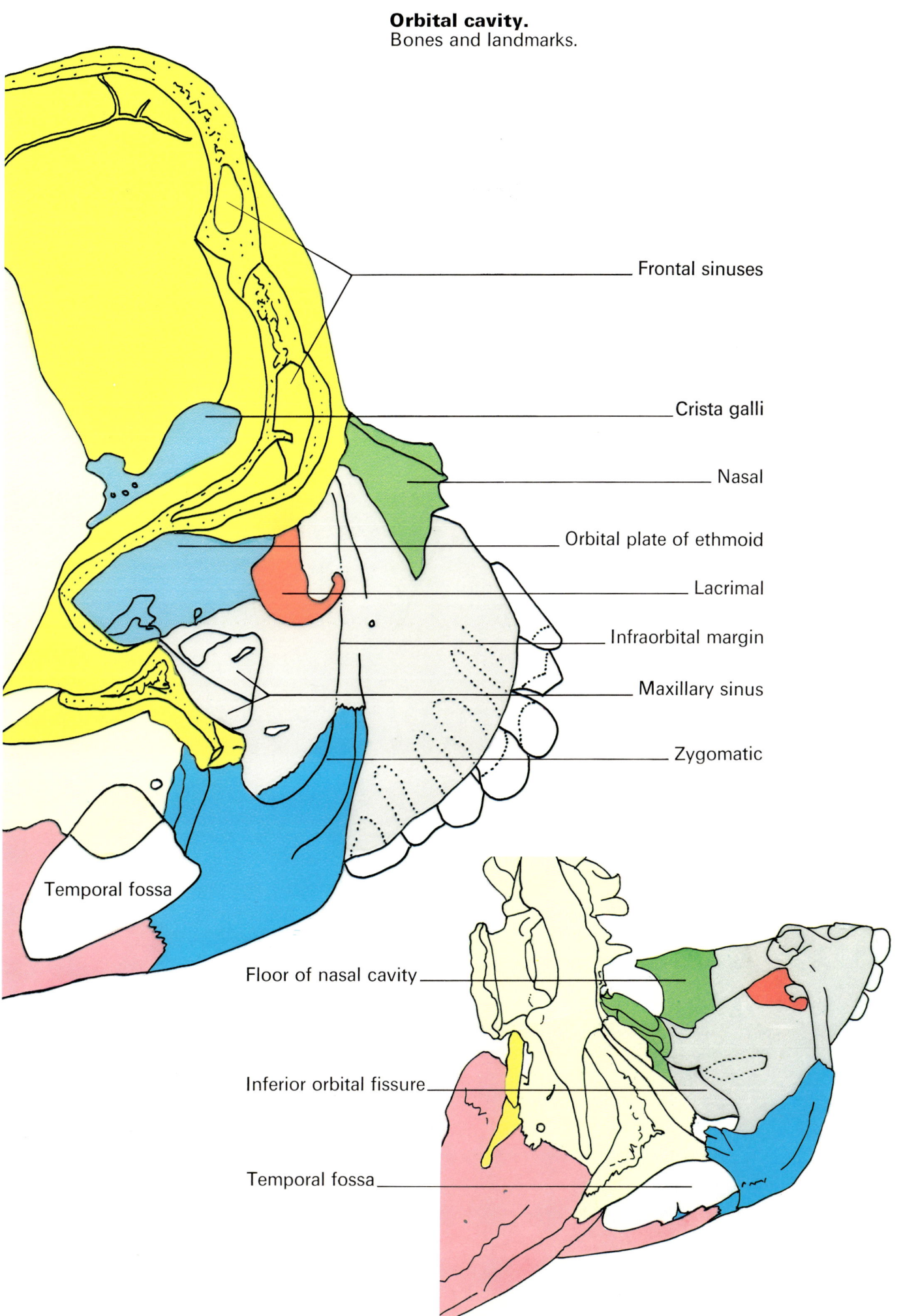

Orbital cavity. Bones and landmarks.

SECTION 10.   **Nasal cavity.**

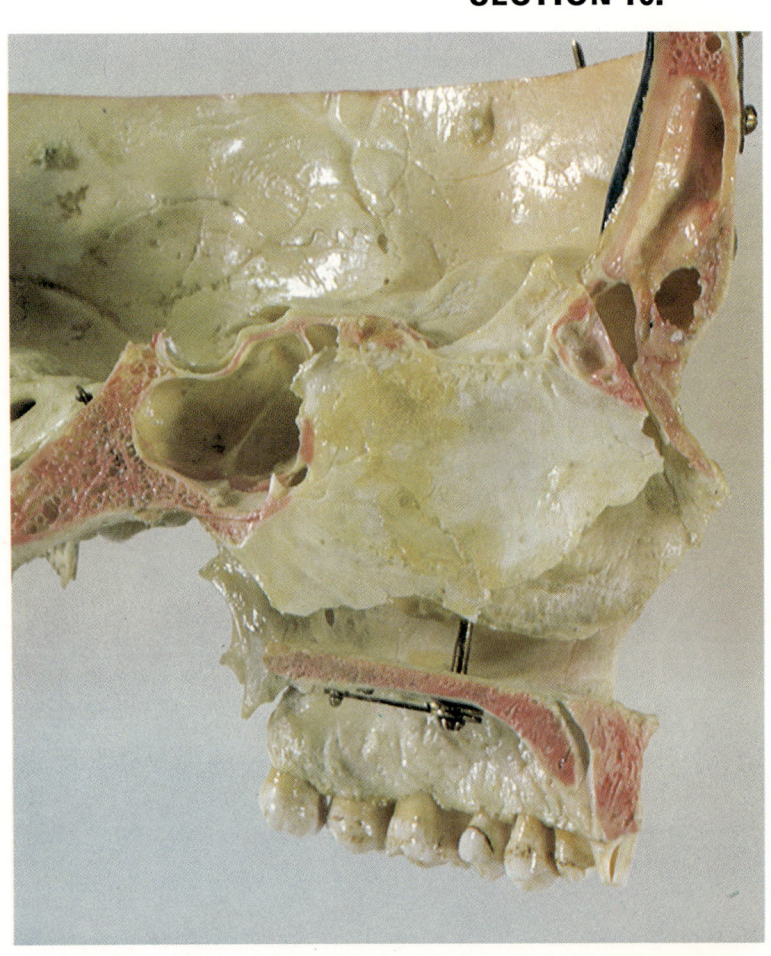

Midsagittal sections of skull. Bones, sinuses and landmarks.

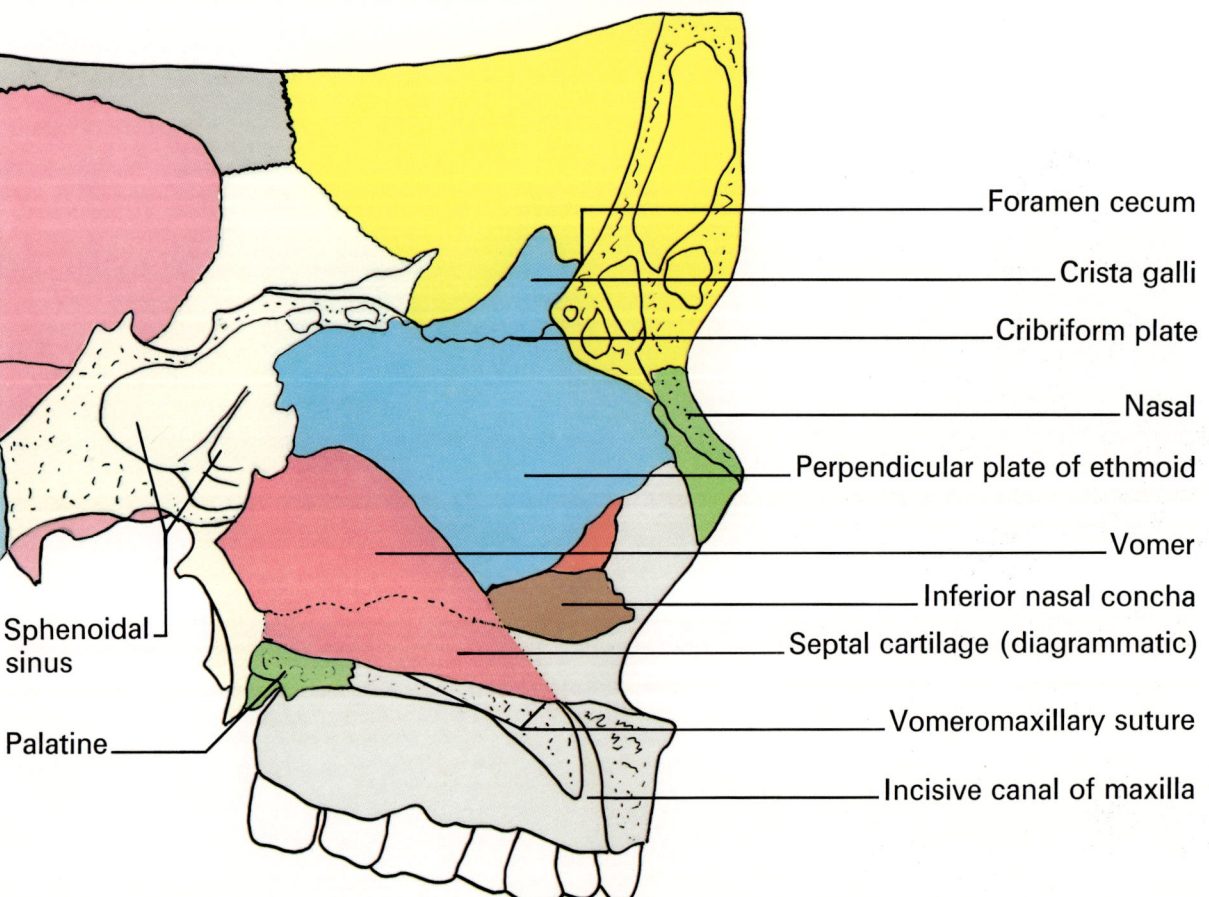

Nasal cavity.

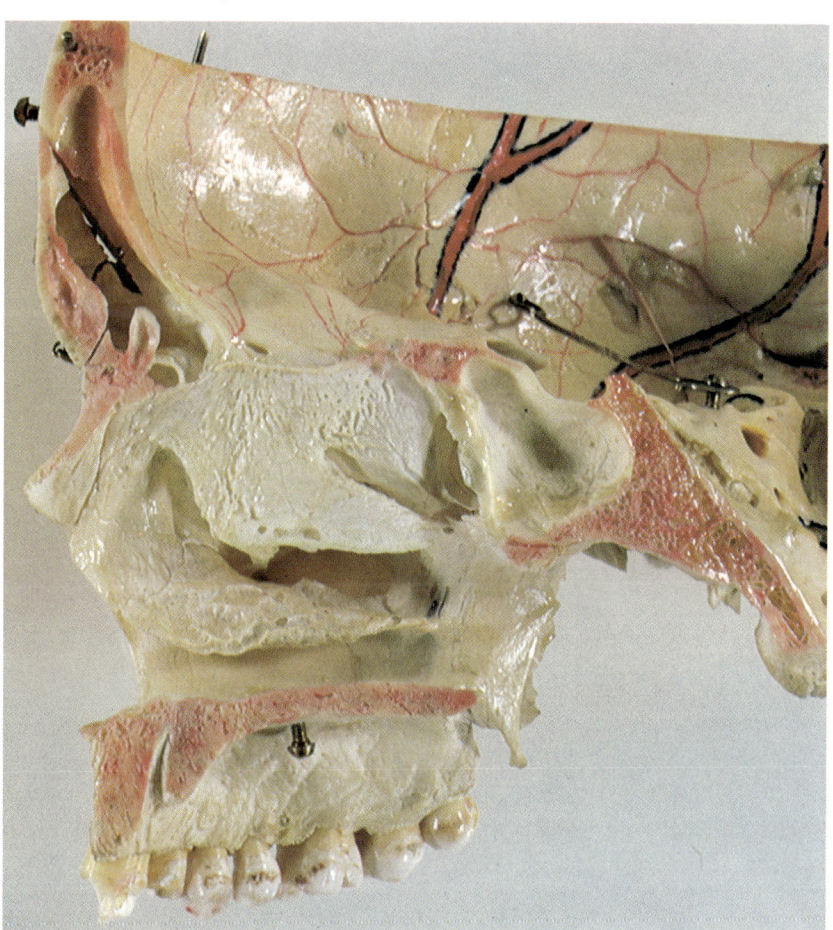

Midsagittal section of skull. Vascular markings.

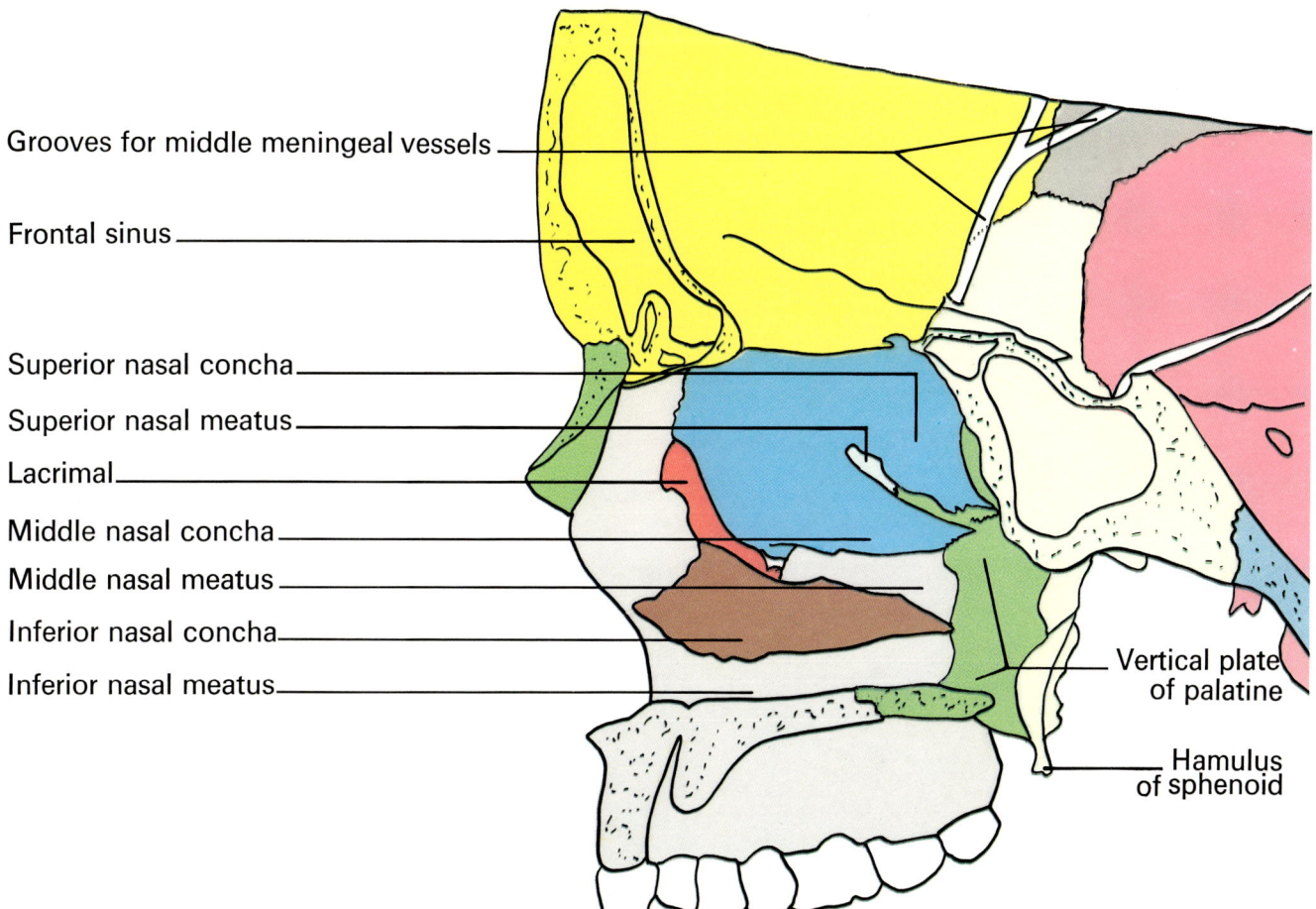

- Grooves for middle meningeal vessels
- Frontal sinus
- Superior nasal concha
- Superior nasal meatus
- Lacrimal
- Middle nasal concha
- Middle nasal meatus
- Inferior nasal concha
- Inferior nasal meatus
- Vertical plate of palatine
- Hamulus of sphenoid

Nasal cavity.

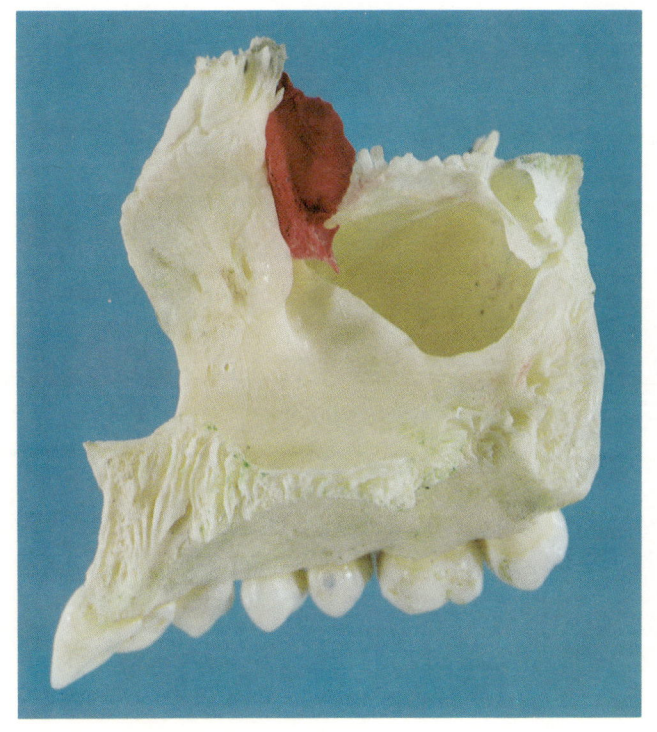
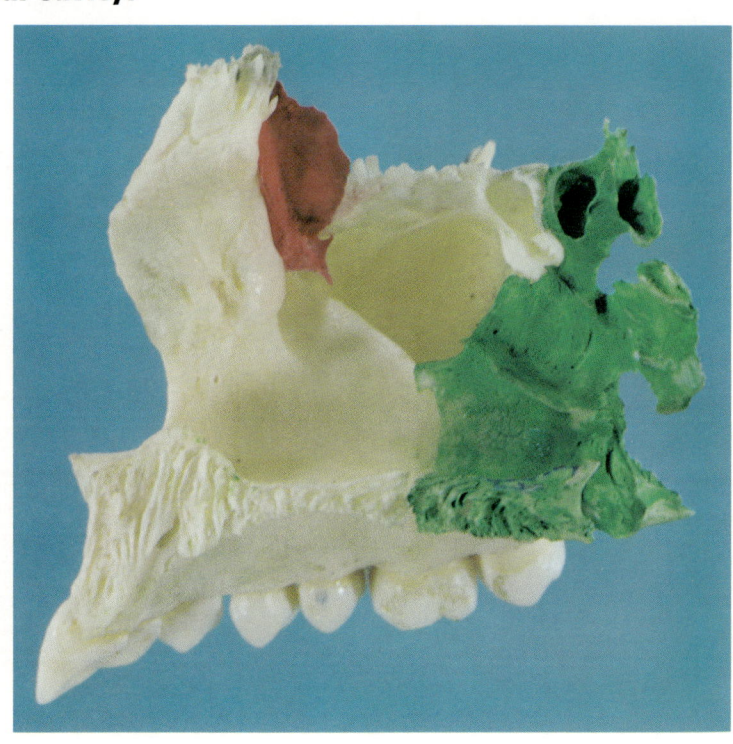

Right nasal cavity, from side. Composite of:
A. Maxilla with lacrimal.
B. With palatine.

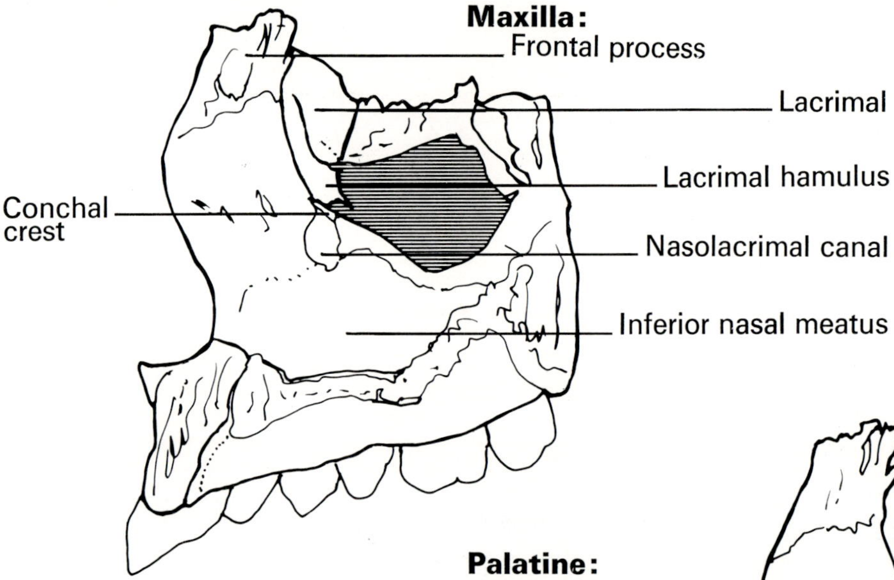

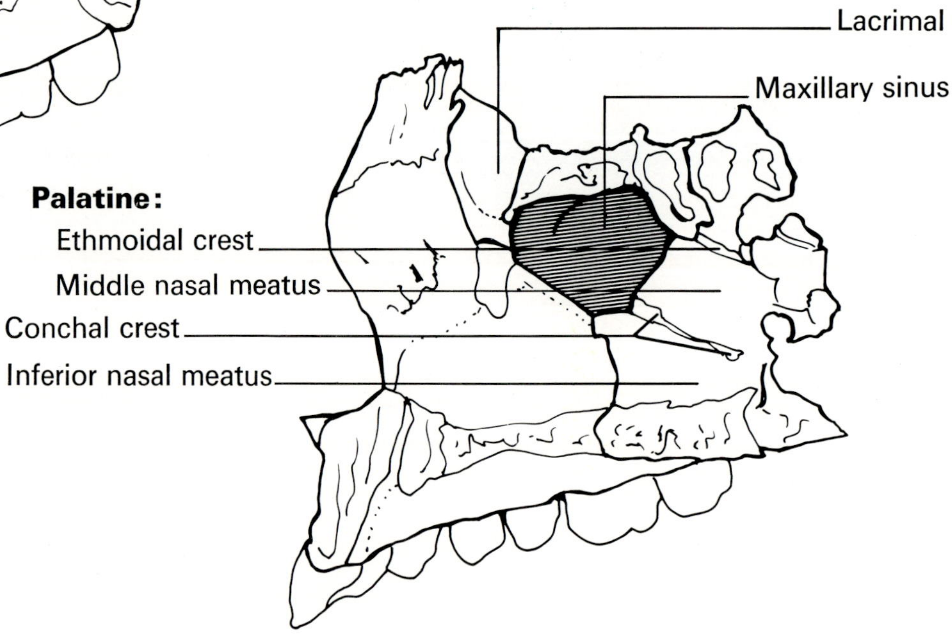

Maxilla:
- Frontal process
- Lacrimal
- Lacrimal hamulus
- Nasolacrimal canal
- Inferior nasal meatus
- Conchal crest

- Lacrimal
- Maxillary sinus

Palatine:
- Ethmoidal crest
- Middle nasal meatus
- Conchal crest
- Inferior nasal meatus

Nasal cavity.

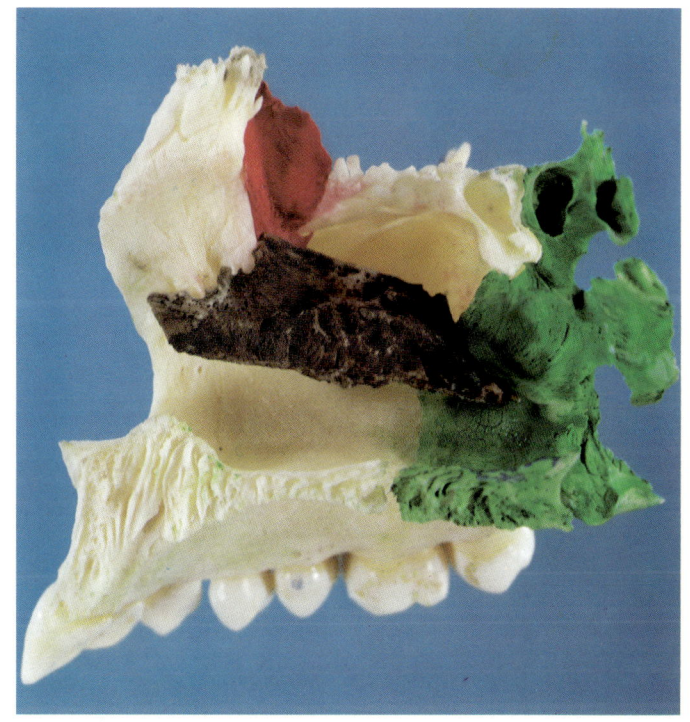

A. With inferior concha.

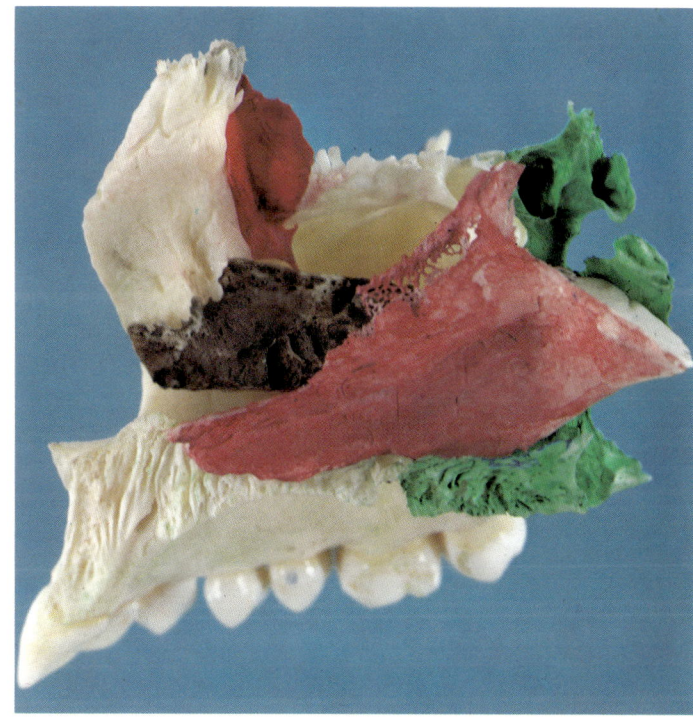

B. With vomer.

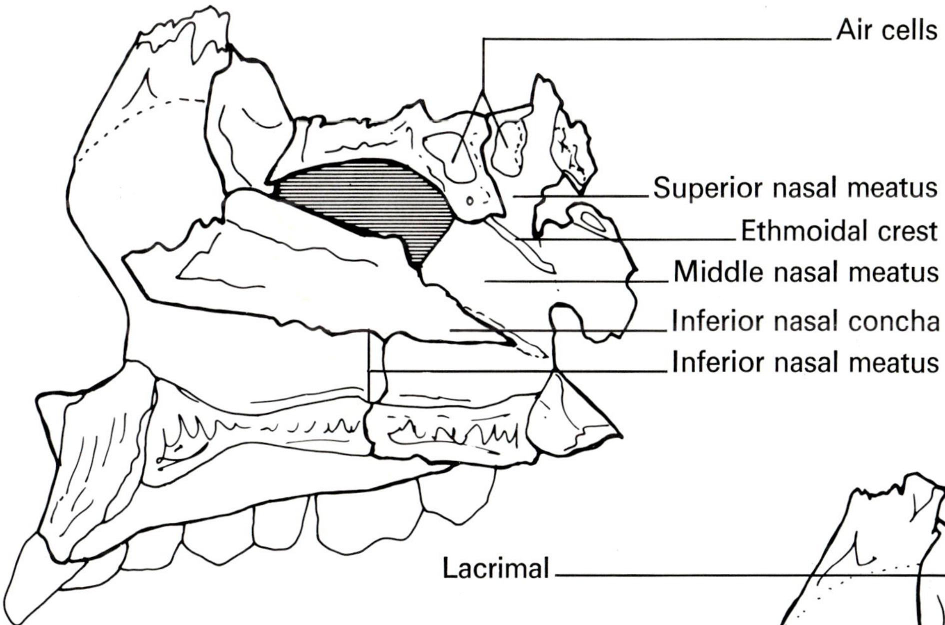

- Air cells
- Superior nasal meatus
- Ethmoidal crest
- Middle nasal meatus
- Inferior nasal concha
- Inferior nasal meatus

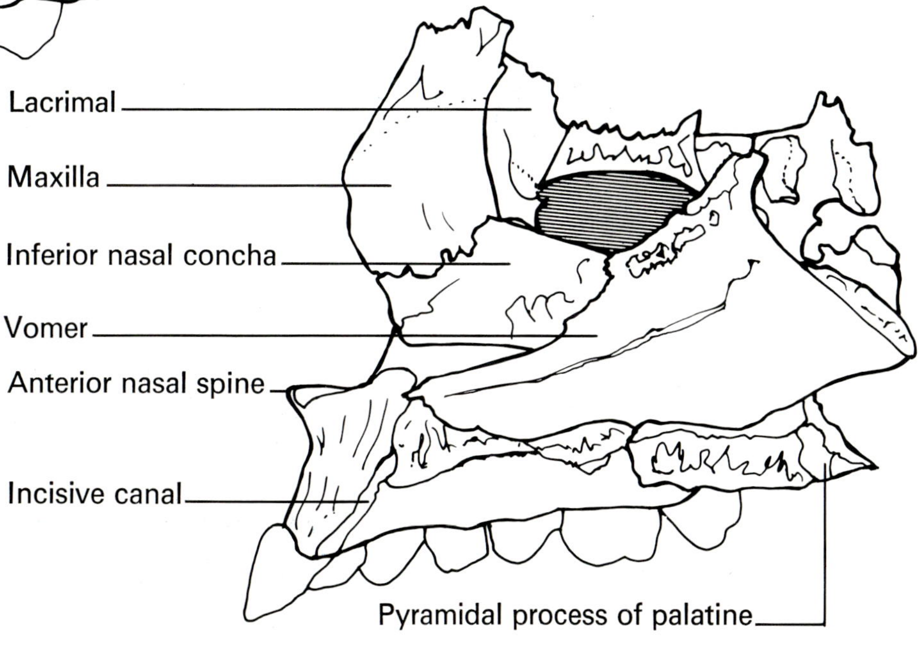

- Lacrimal
- Maxilla
- Inferior nasal concha
- Vomer
- Anterior nasal spine
- Incisive canal
- Pyramidal process of palatine

Nasal cavity.

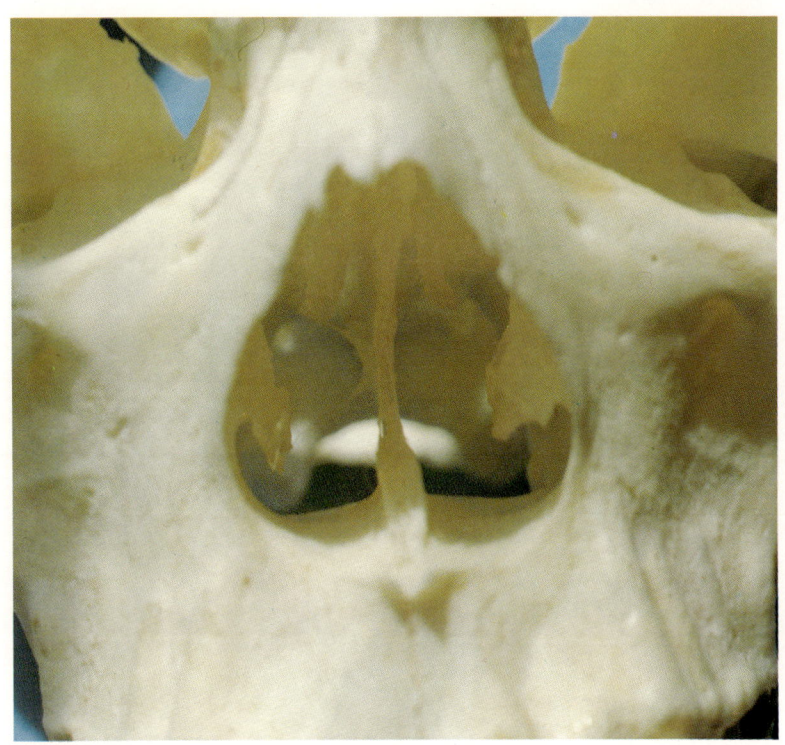

From front. A. Skull.

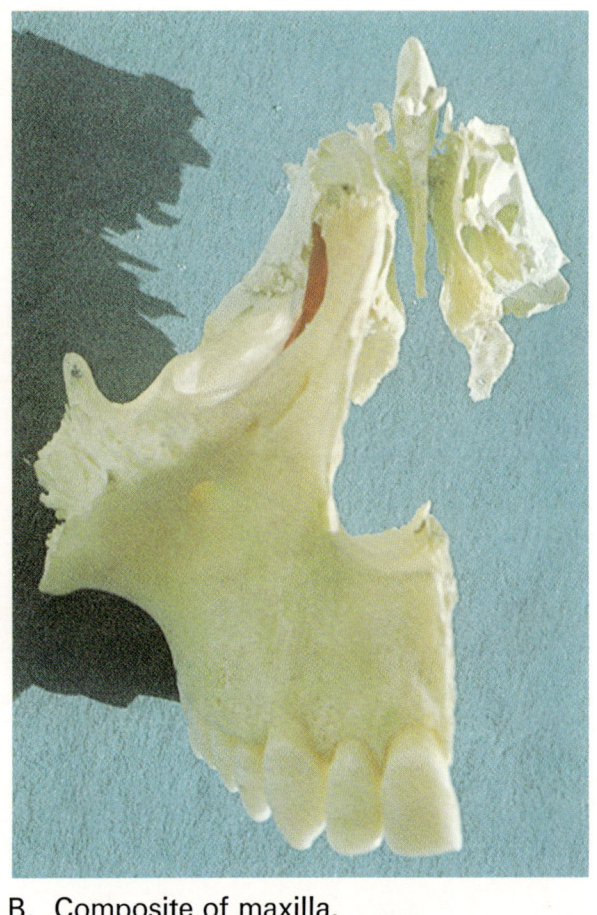

B. Composite of maxilla, ethmoid and lacrimal.

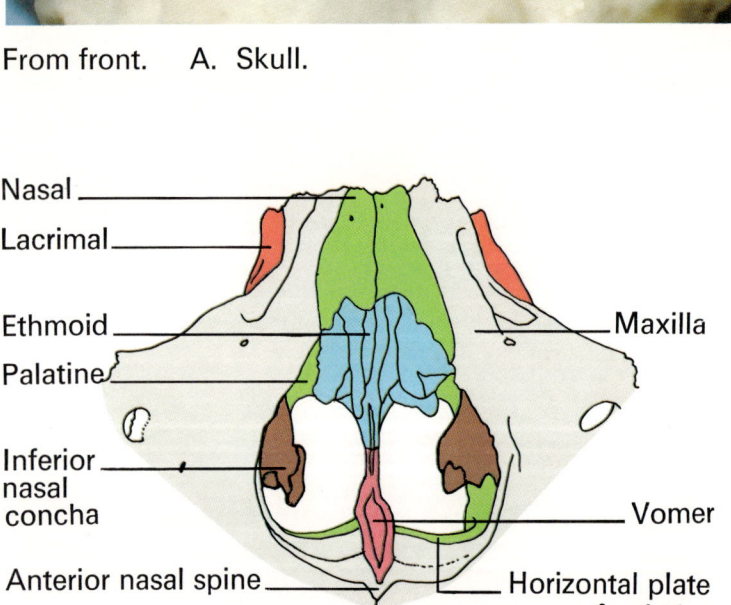

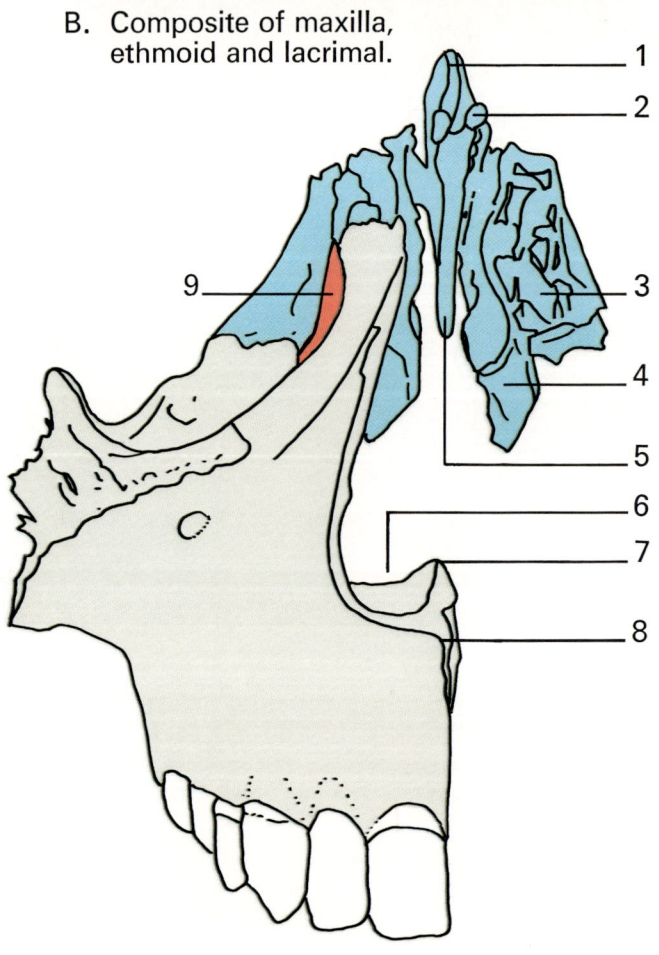

Key to numbers, Fig. B
Ethmoid:
1. Crista galli
2. Ala cristae galli
3. Uncinate process
4. Middle nasal concha
5. Perpendicular plate
6. Nasal notch

Maxilla, nasal spines:
7. Posterior
8. Anterior
9. Lacrimal

Nasal cavity.

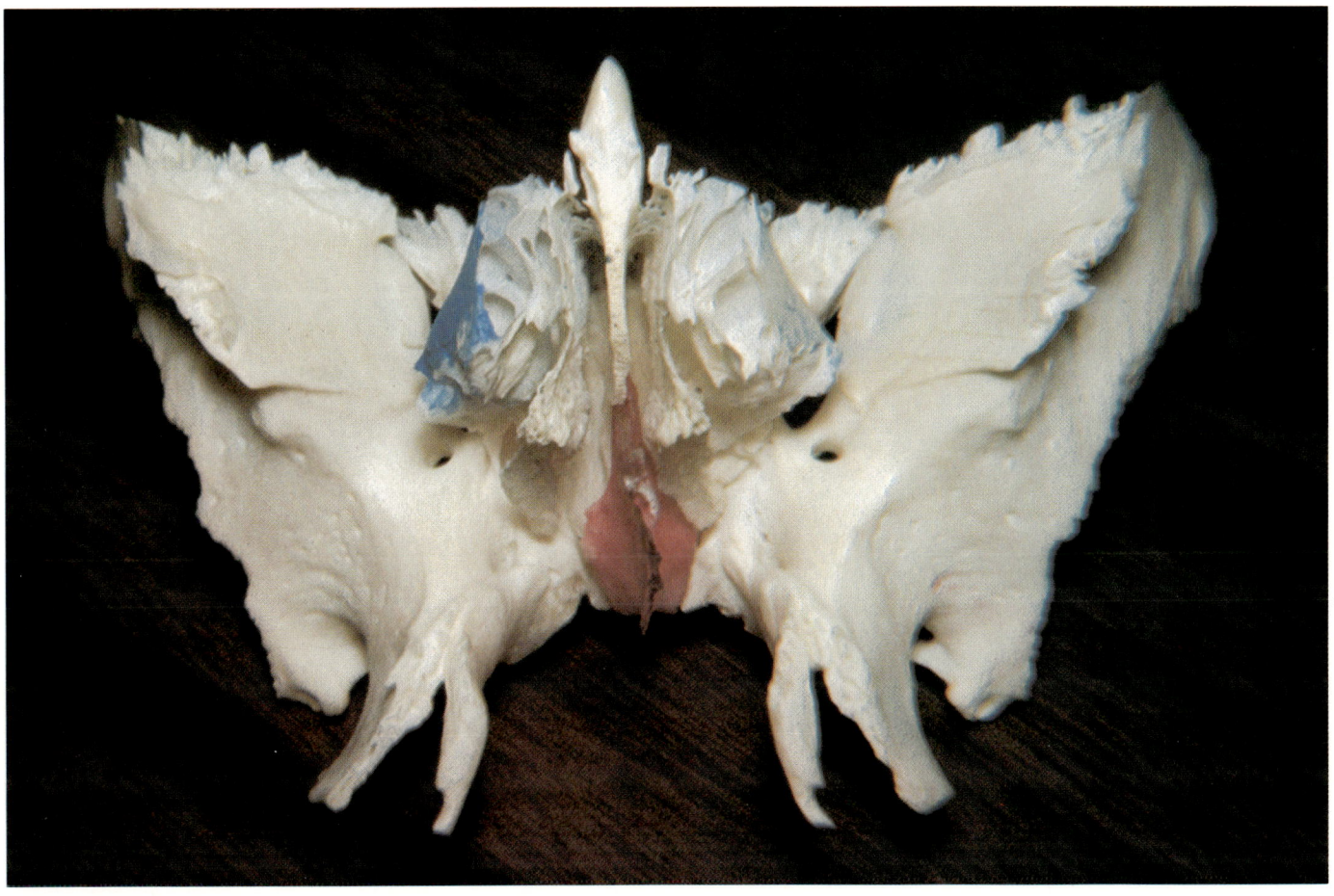

Composite of sphenoid, ethmoid and vomer.

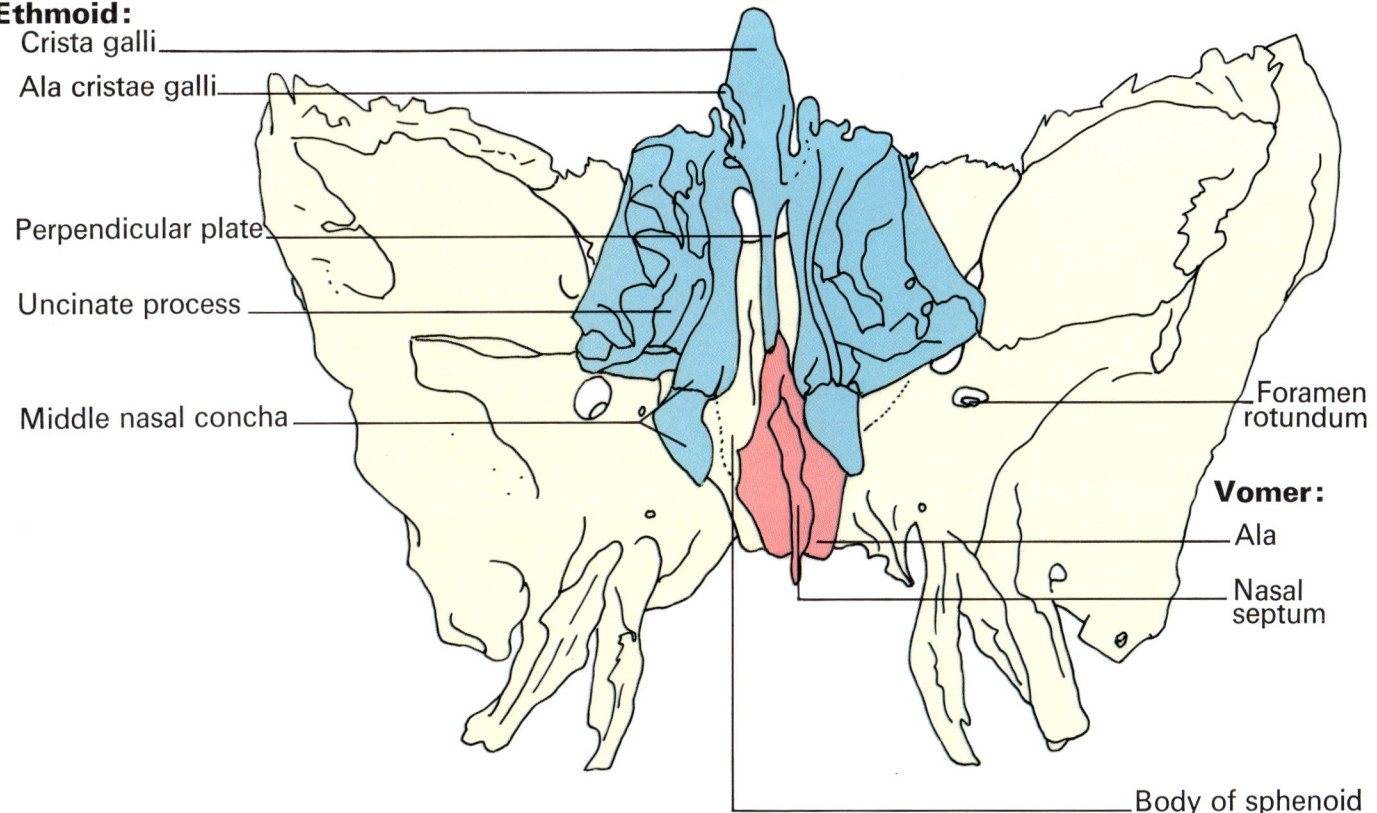

Ethmoid:
- Crista galli
- Ala cristae galli
- Perpendicular plate
- Uncinate process
- Middle nasal concha

- Foramen rotundum

Vomer:
- Ala
- Nasal septum

- Body of sphenoid

Nasal cavity.

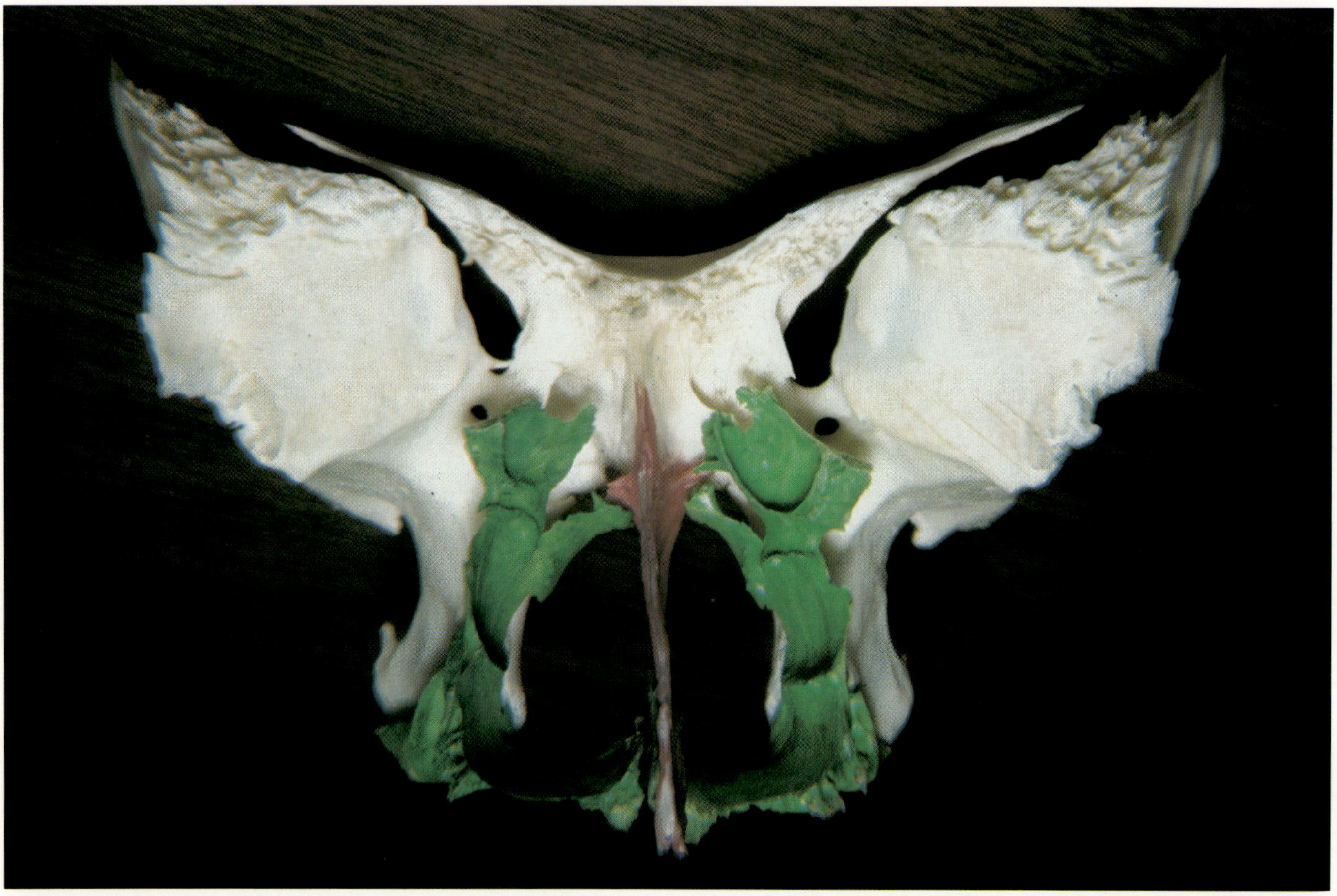

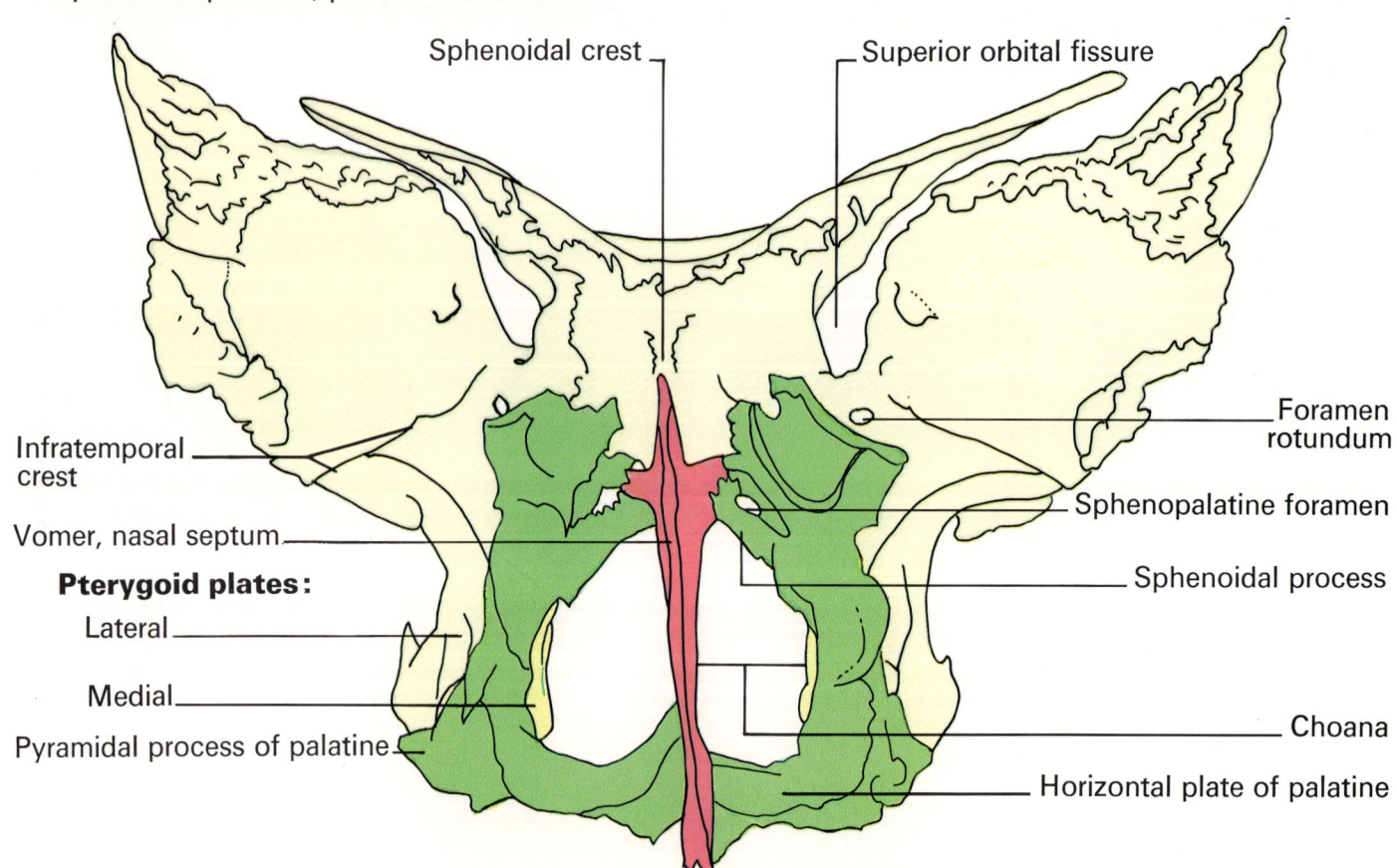

Composite of sphenoid, palatines and vomer.

- Sphenoidal crest
- Superior orbital fissure
- Infratemporal crest
- Vomer, nasal septum
- **Pterygoid plates:**
  - Lateral
  - Medial
- Pyramidal process of palatine
- Foramen rotundum
- Sphenopalatine foramen
- Sphenoidal process
- Choana
- Horizontal plate of palatine

Nasal cavity.

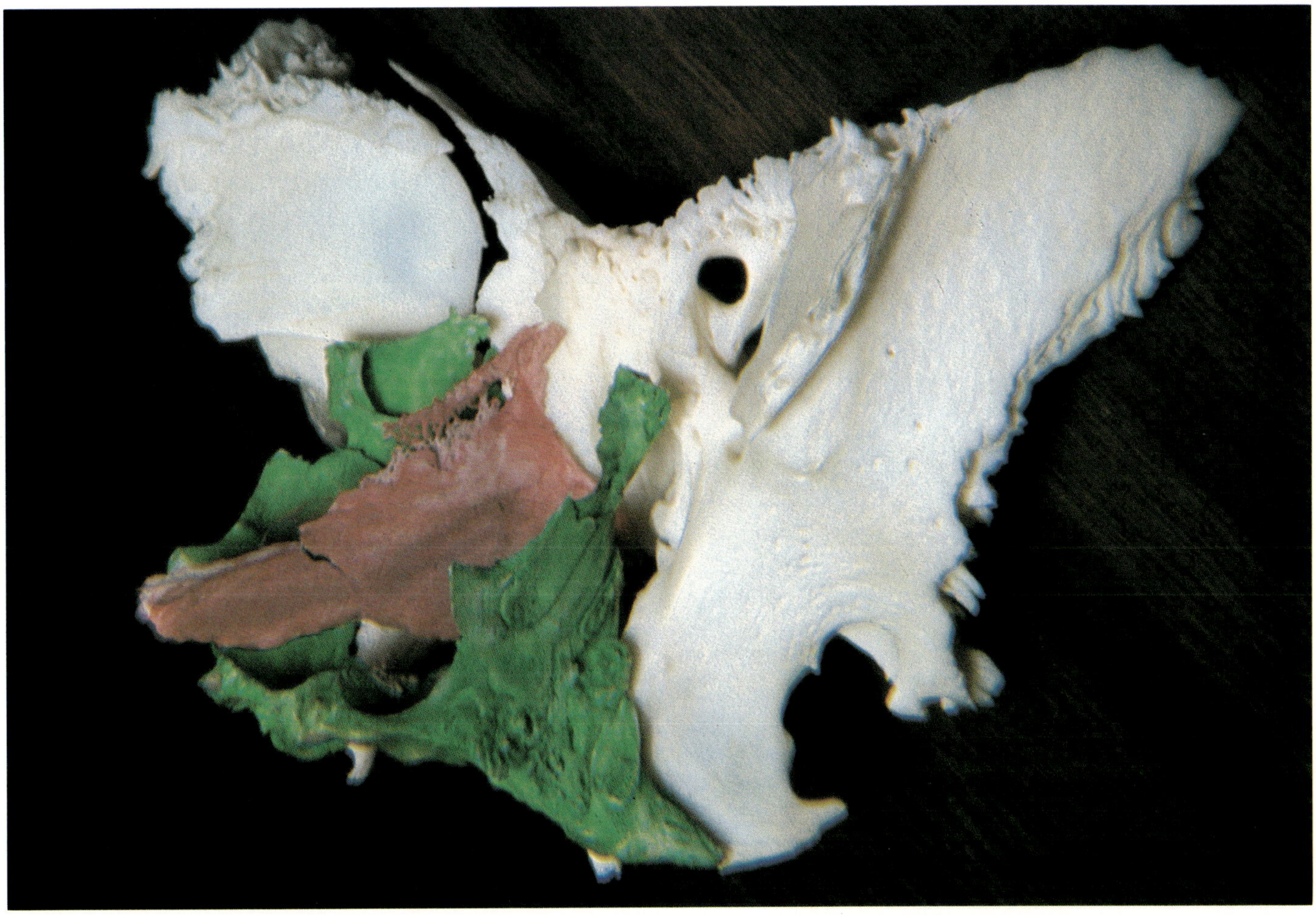

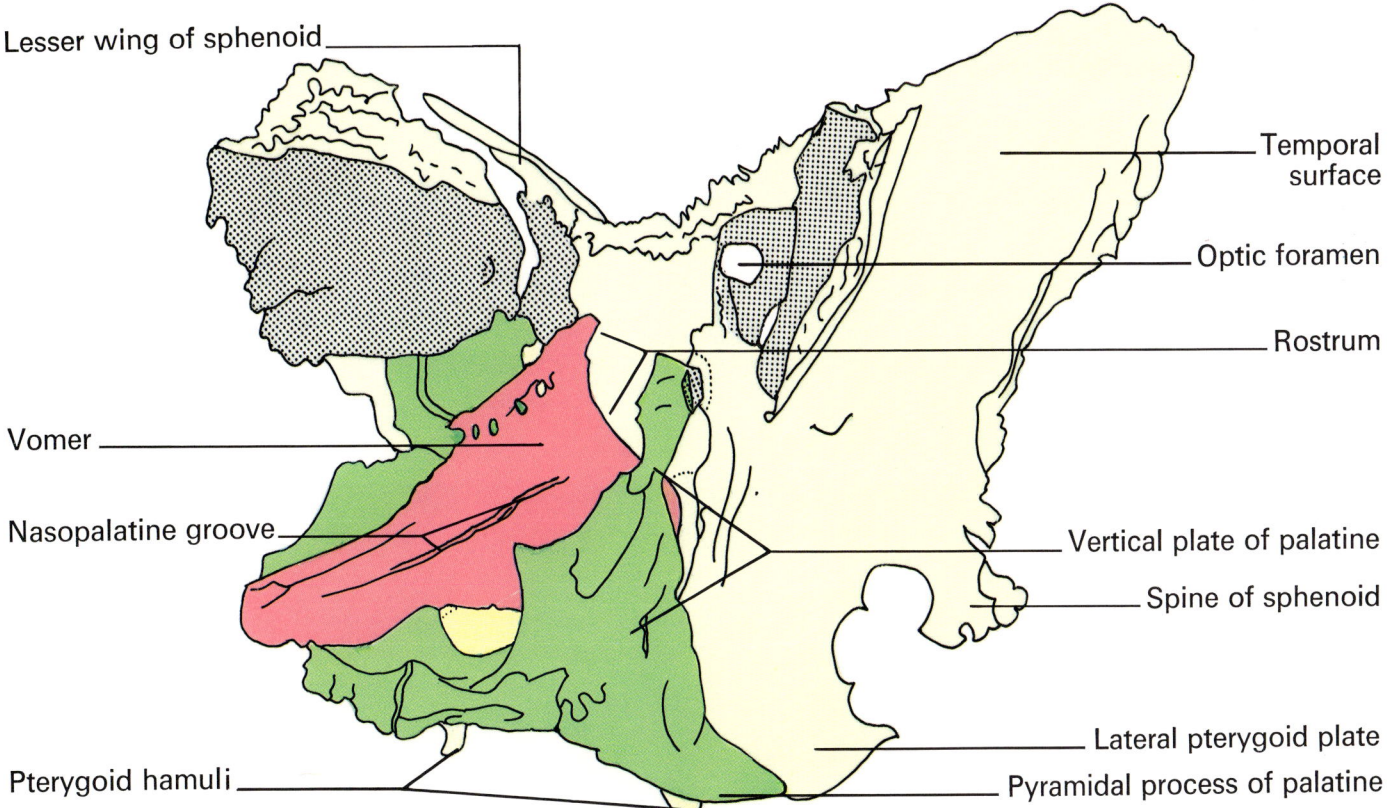

Oblique view. Composite of sphenoid, palatines and vomer. Orbital surfaces stippled.

Nasal cavity.

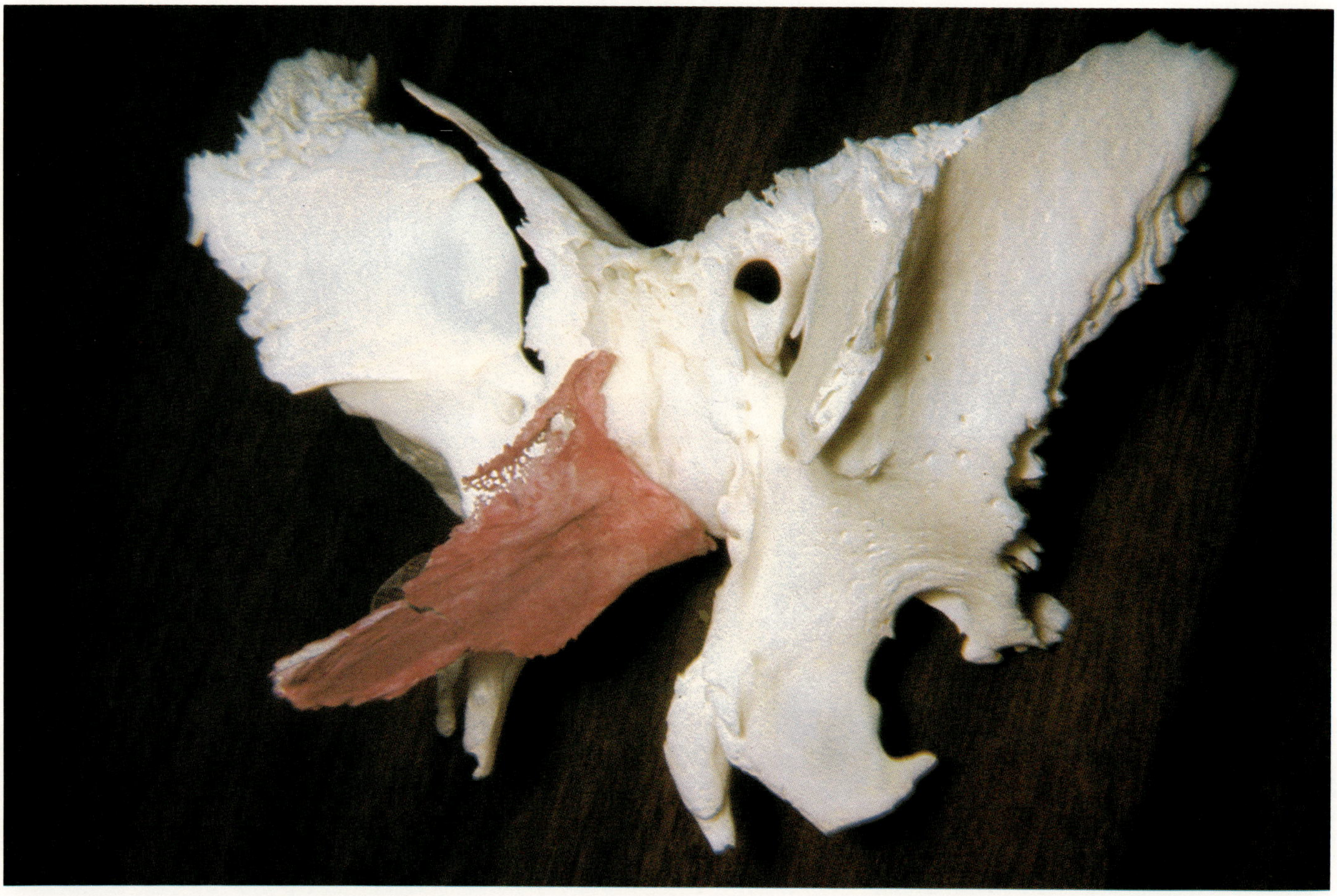

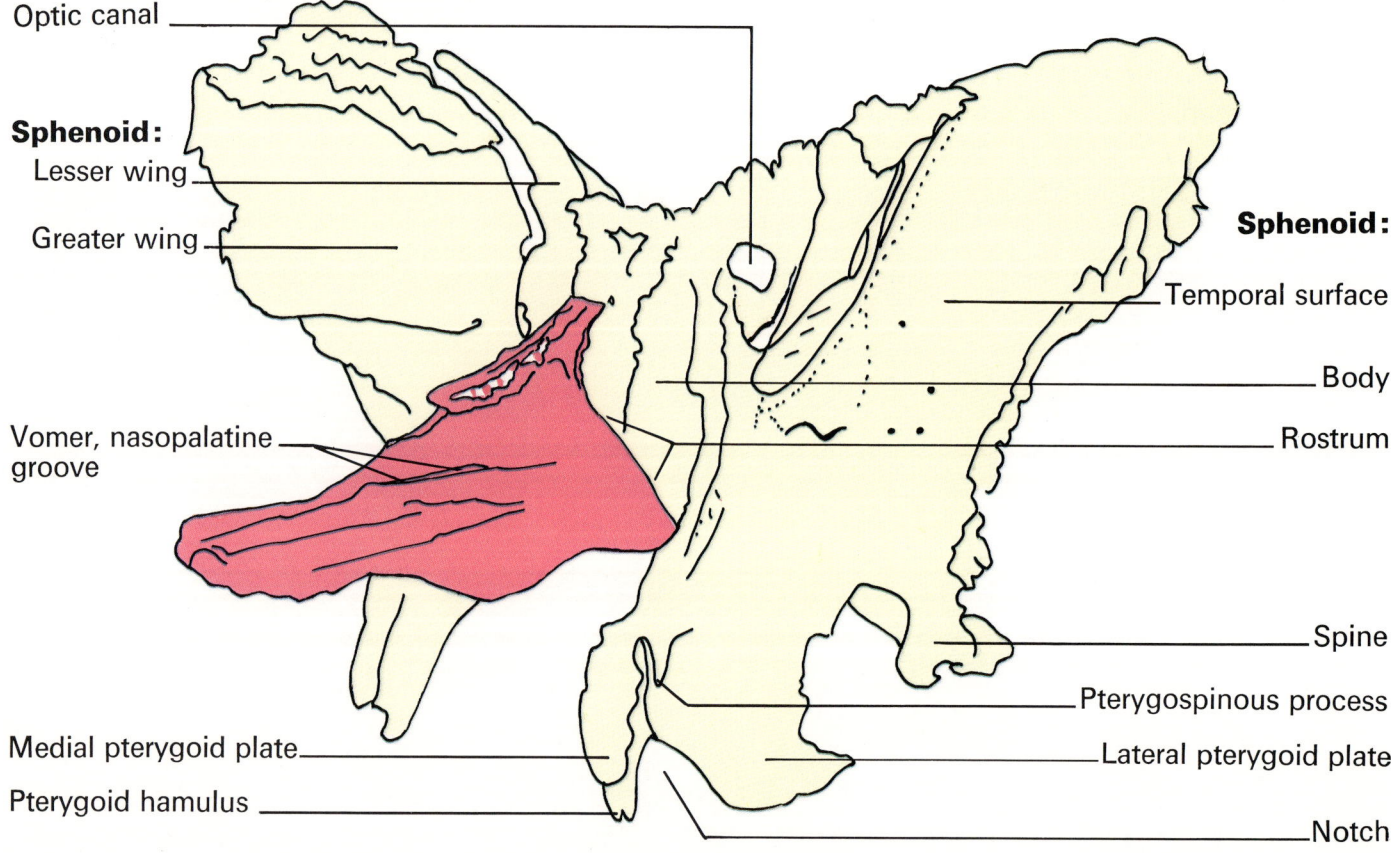

Composite of sphenoid and vomer.

Nasal cavity.

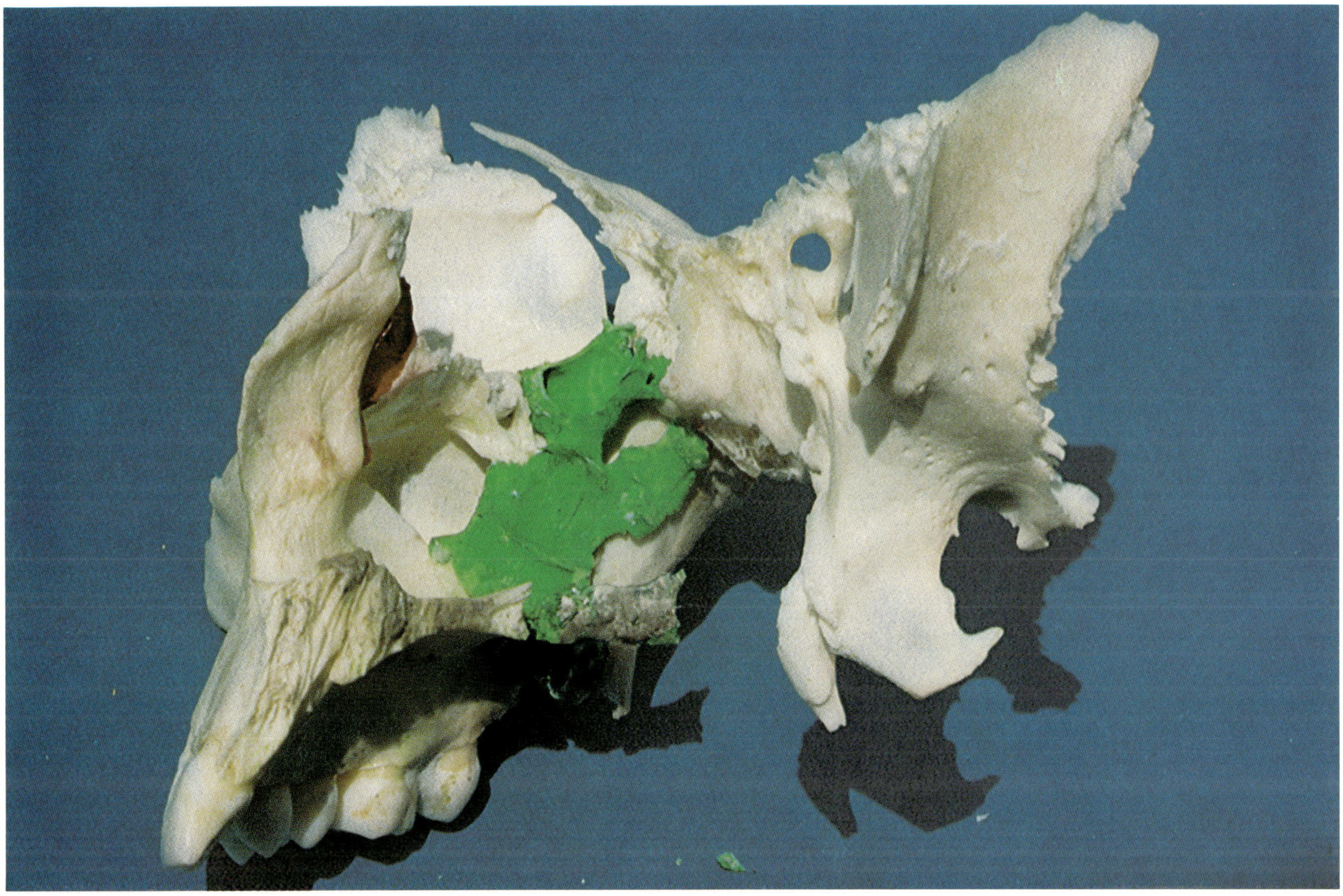

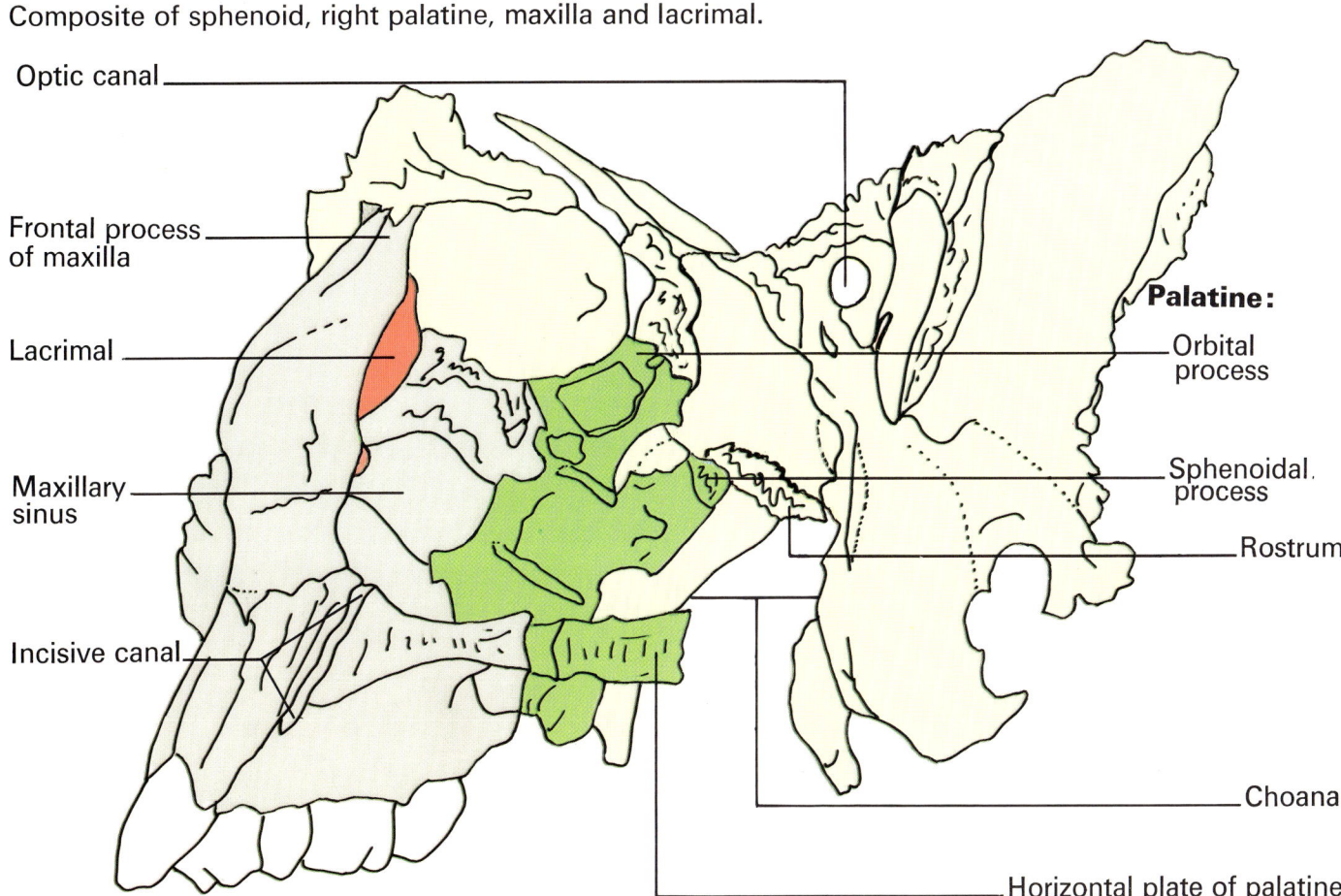

Composite of sphenoid, right palatine, maxilla and lacrimal.

Nasal cavity.

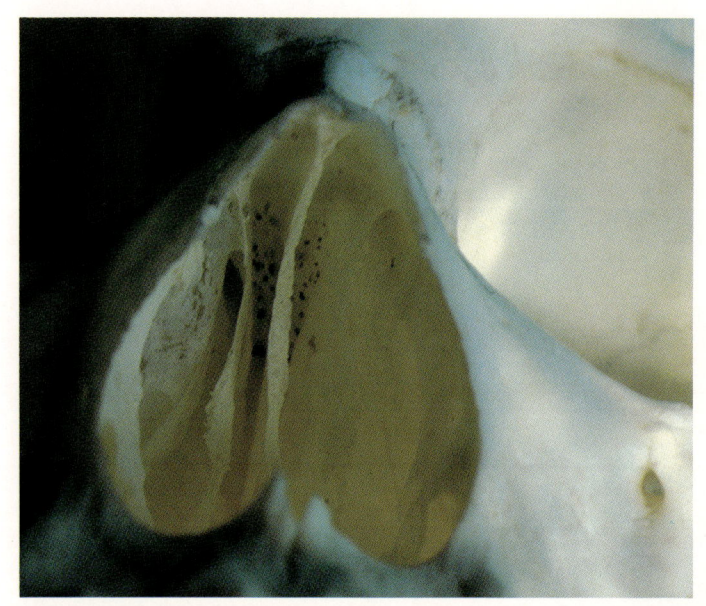
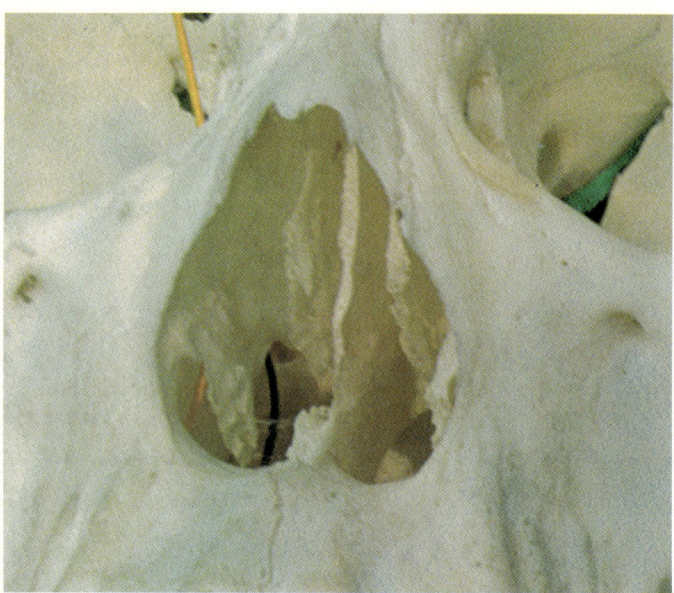

From below.

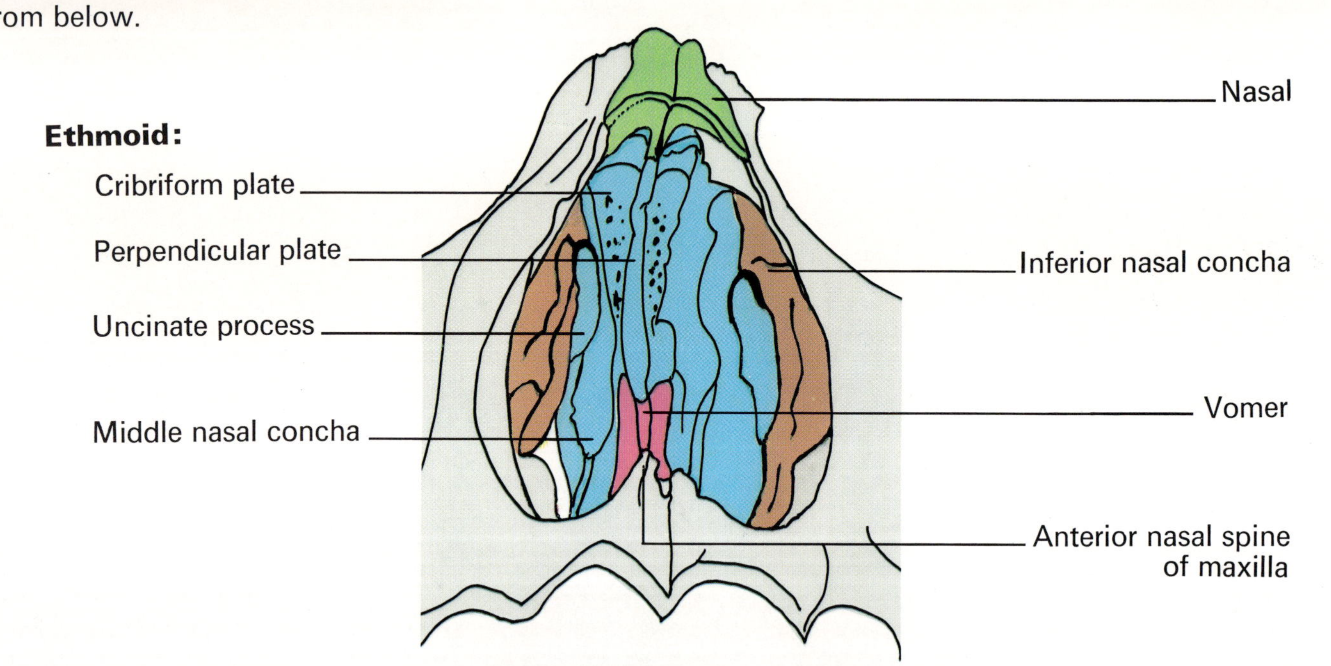

Ethmoid:
- Cribriform plate
- Perpendicular plate
- Uncinate process
- Middle nasal concha

- Nasal
- Inferior nasal concha
- Vomer
- Anterior nasal spine of maxilla

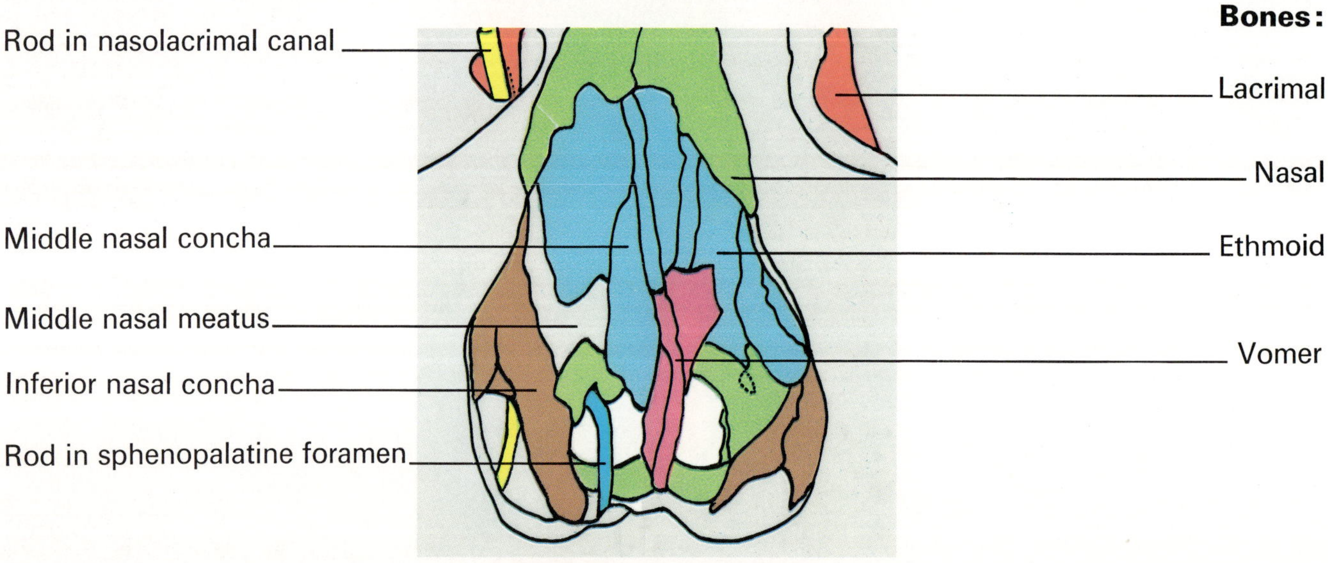

- Rod in nasolacrimal canal
- Middle nasal concha
- Middle nasal meatus
- Inferior nasal concha
- Rod in sphenopalatine foramen

Bones:
- Lacrimal
- Nasal
- Ethmoid
- Vomer

Nasal cavity.

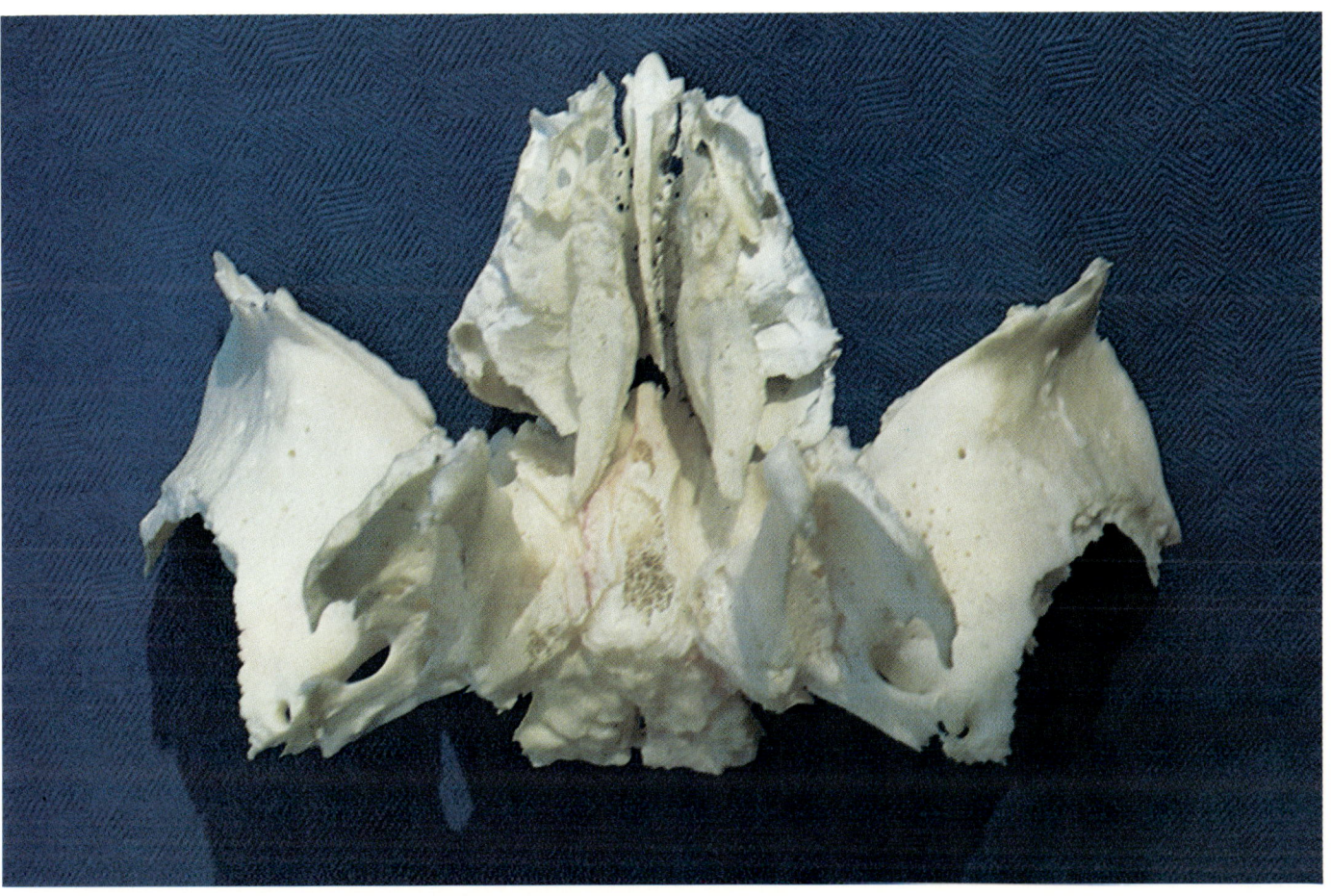

From below. Composite of sphenoid and ethmoid.

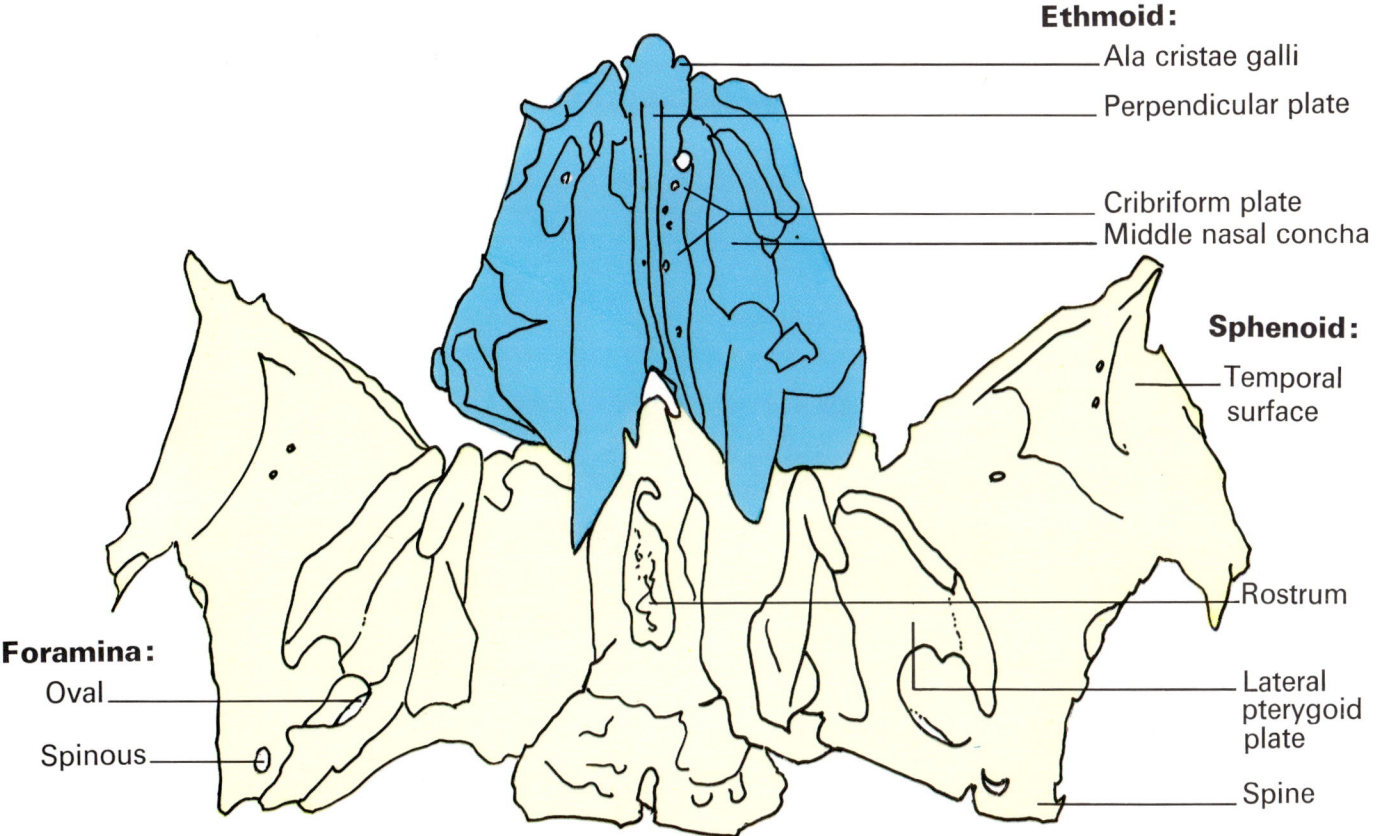

Ethmoid:
- Ala cristae galli
- Perpendicular plate
- Cribriform plate
- Middle nasal concha

Sphenoid:
- Temporal surface
- Rostrum
- Lateral pterygoid plate
- Spine

Foramina:
- Oval
- Spinous

Nasal cavity.

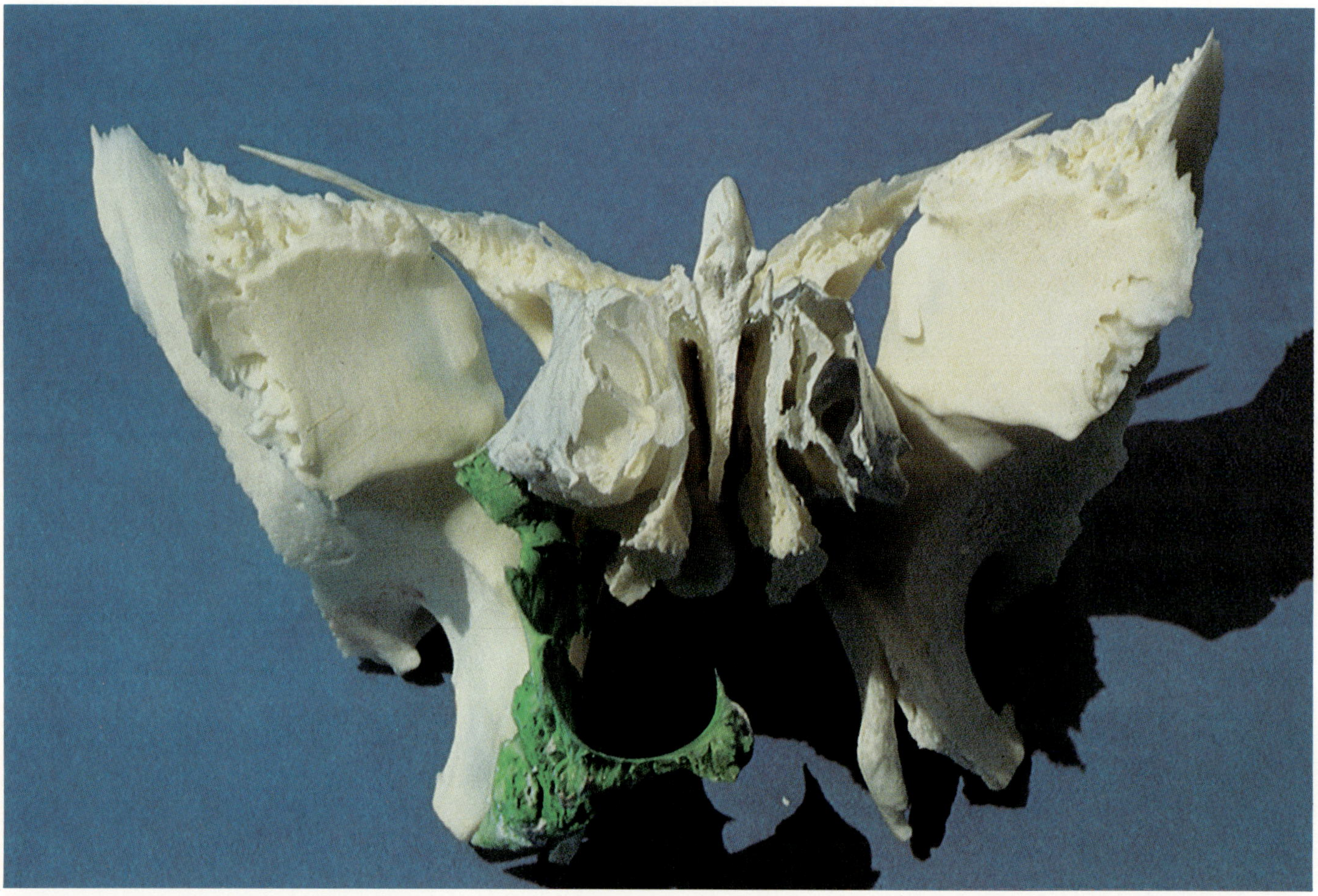

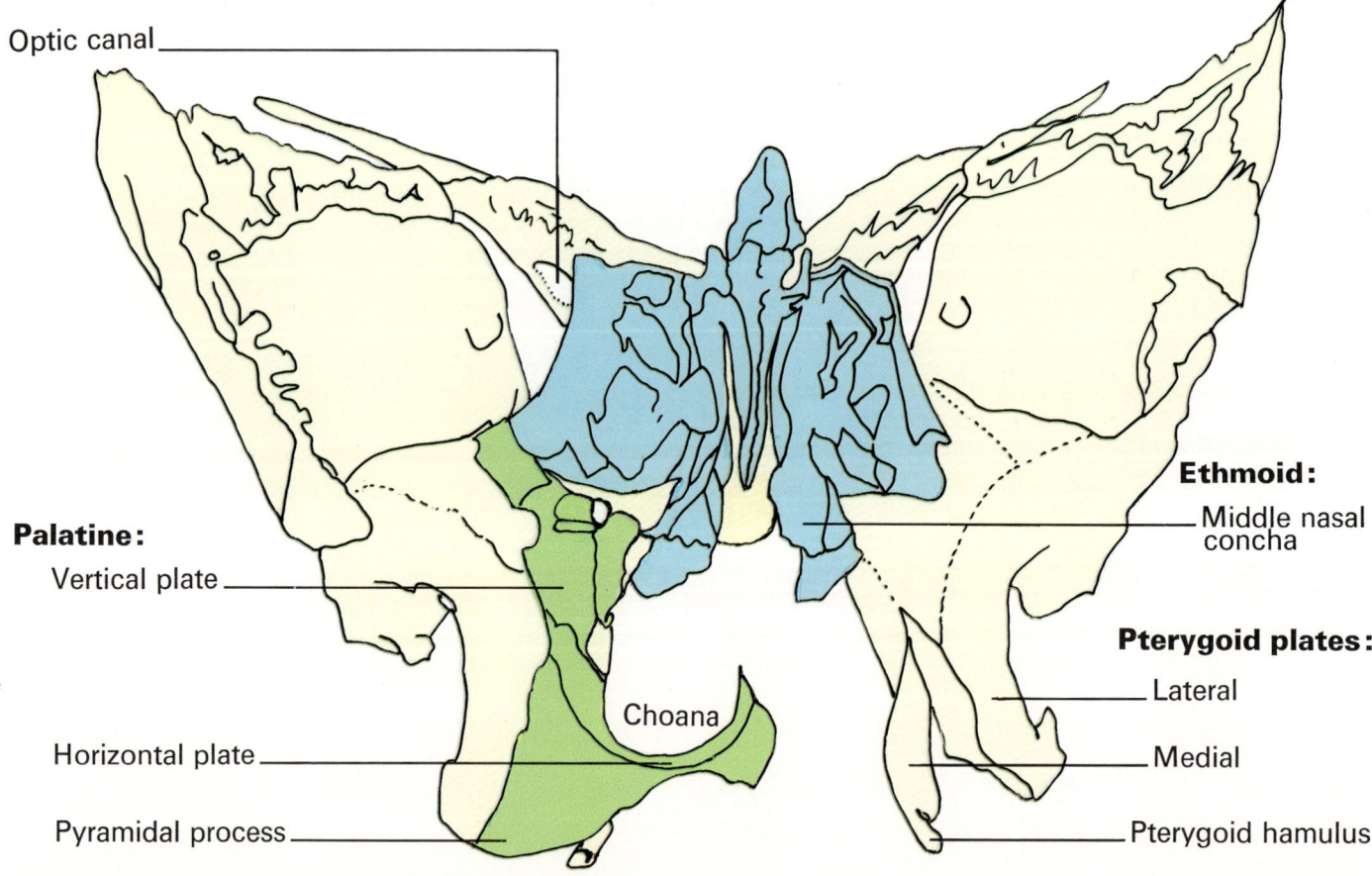

From front. Composite of sphenoid, ethmoid and right palatine.

- Optic canal

Palatine:
- Vertical plate
- Horizontal plate
- Pyramidal process

- Choana

Ethmoid:
- Middle nasal concha

Pterygoid plates:
- Lateral
- Medial
- Pterygoid hamulus

Nasal cavity.

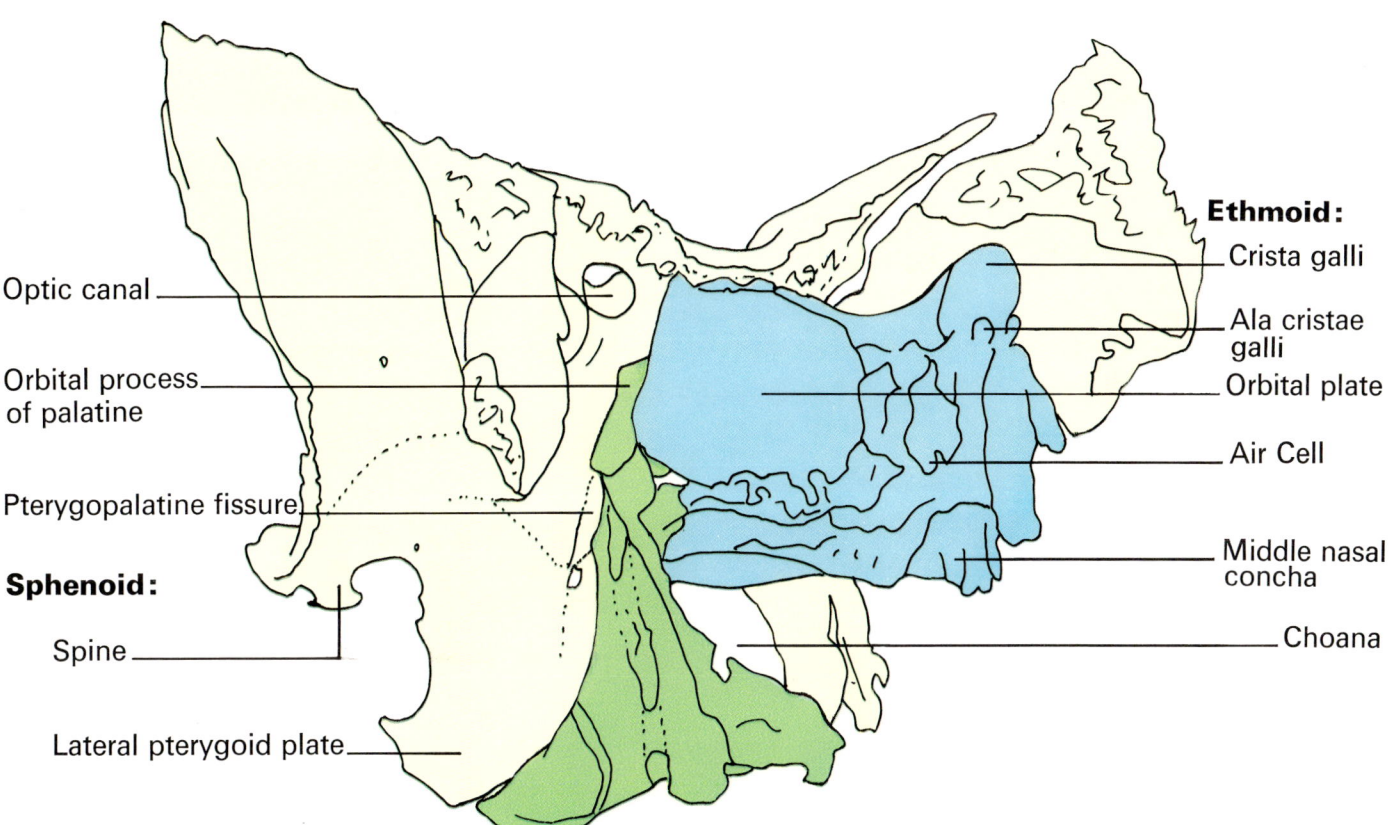

Oblique view. Composite of sphenoid, ethmoid and right palatine.

Sphenoid:
- Optic canal
- Orbital process of palatine
- Pterygopalatine fissure
- Spine
- Lateral pterygoid plate

Ethmoid:
- Crista galli
- Ala cristae galli
- Orbital plate
- Air Cell
- Middle nasal concha
- Choana

Nasal cavity.

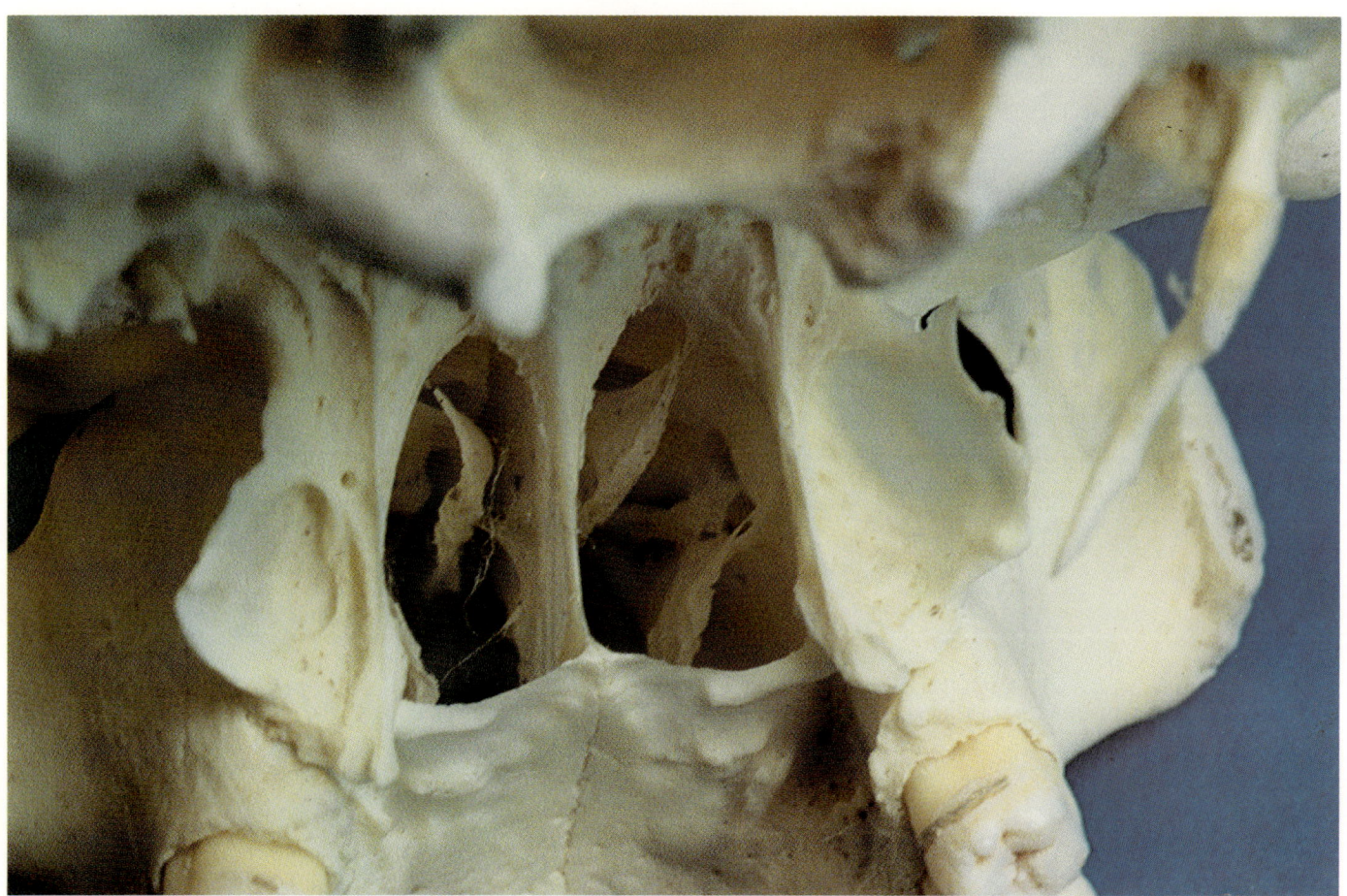

From behind, and slightly below. Skull.

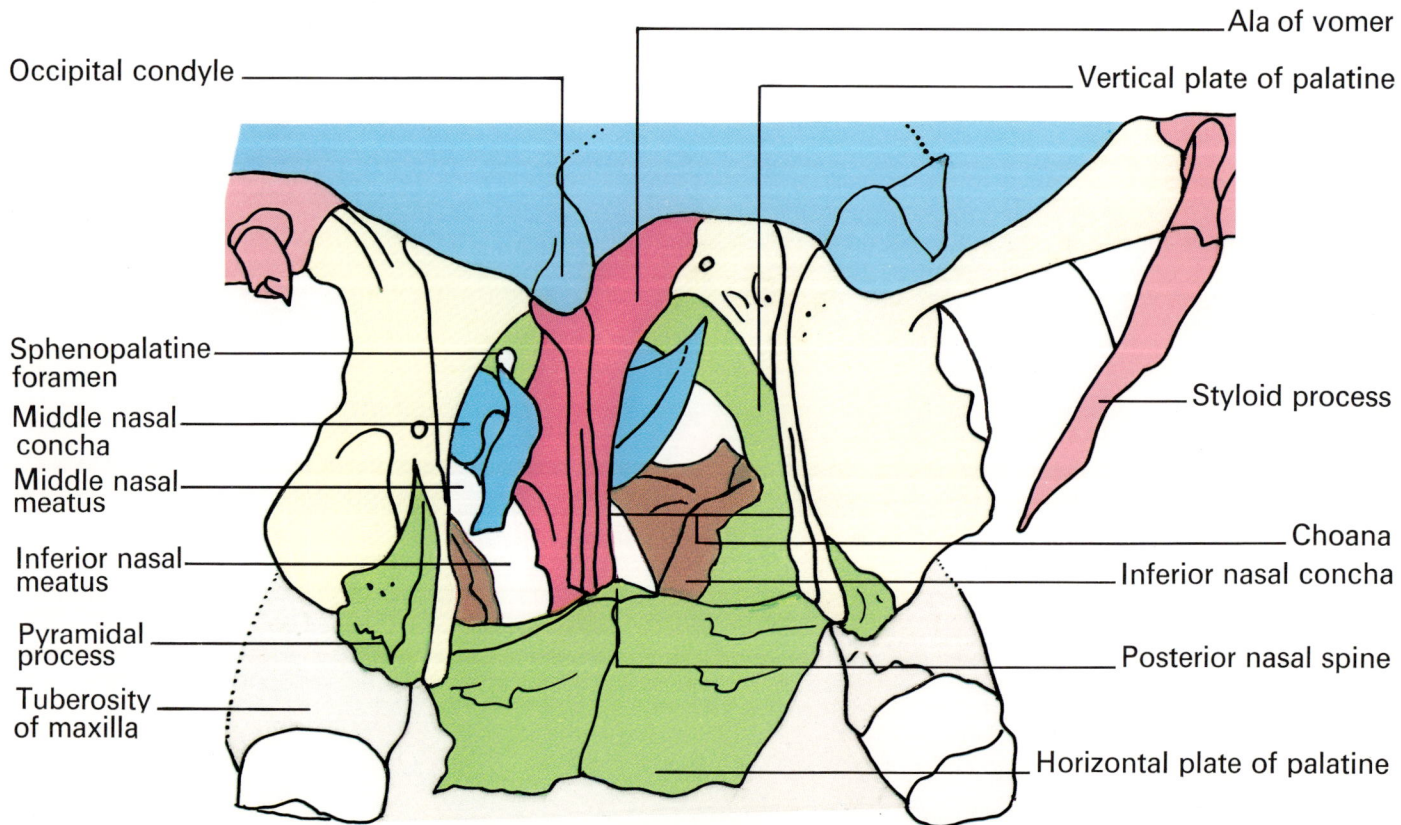

Nasal cavity.

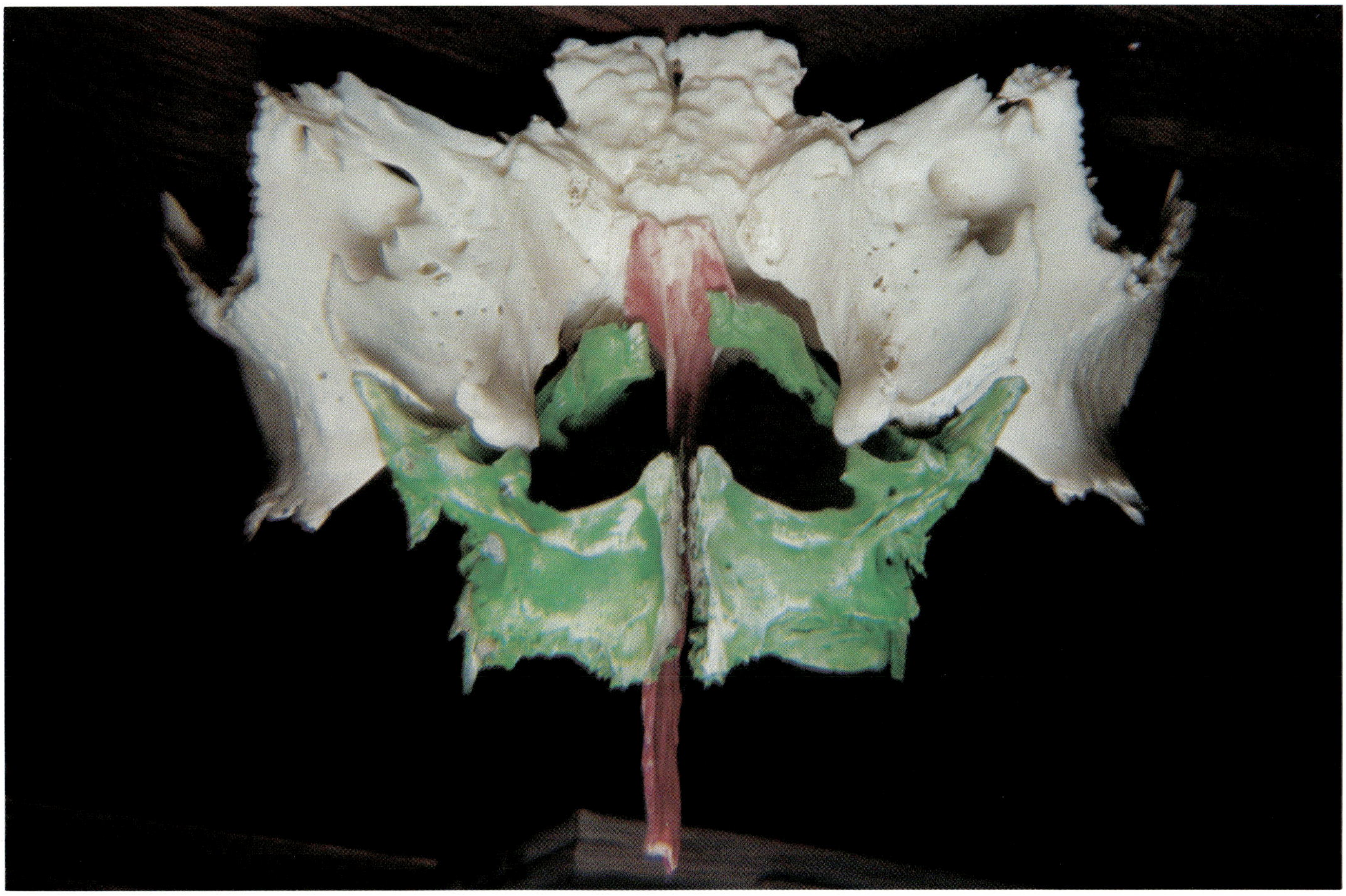

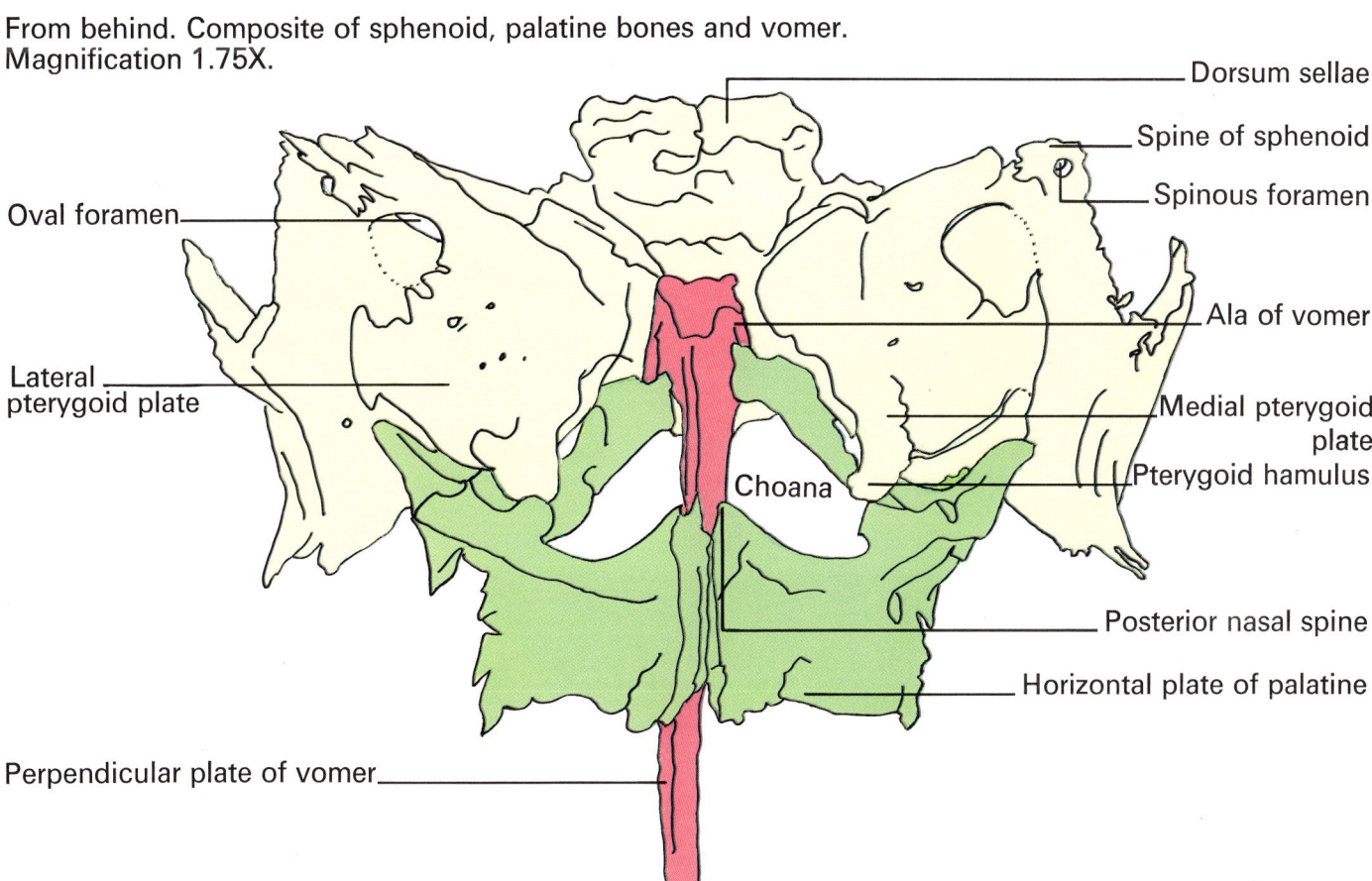

From behind. Composite of sphenoid, palatine bones and vomer. Magnification 1.75X.

Nasal cavity.

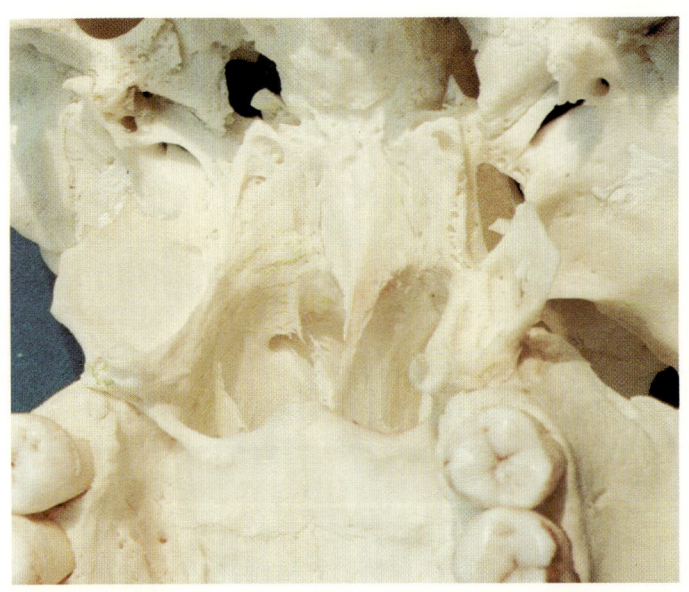
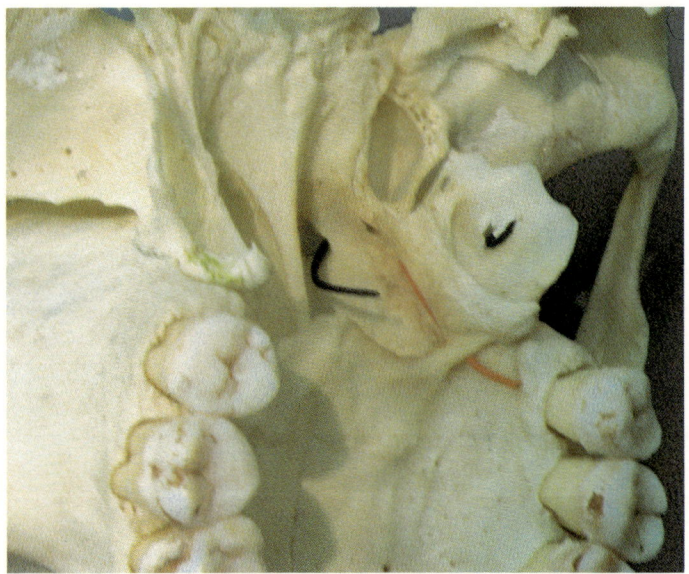

From behind.

Foramina:
- Spinous
- Oval
- Rotundum

- Ala of vomer
- Sphenopalatine foramen
- Maxilla
- Inferior orbital fissure
- Pterygoid hamulus
- Horizontal plate of palatine

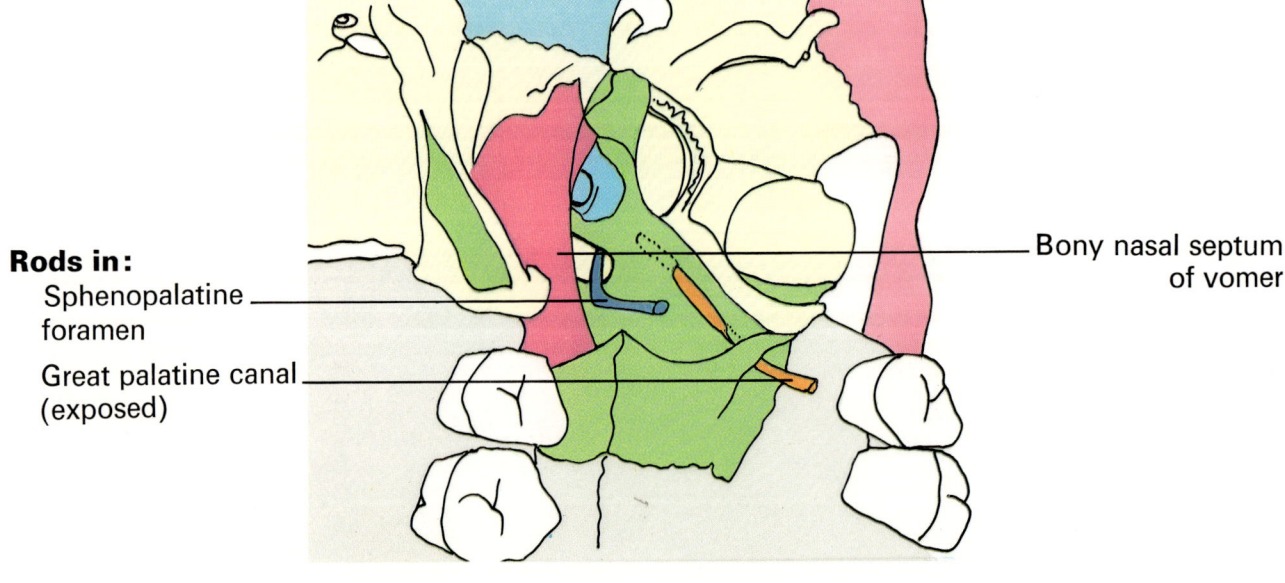

Rods in:
- Sphenopalatine foramen
- Great palatine canal (exposed)
- Bony nasal septum of vomer

Nasal cavity.

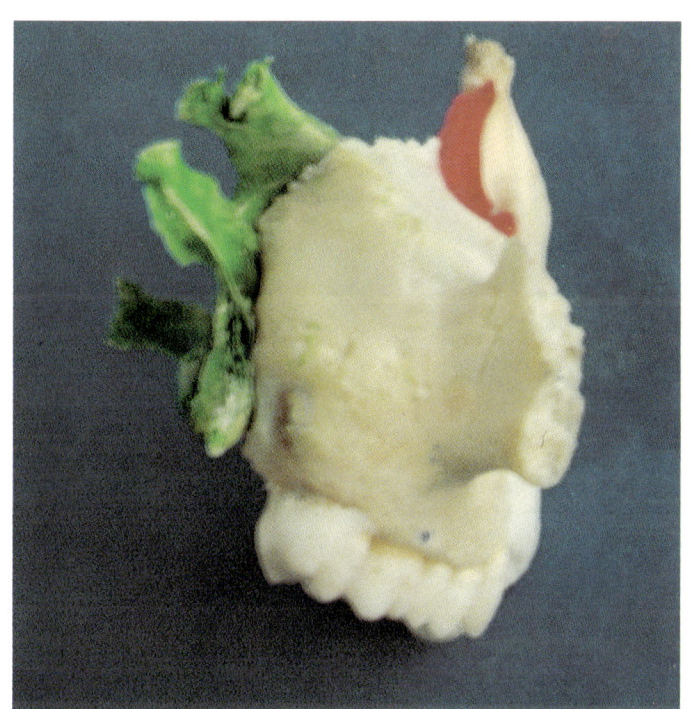
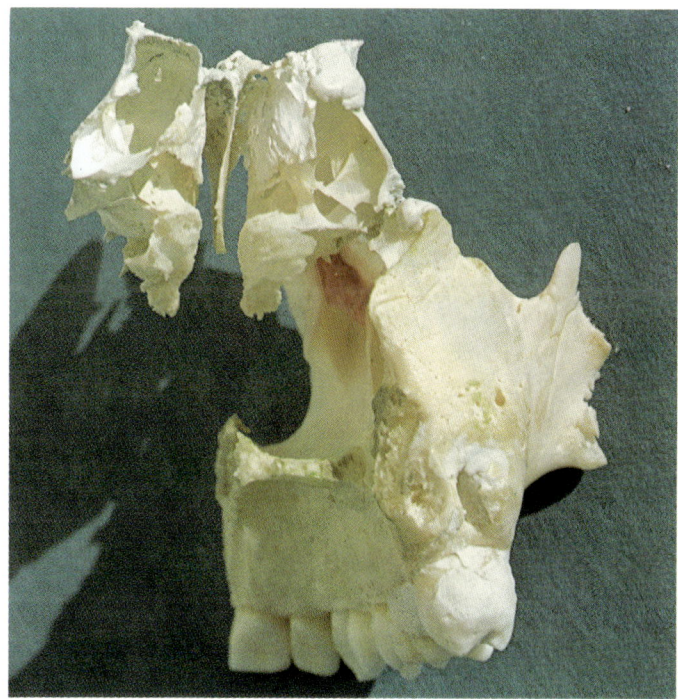

From behind.

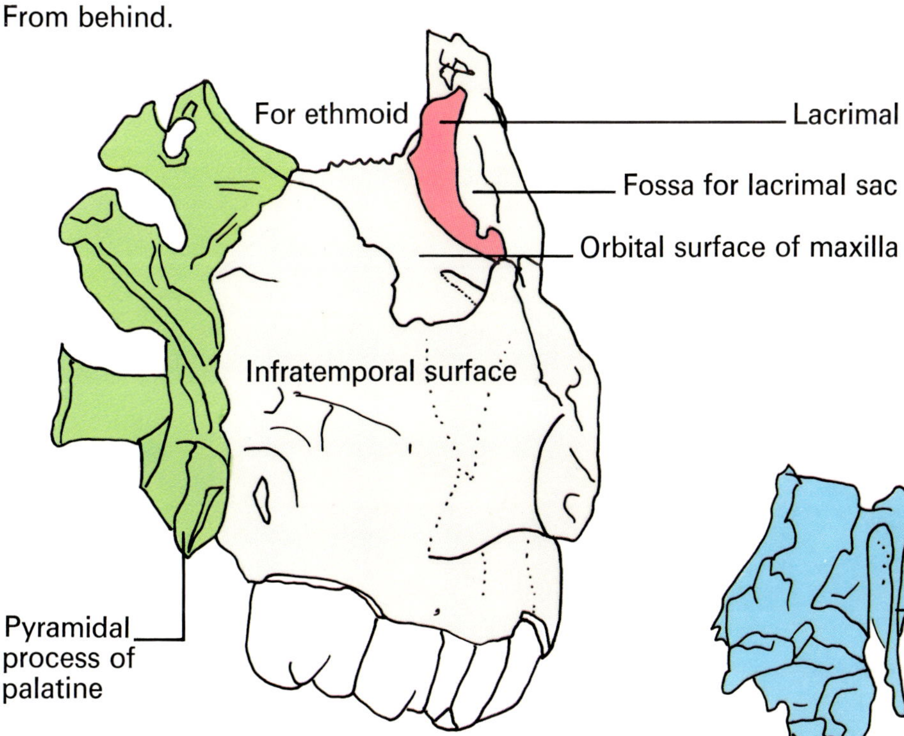

- For ethmoid
- Lacrimal
- Fossa for lacrimal sac
- Orbital surface of maxilla
- Infratemporal surface
- Pyramidal process of palatine

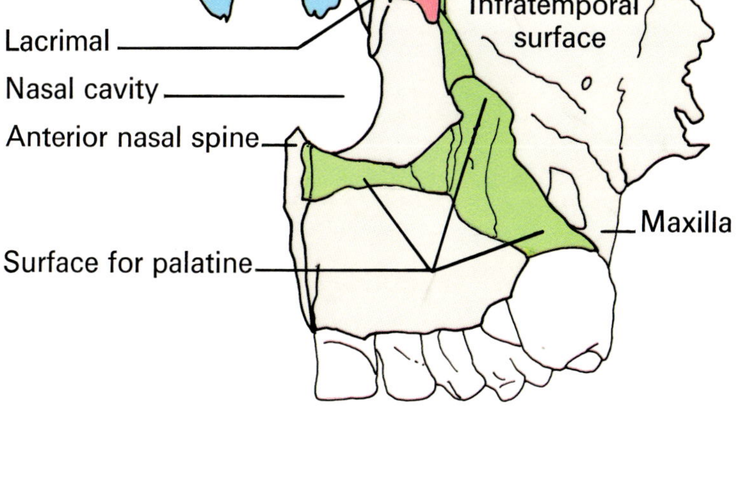

Ethmoid:
- Orbital plate
- Perpendicular plate
- Lacrimal
- Nasal cavity
- Anterior nasal spine
- Infratemporal surface
- Maxilla
- Surface for palatine

SECTION 11. Pterygopalatine fossa.

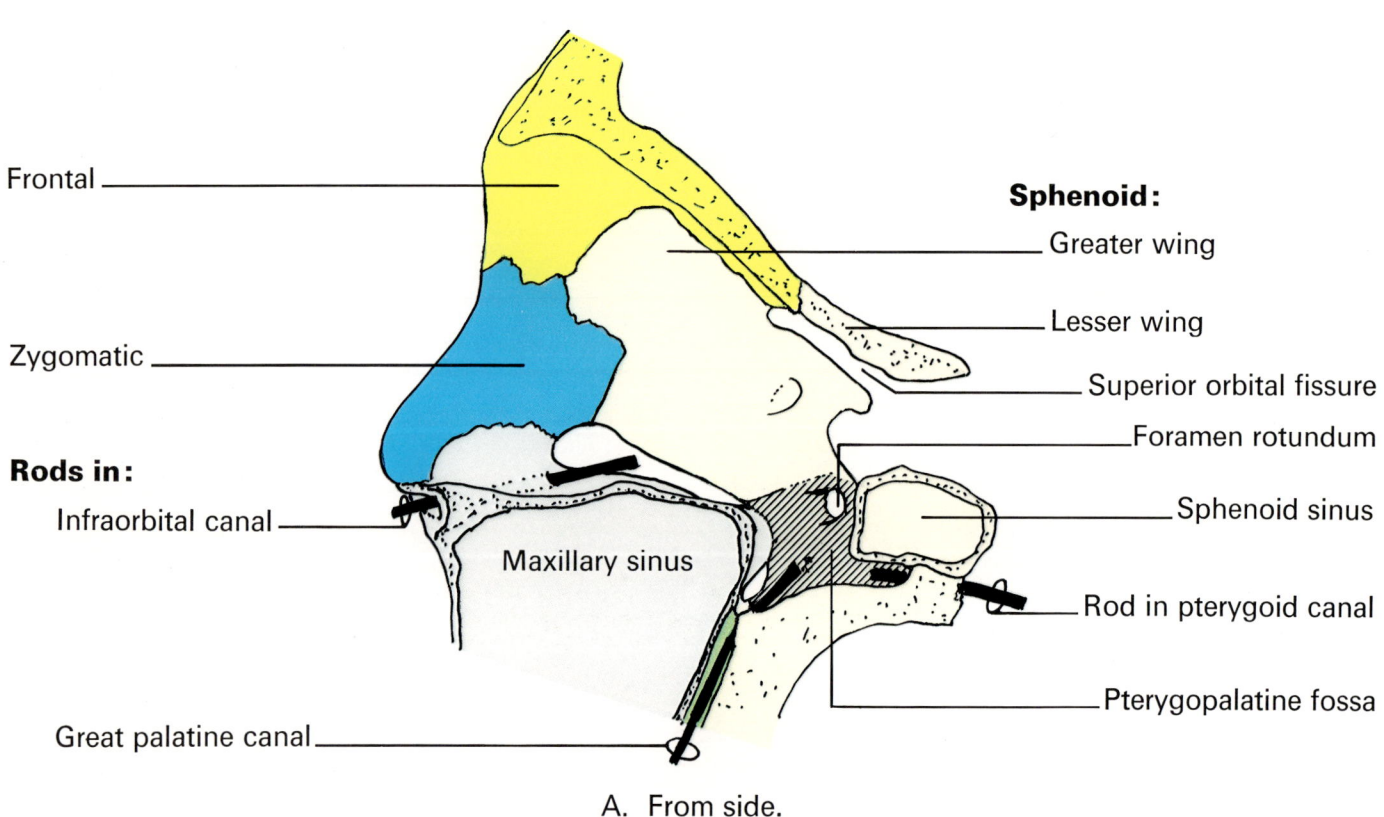

A. From side.

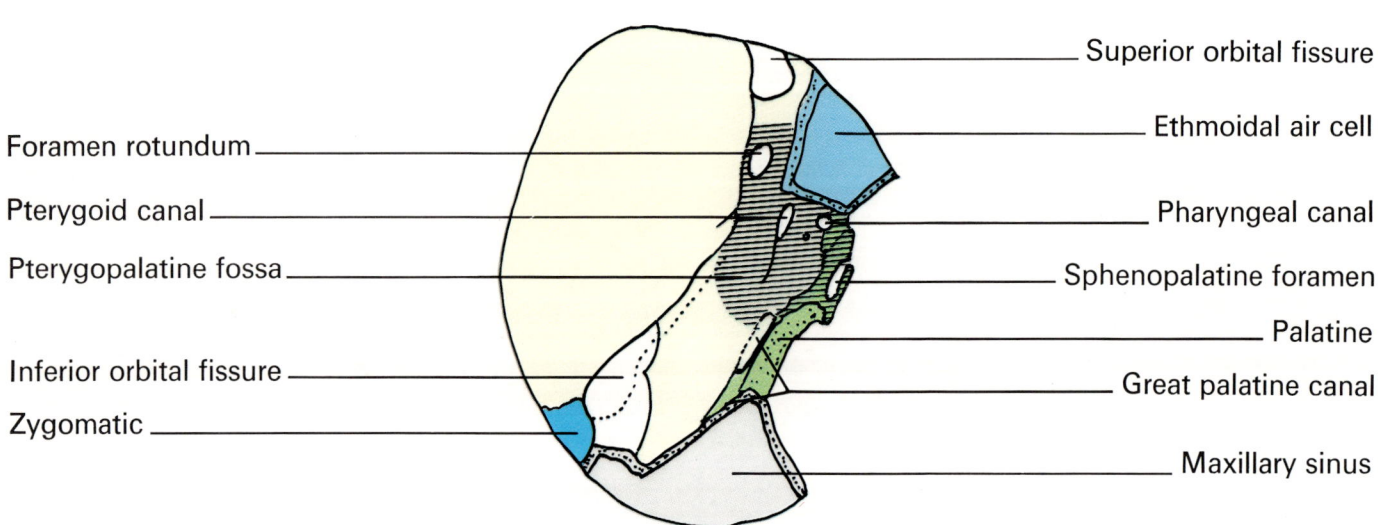

Pterygopalatine fossa.

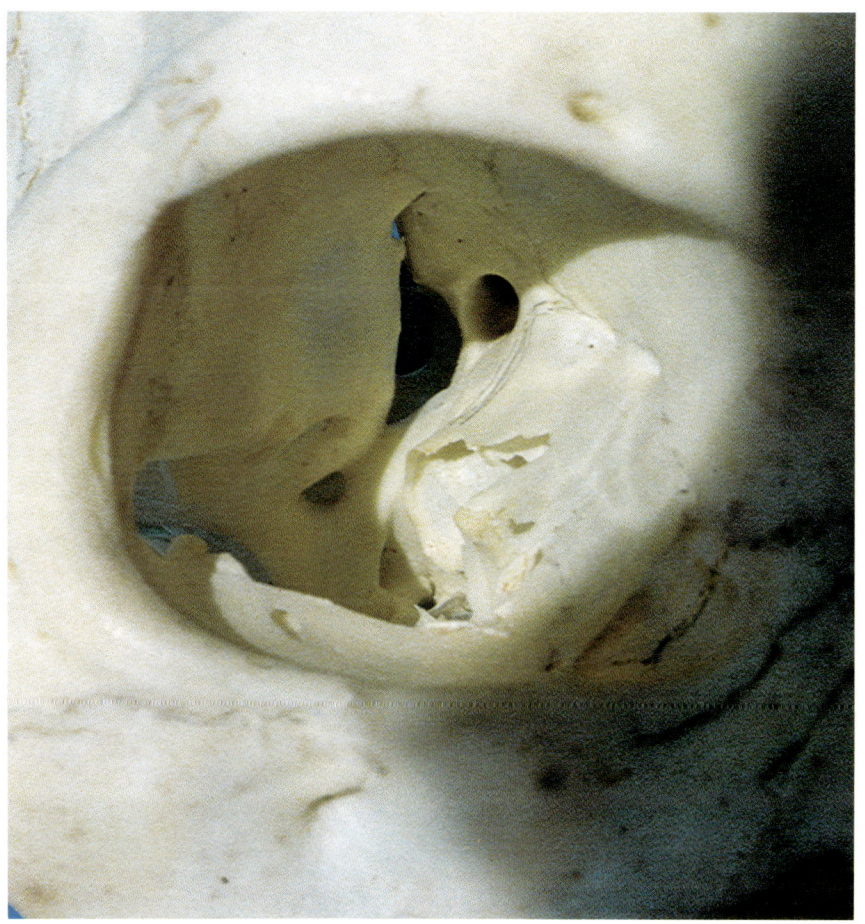

View through right orbital cavity with partial resection of maxilla, ethmoid and palatine.

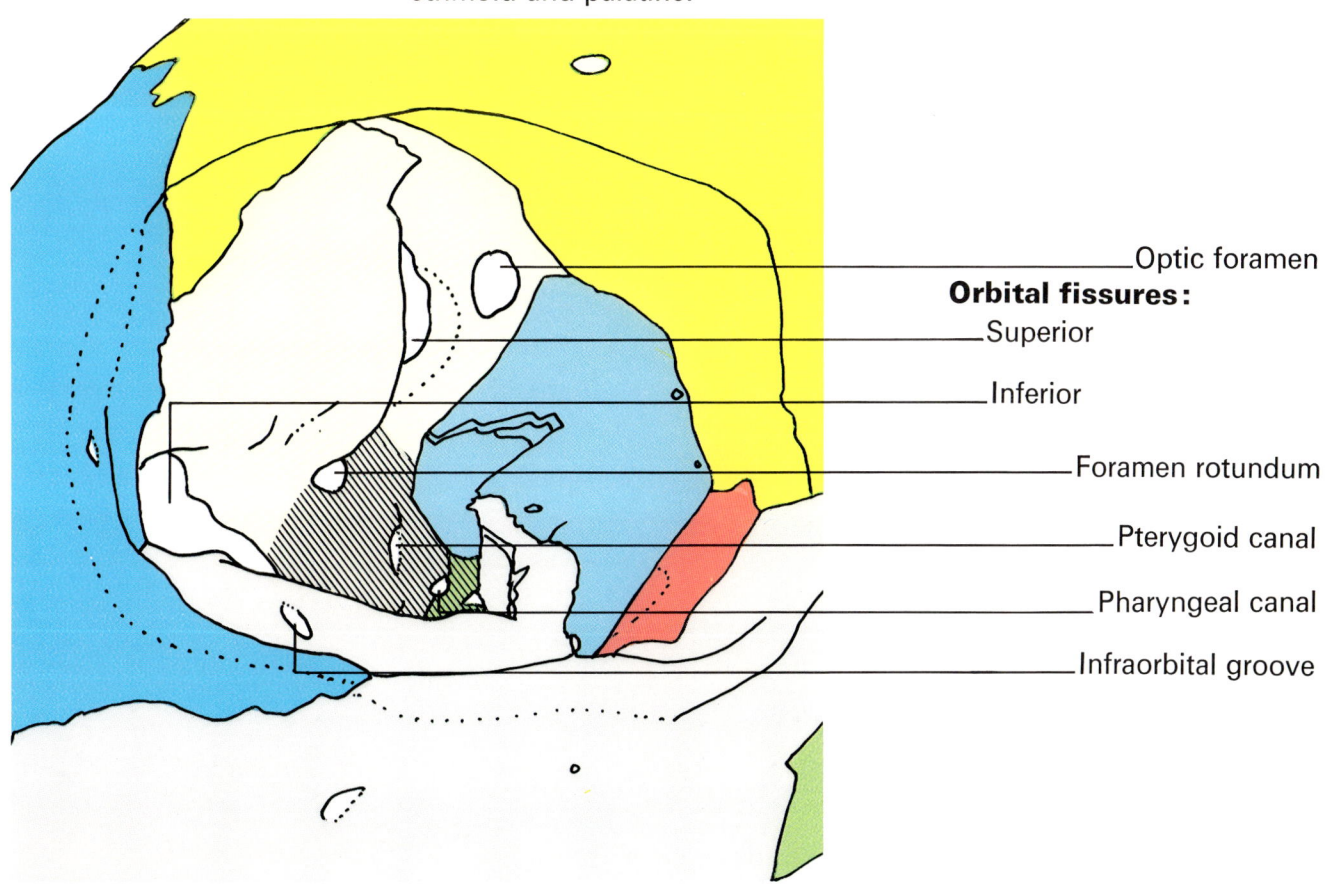

- Optic foramen
- **Orbital fissures:**
  - Superior
  - Inferior
- Foramen rotundum
- Pterygoid canal
- Pharyngeal canal
- Infraorbital groove

Pterygopalatine fossa.

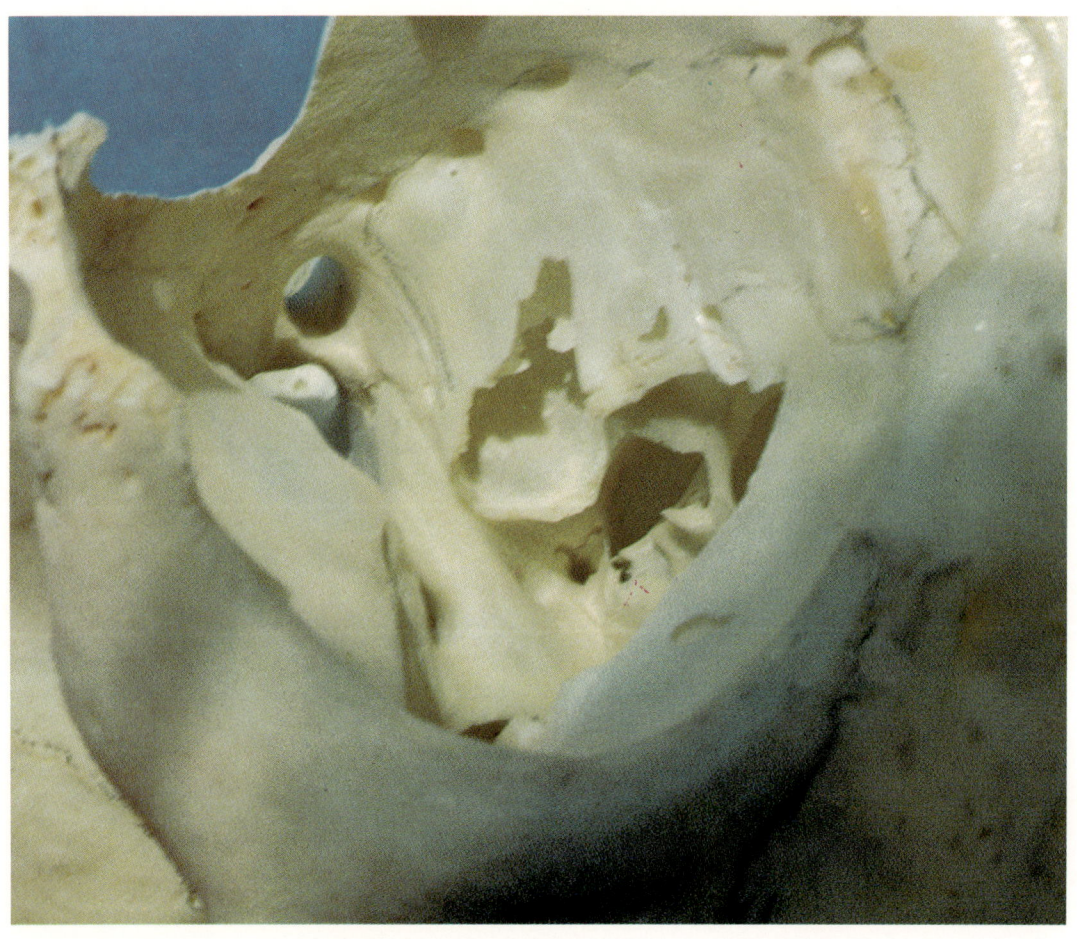

Further resection. View slightly more lateral.

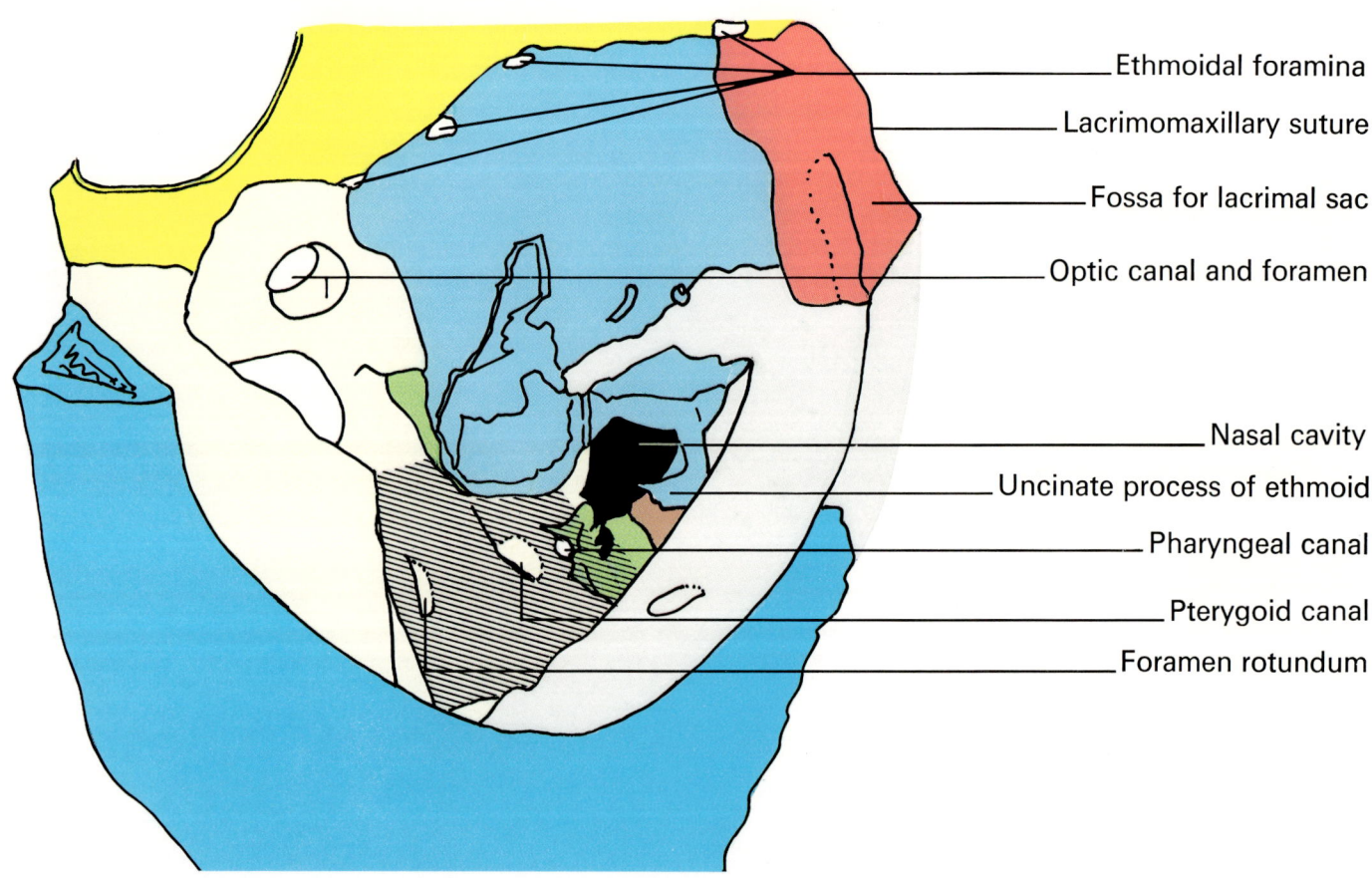

- Ethmoidal foramina
- Lacrimomaxillary suture
- Fossa for lacrimal sac
- Optic canal and foramen
- Nasal cavity
- Uncinate process of ethmoid
- Pharyngeal canal
- Pterygoid canal
- Foramen rotundum

Pterygopalatine fossa.

A. Rods in foramina.

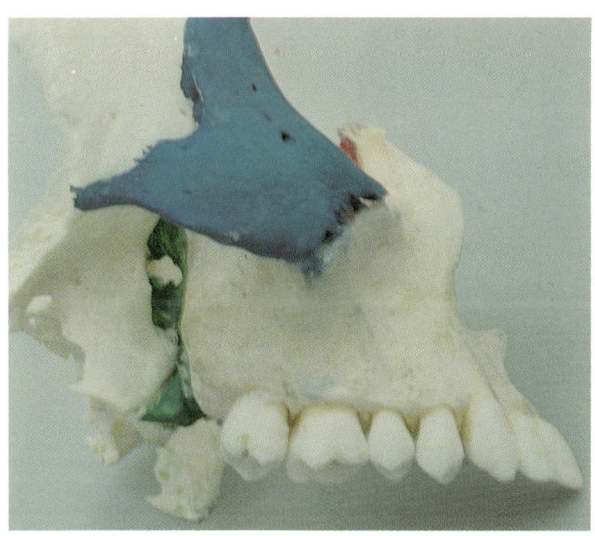

B. From side. Composite of sphenoid right maxilla, palatine and zygomatic.

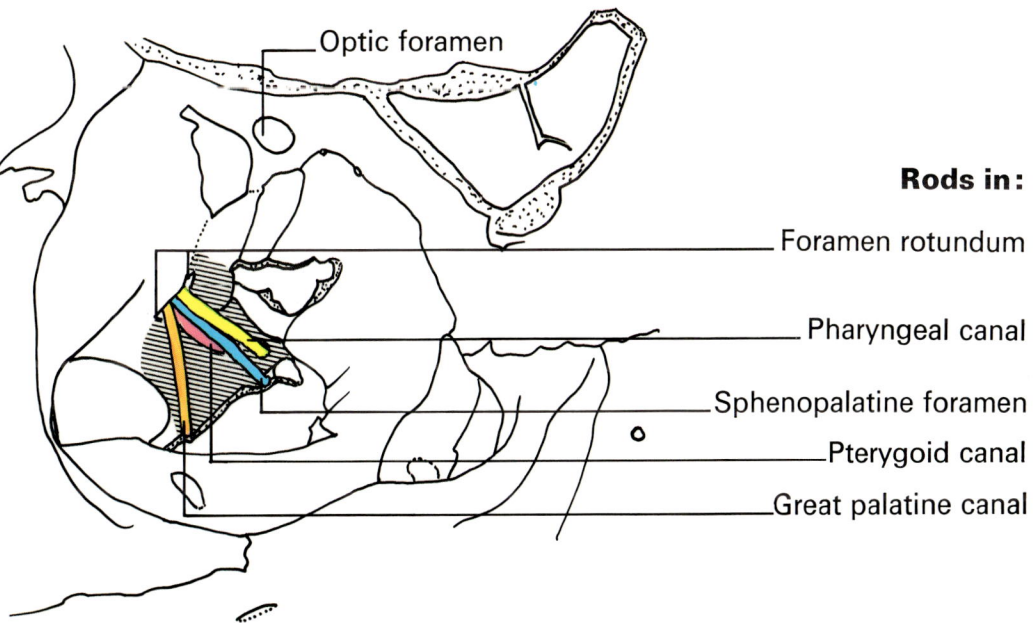

Rods in:

Foramen rotundum
Pharyngeal canal
Sphenopalatine foramen
Pterygoid canal
Great palatine canal

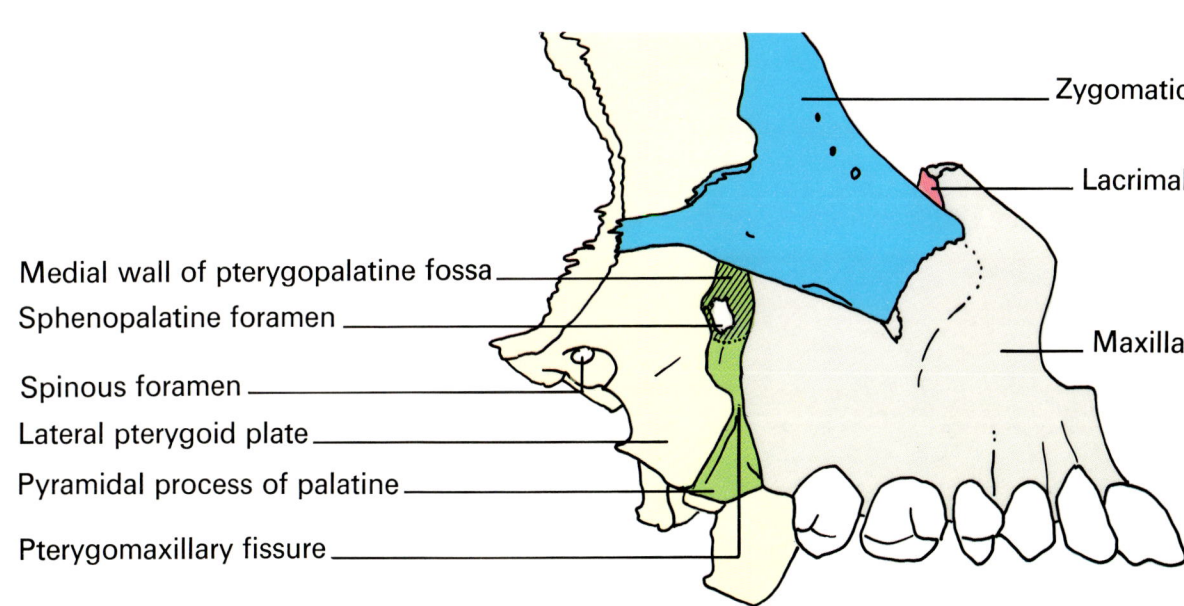

Pterygopalatine fossa.

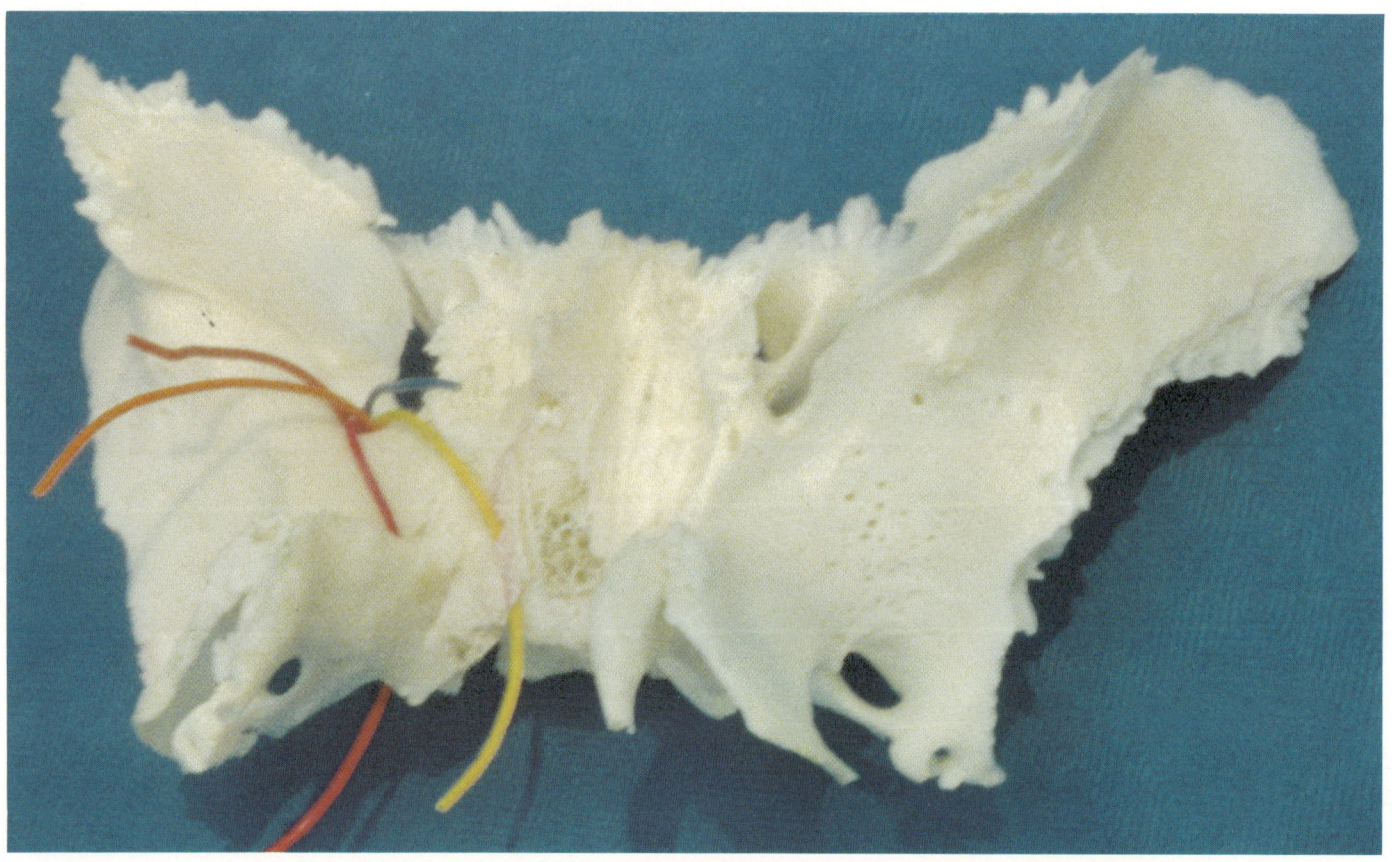

Posterior wall. Sphenoid. Rods in foramen rotundum, pharyngeal and pterygoid canal.

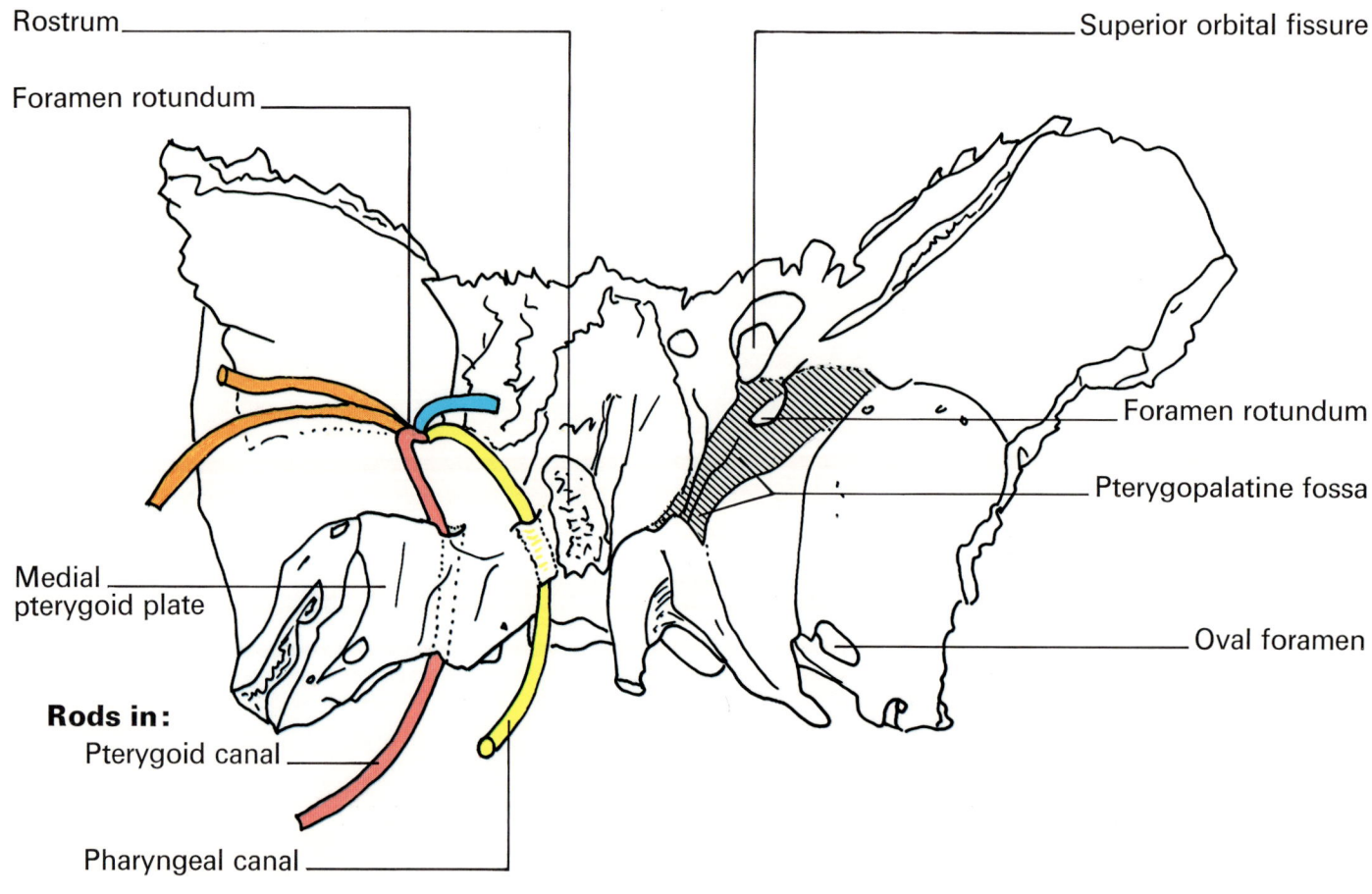

Pterygopalatine fossa.

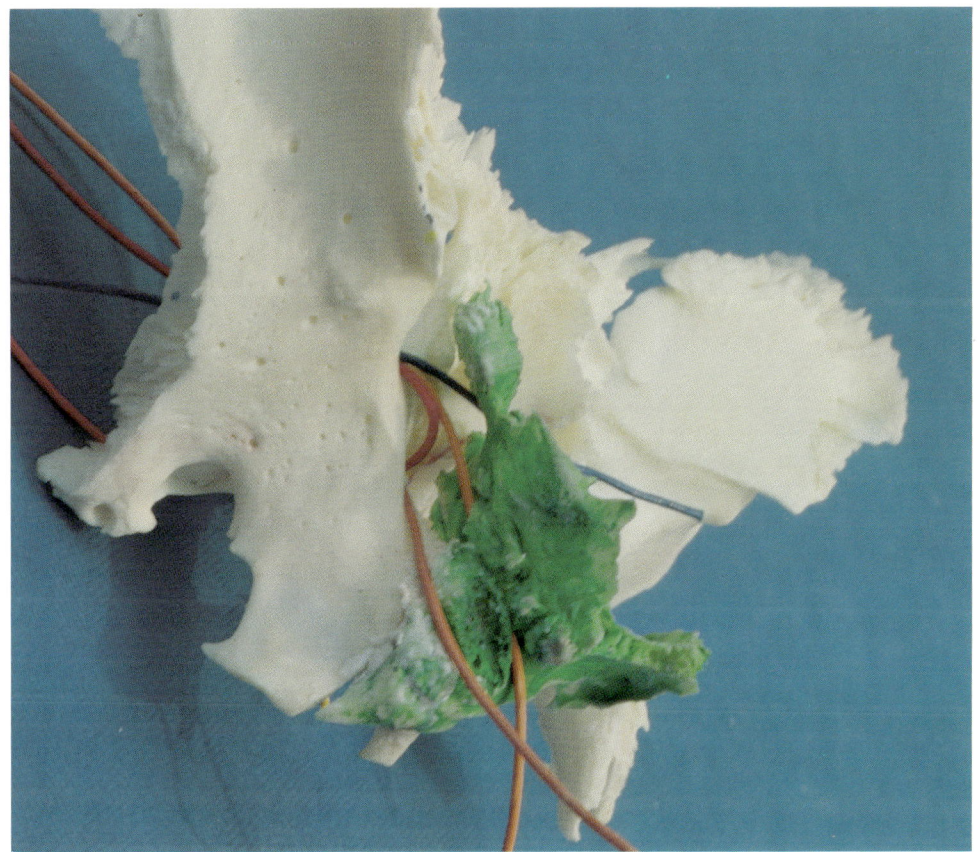

Medial wall. Composite of sphenoid and right palatine.

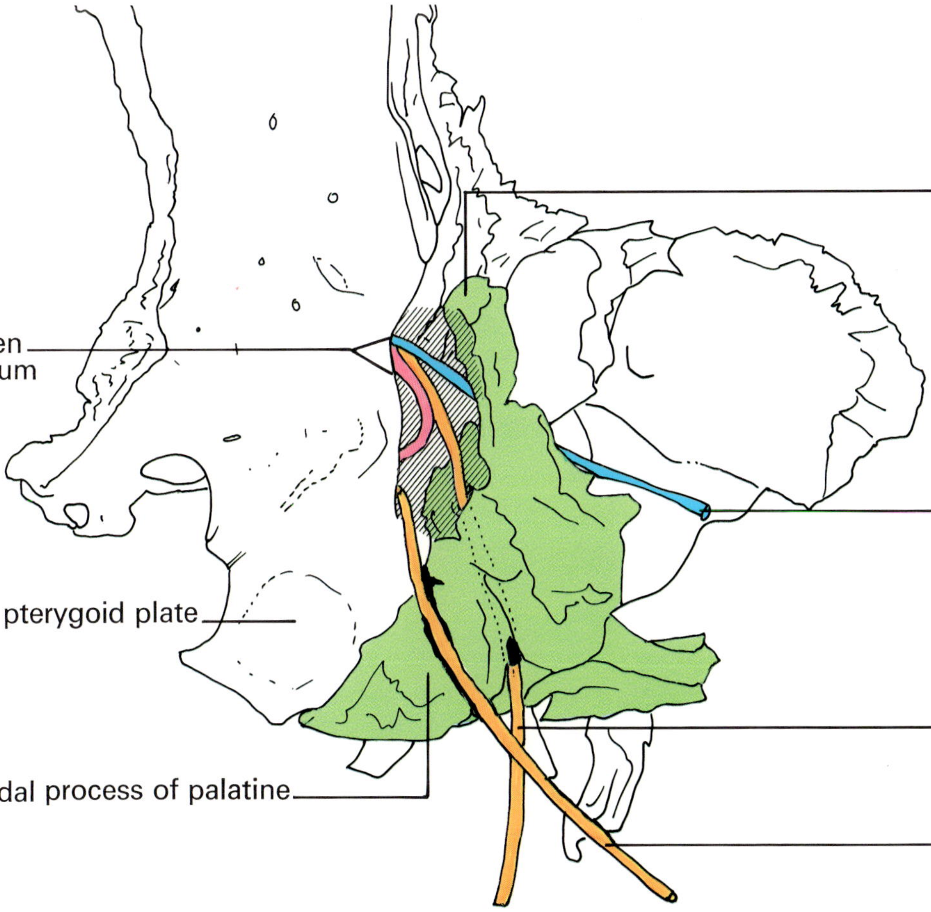

- Orbital process of palatine
- Foramen rotundum
- Lateral pterygoid plate
- Pyramidal process of palatine

Rods in:
- Sphenopalatine foramen
- Great palatine canal
- Lesser palatine canal

SECTION 12.   Infratemporal fossa.

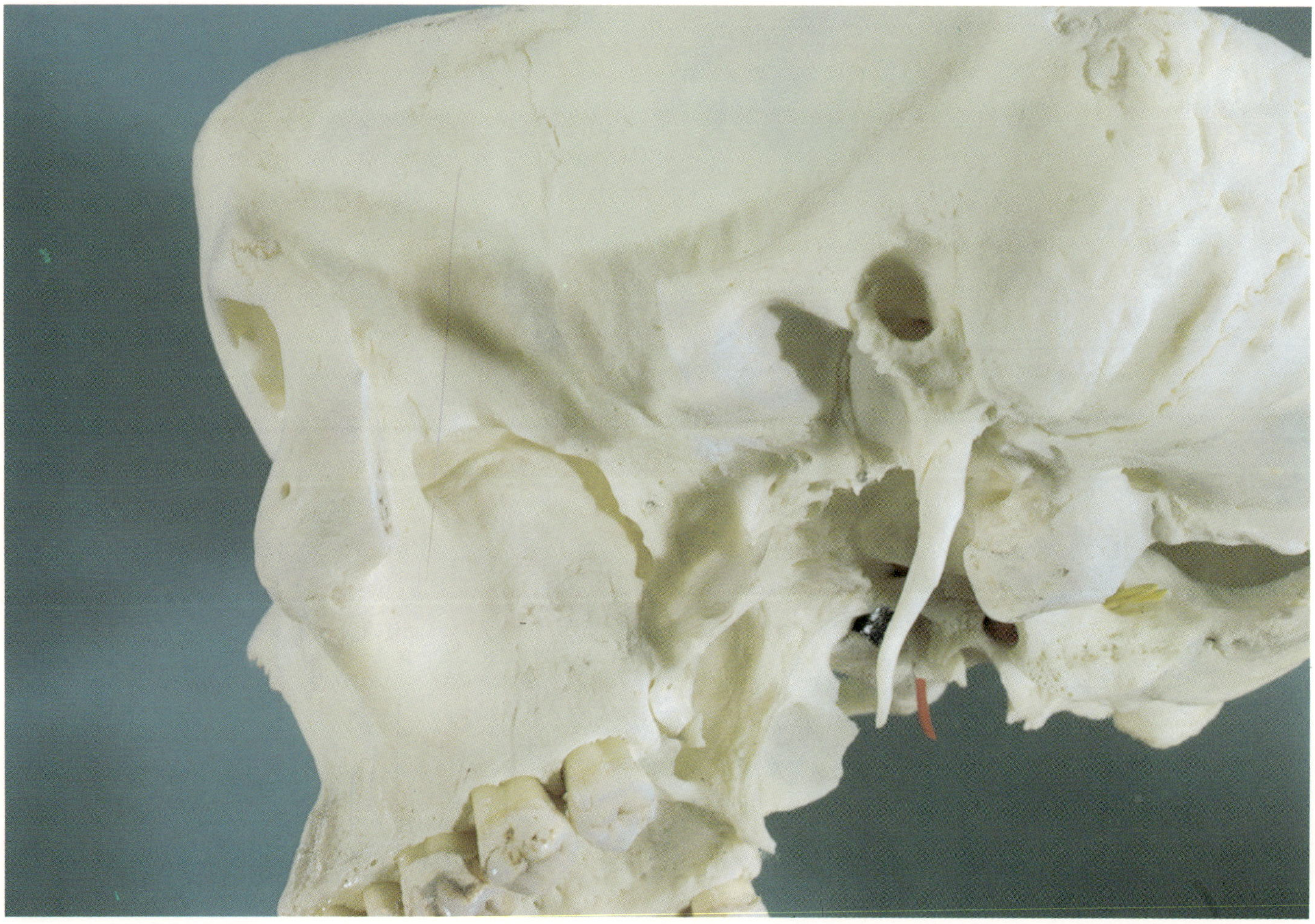

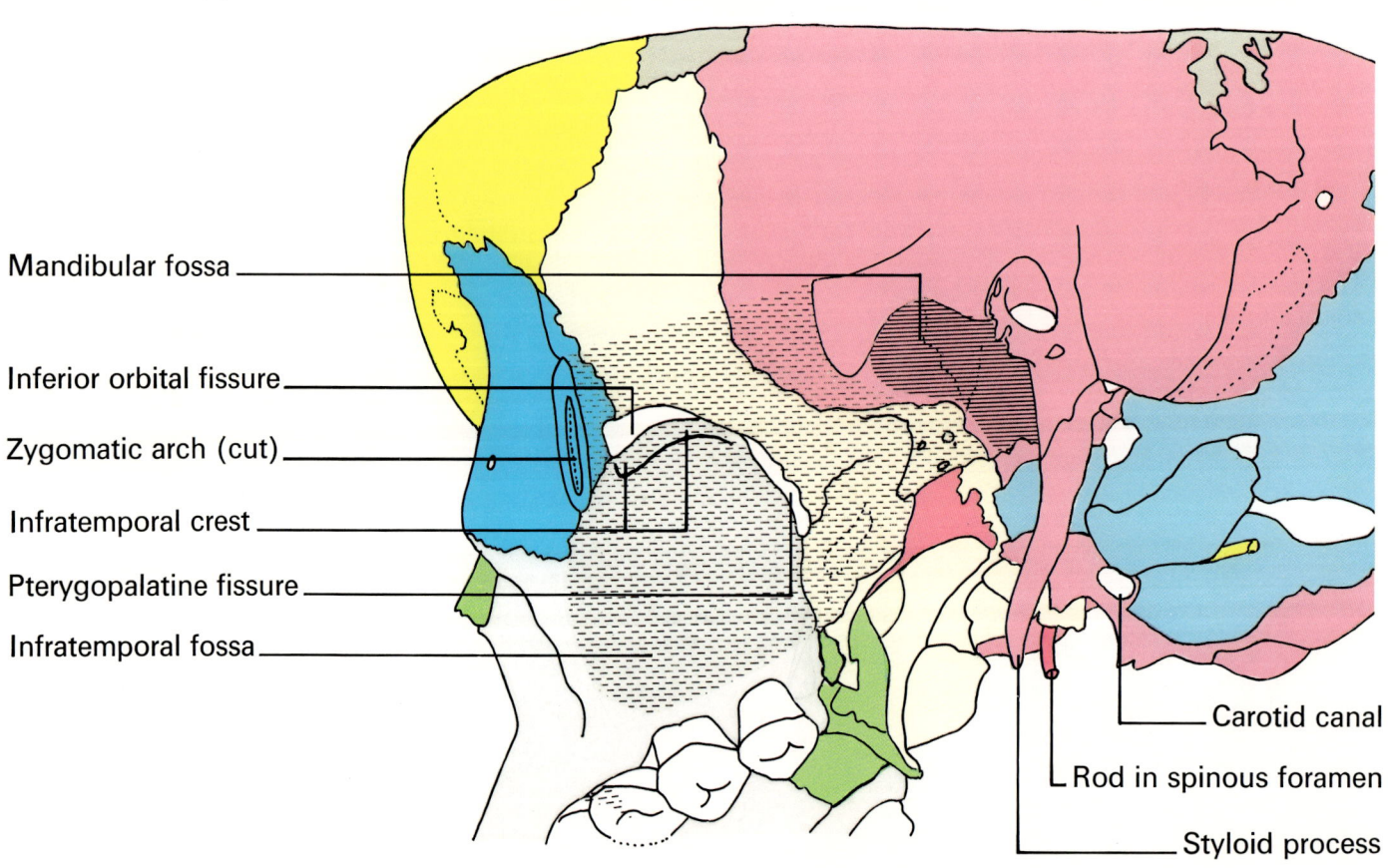

Side of skull. Zygomatic arch resected. Pertinent landmarks.

- Mandibular fossa
- Inferior orbital fissure
- Zygomatic arch (cut)
- Infratemporal crest
- Pterygopalatine fissure
- Infratemporal fossa
- Carotid canal
- Rod in spinous foramen
- Styloid process

Infratemporal fossa.

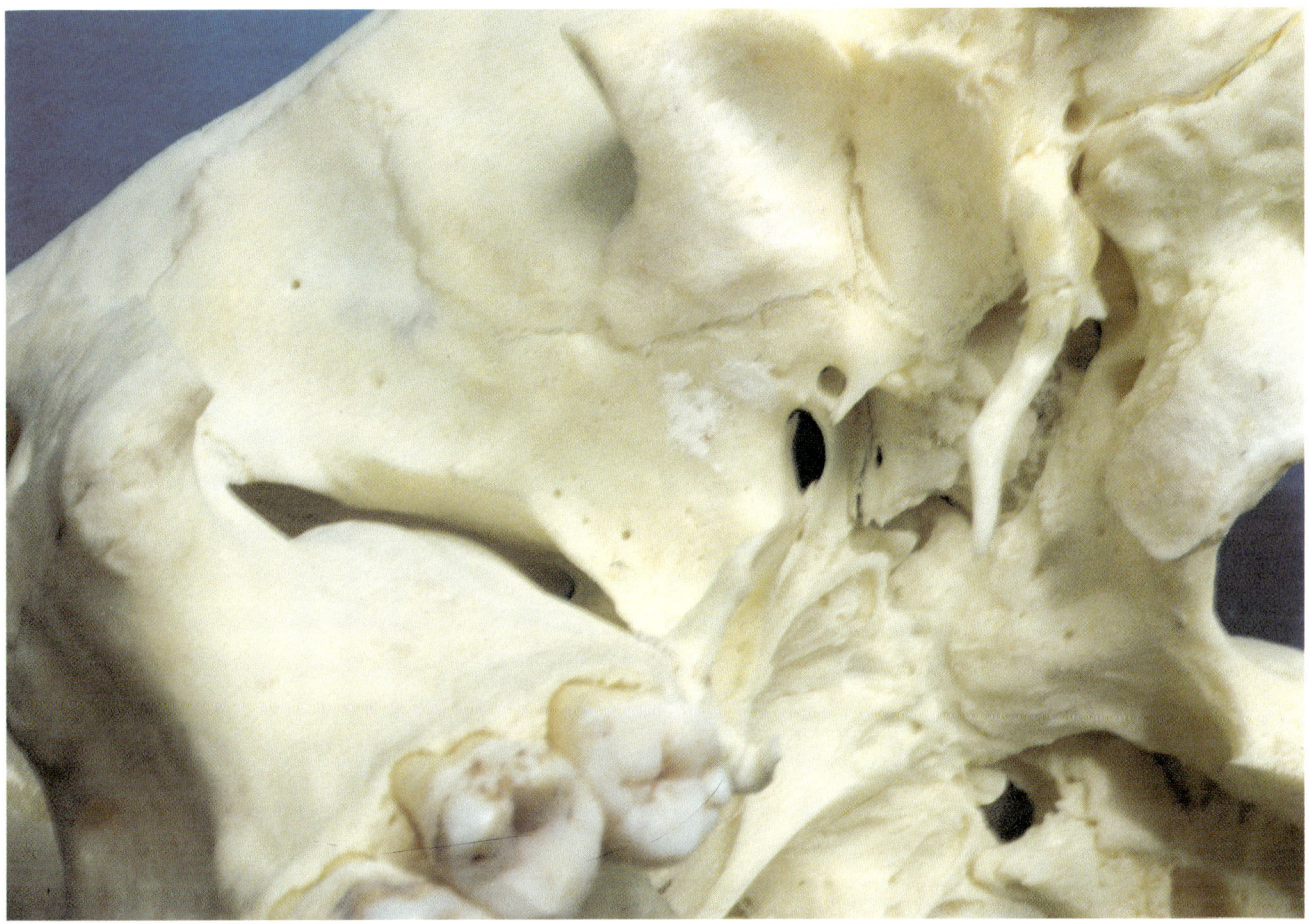

Skull more oblique, with view of optic canal through inferior orbital fissure.

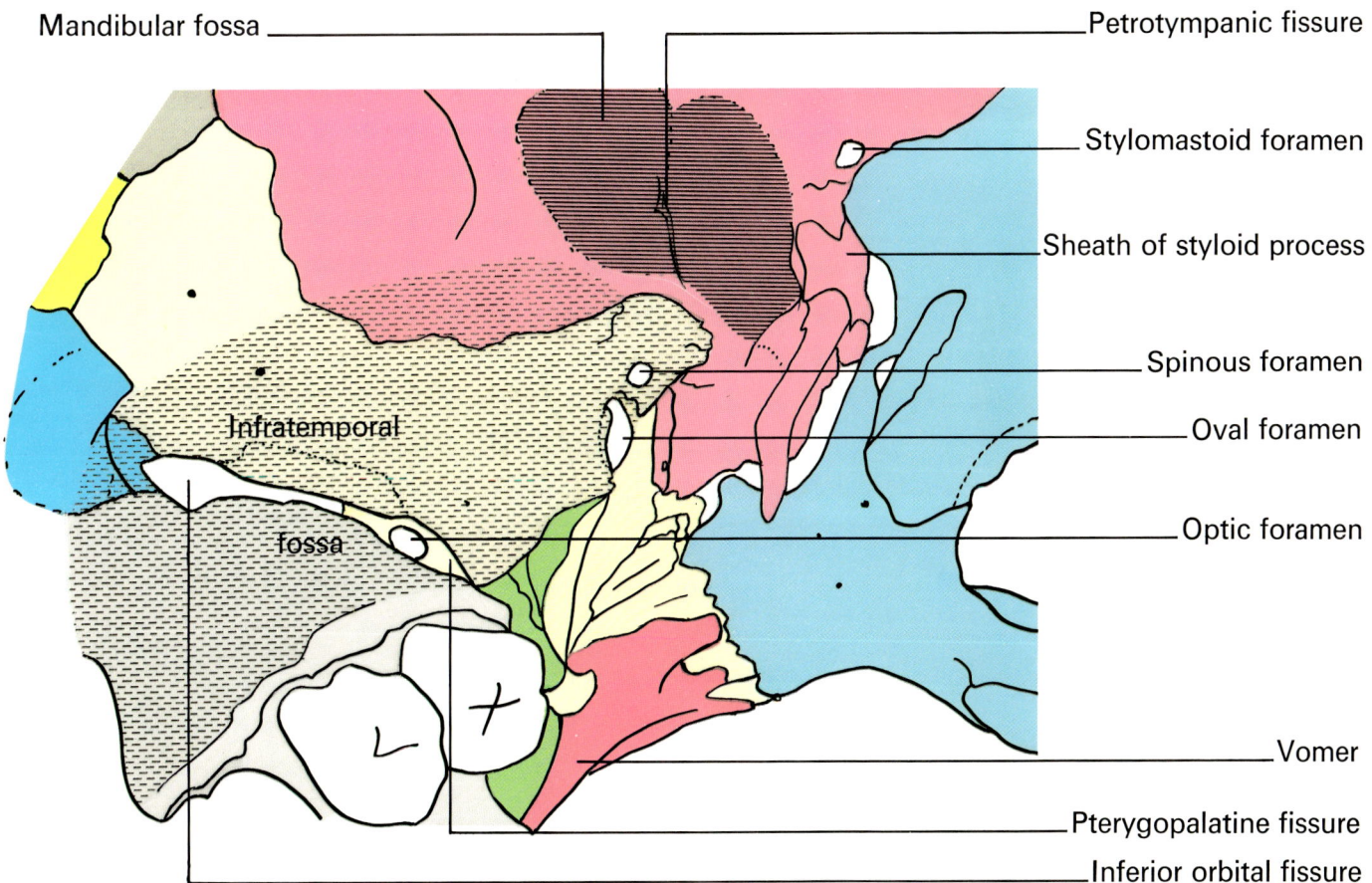

SECTION 13.   Temporomandibular joint.

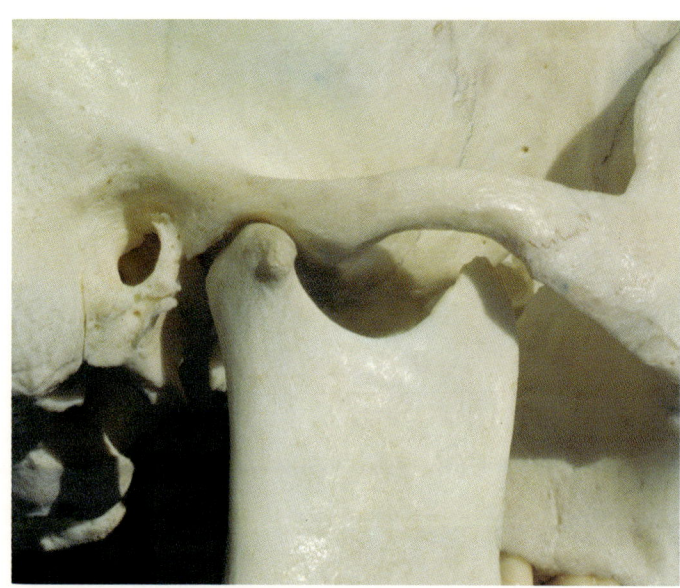
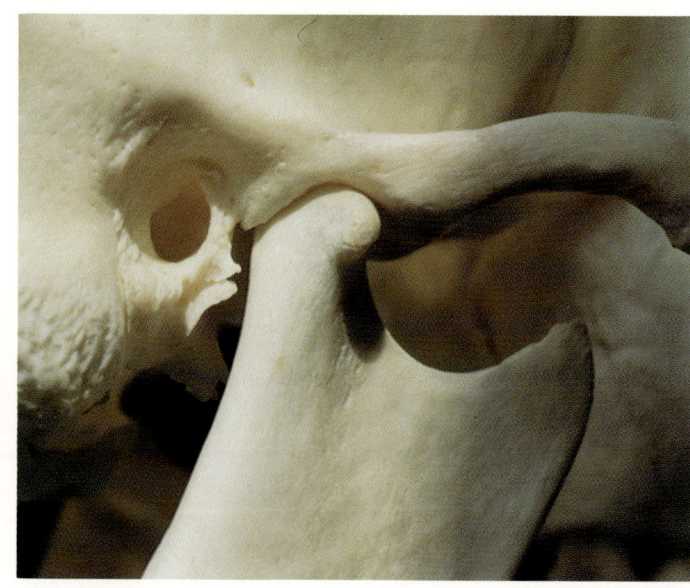

A. From side, closed.

B. Open.

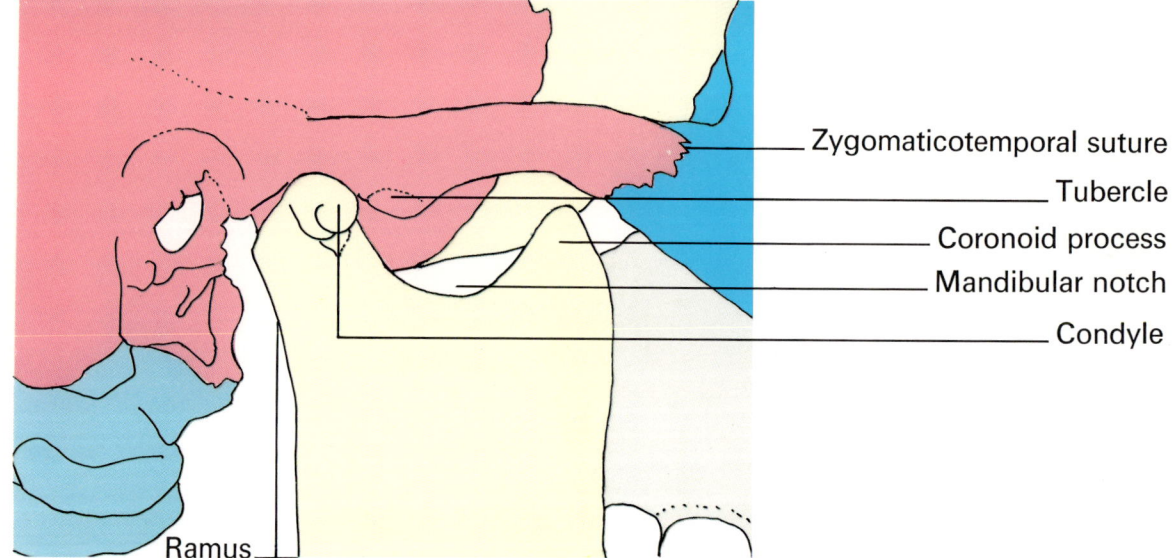

- Zygomaticotemporal suture
- Tubercle
- Coronoid process
- Mandibular notch
- Condyle

Ramus

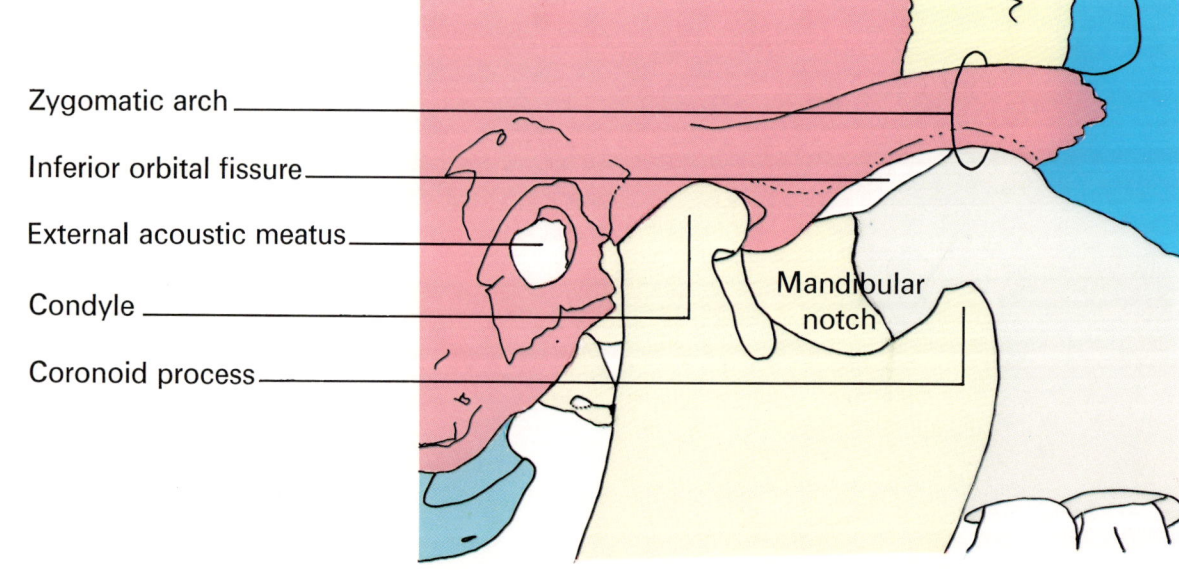

- Zygomatic arch
- Inferior orbital fissure
- External acoustic meatus
- Condyle
- Coronoid process
- Mandibular notch

Temporomandibular joint.

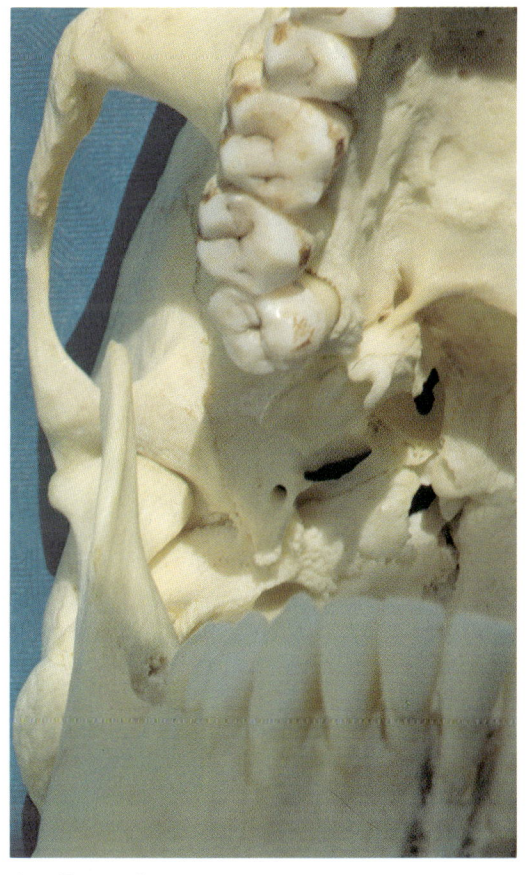

A. From front.

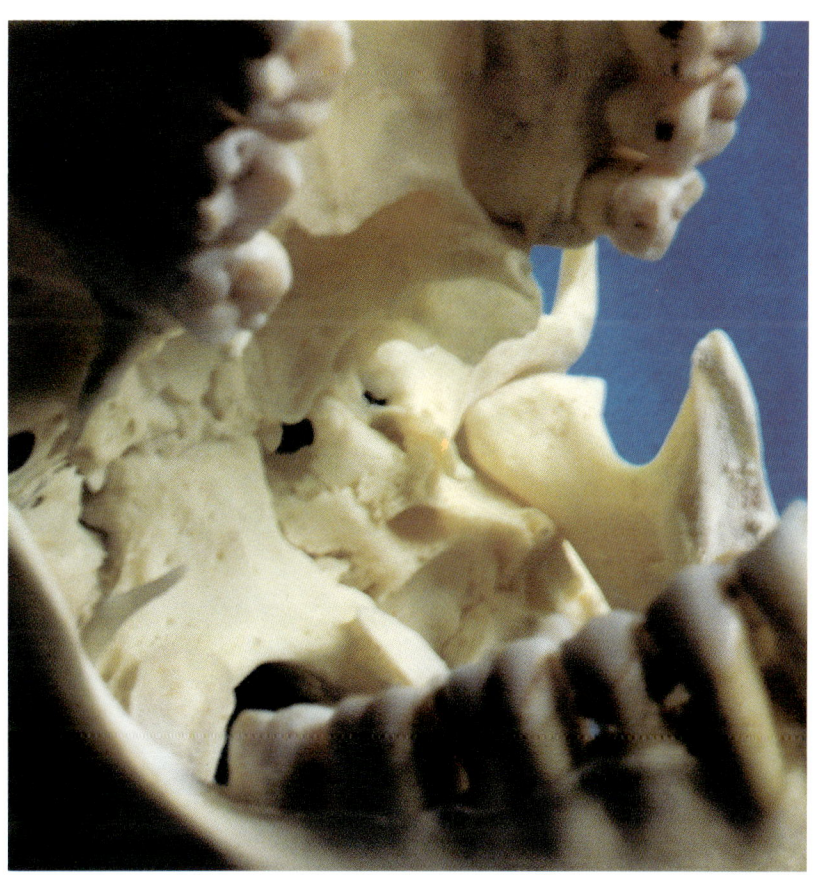

B. Through oral cavity, medial side.

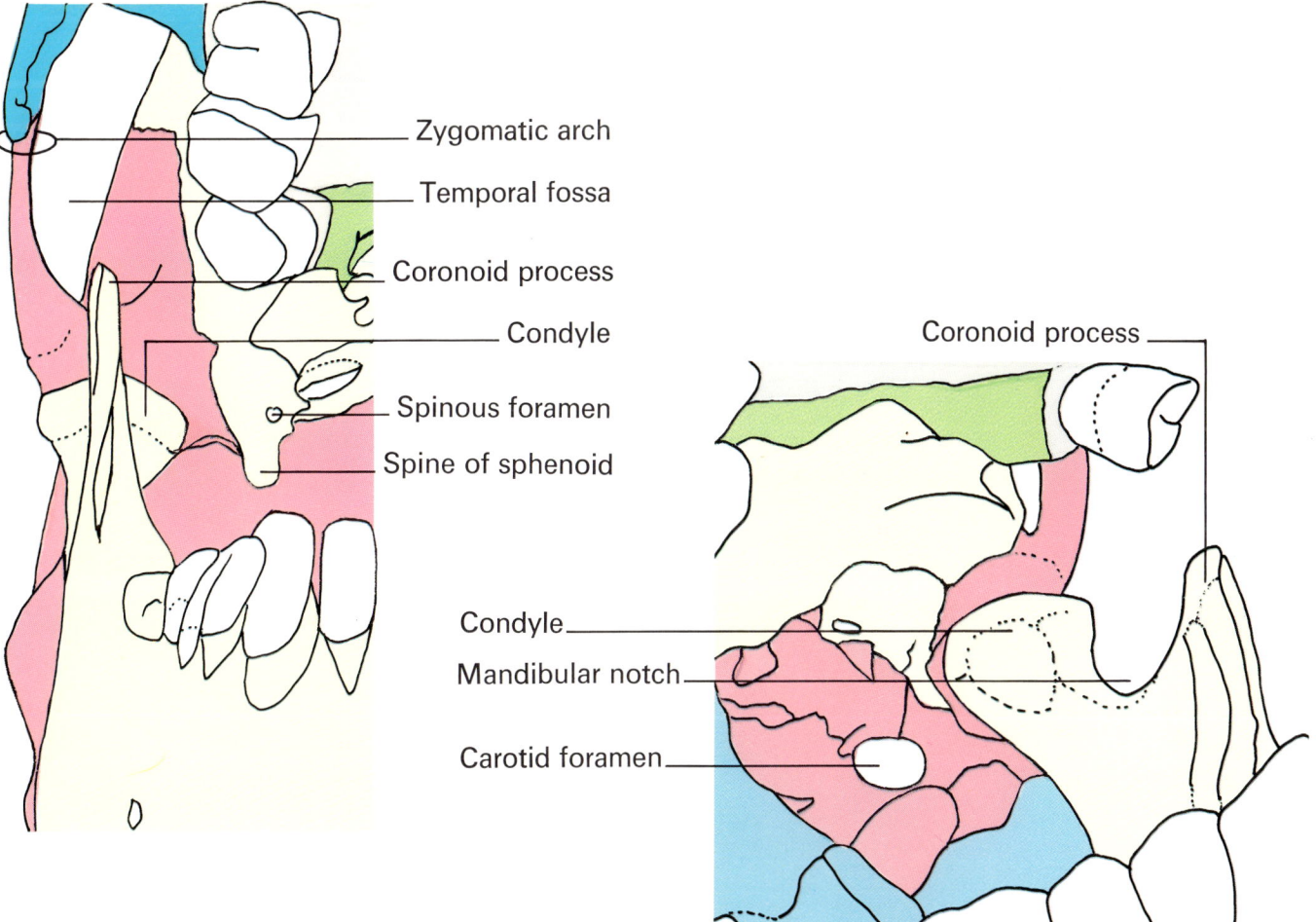

PART II. FACIAL BONES

SECTION 1. Mandible.

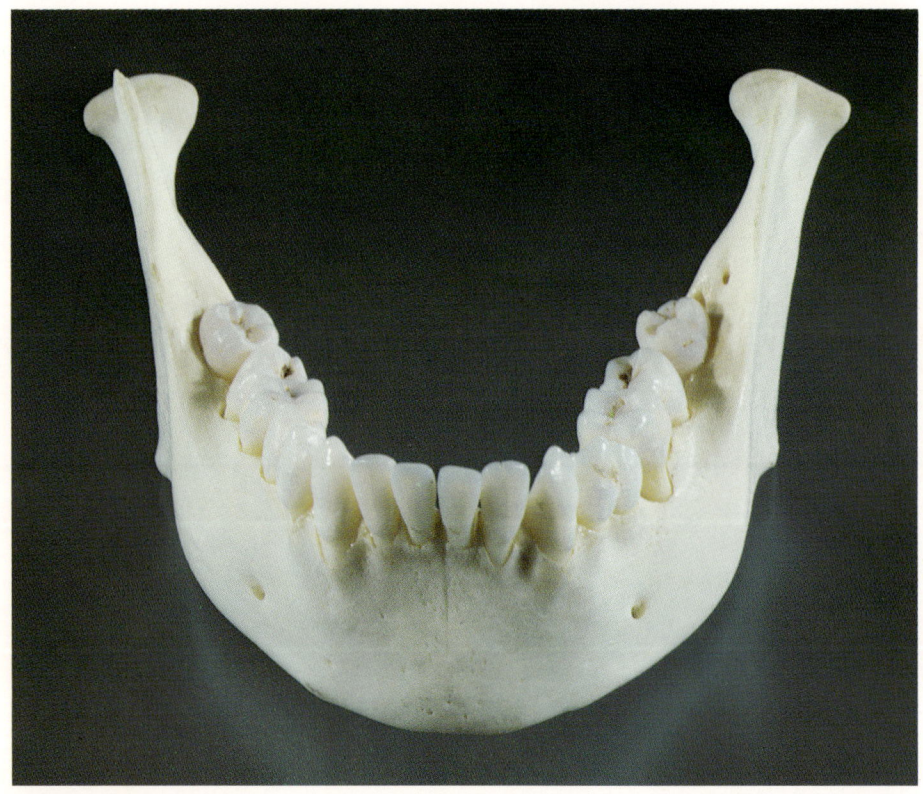

From front.

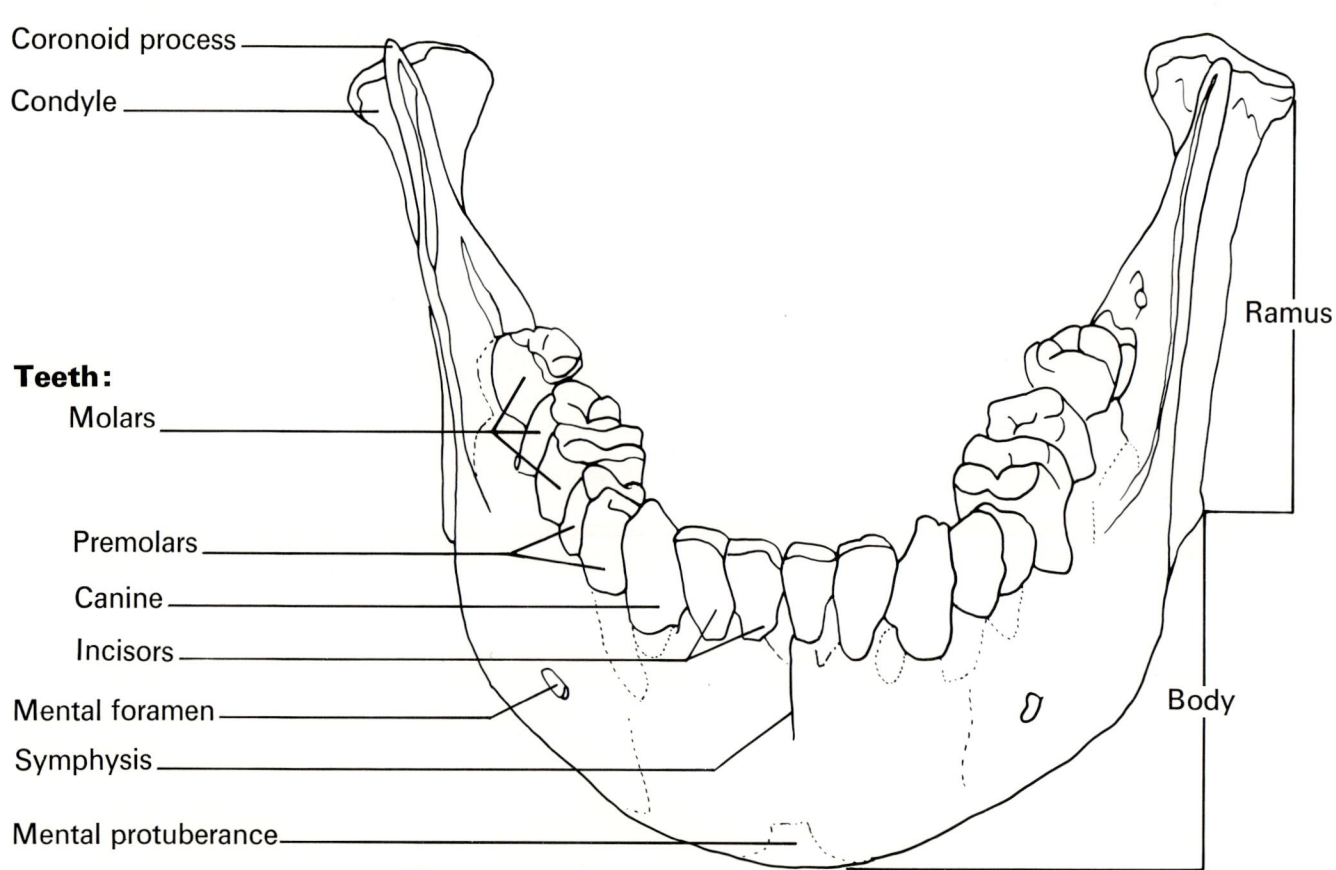

Mandible.

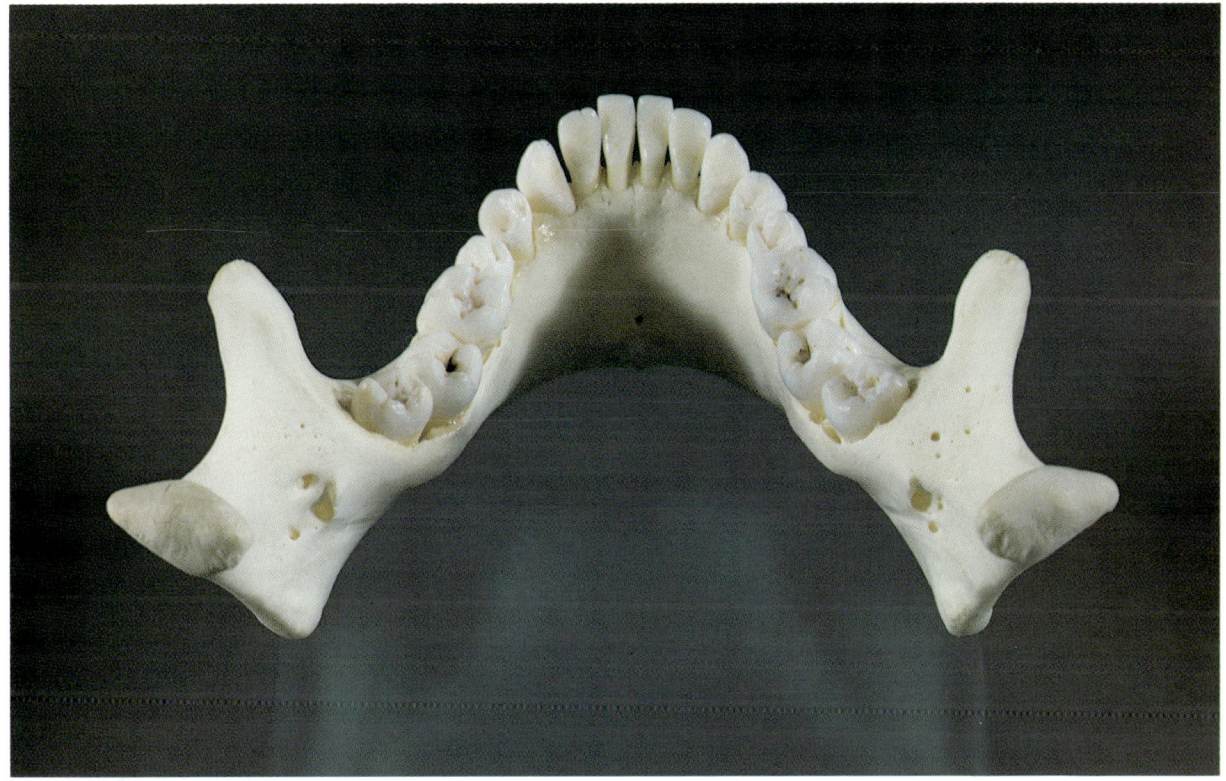

From behind, and above.

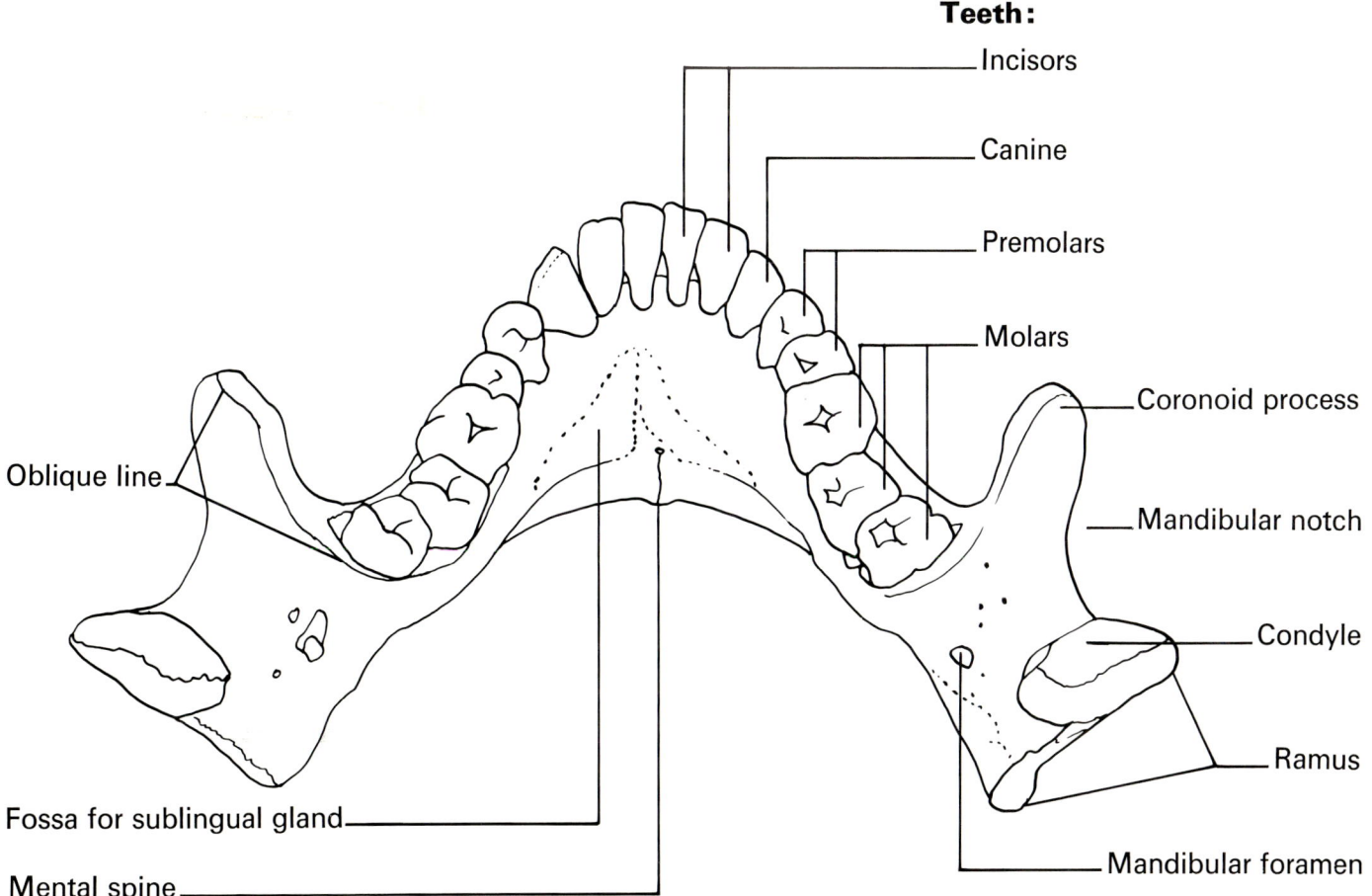

Teeth:
- Incisors
- Canine
- Premolars
- Molars
- Coronoid process
- Mandibular notch
- Condyle
- Ramus
- Mandibular foramen
- Mental spine
- Fossa for sublingual gland
- Oblique line

Mandible.
From side.

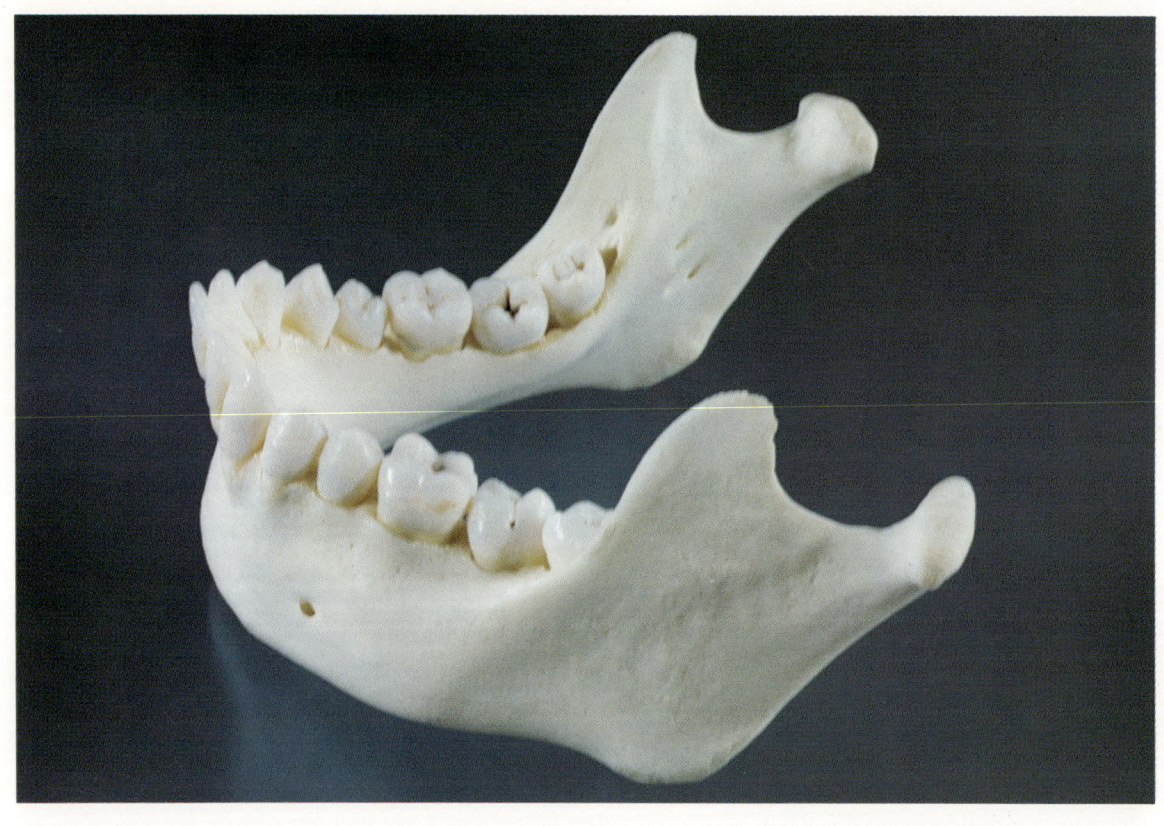

Mandible.
Landmarks and muscle attachments.

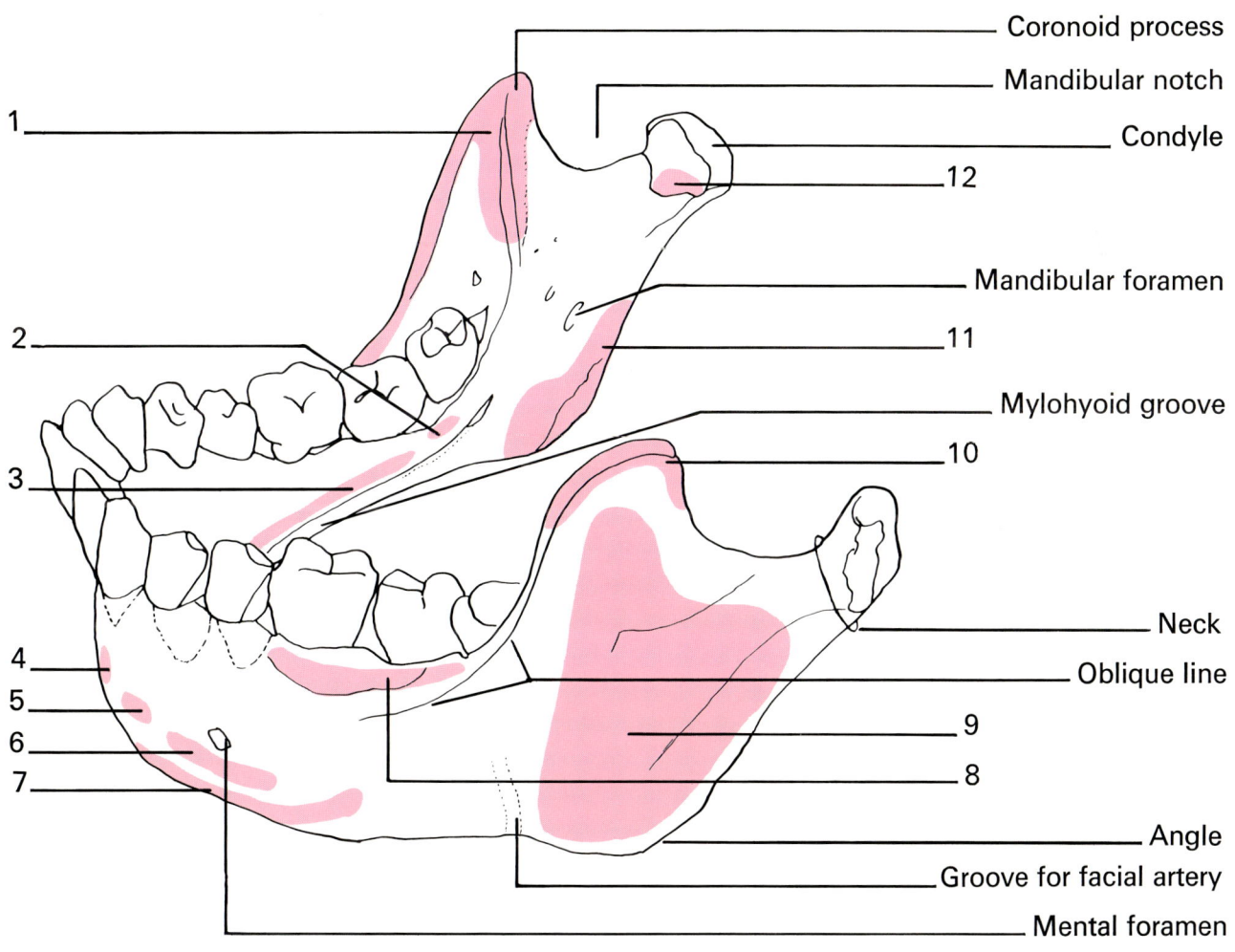

Muscle attachments:

1. Temporal
2. Superior constrictor of pharynx
3. Mylohyoid
4. Mental
5. Depressor of lower lip
6. Depressor of angle of mouth
7. Platysma
8. Buccinator
9. Masseter
10. Temporal
11. Medial pterygoid
12. Lateral pterygoid

Mandible.
From behind and below.

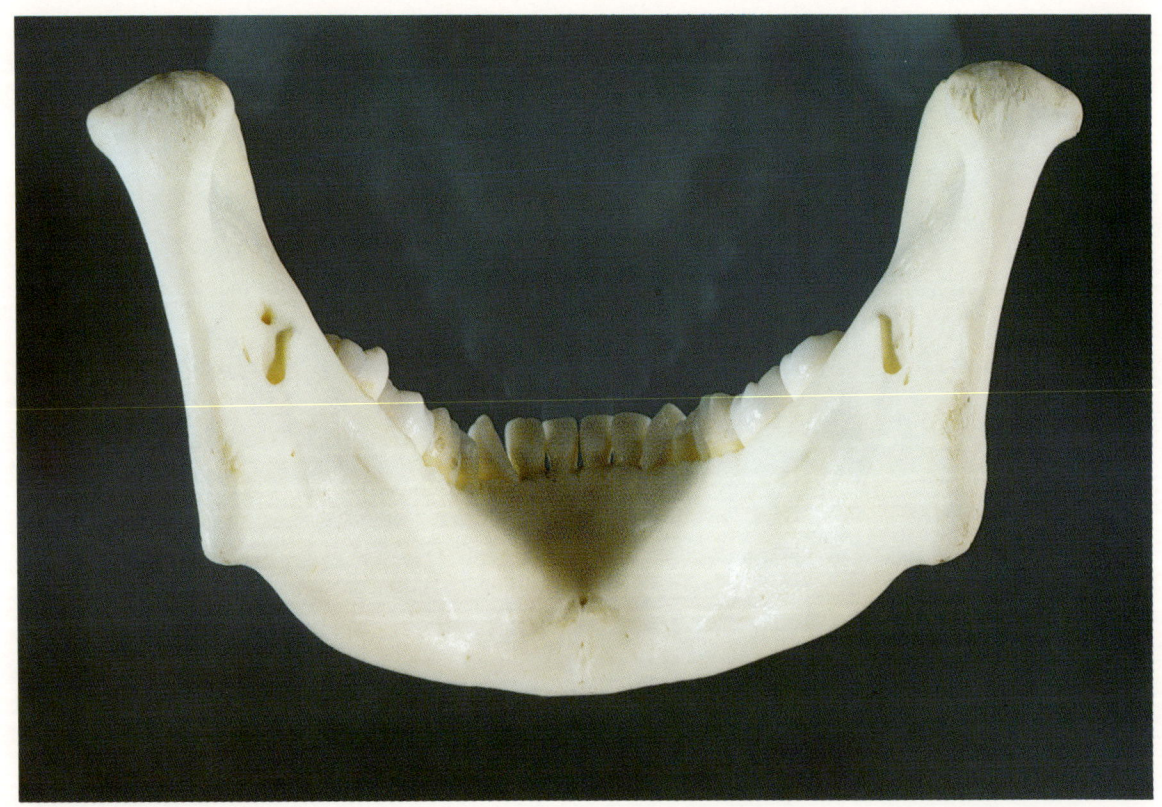

Mandible.
Landmarks and muscle attachments.

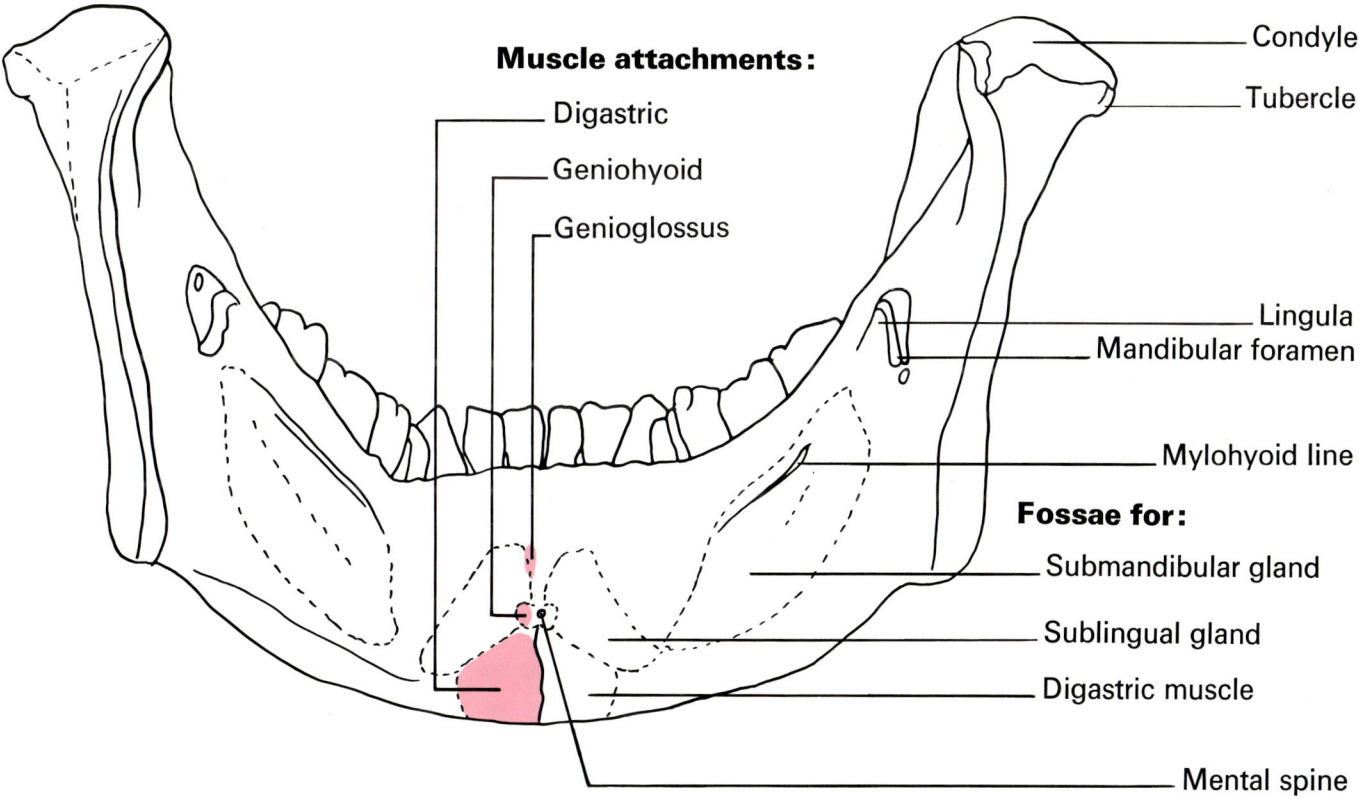

Mandible.

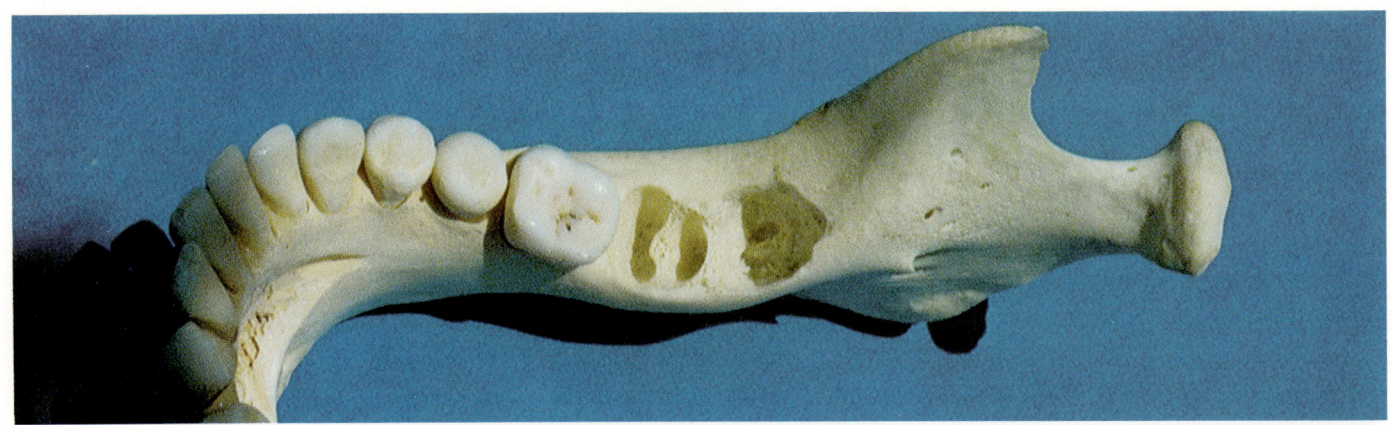

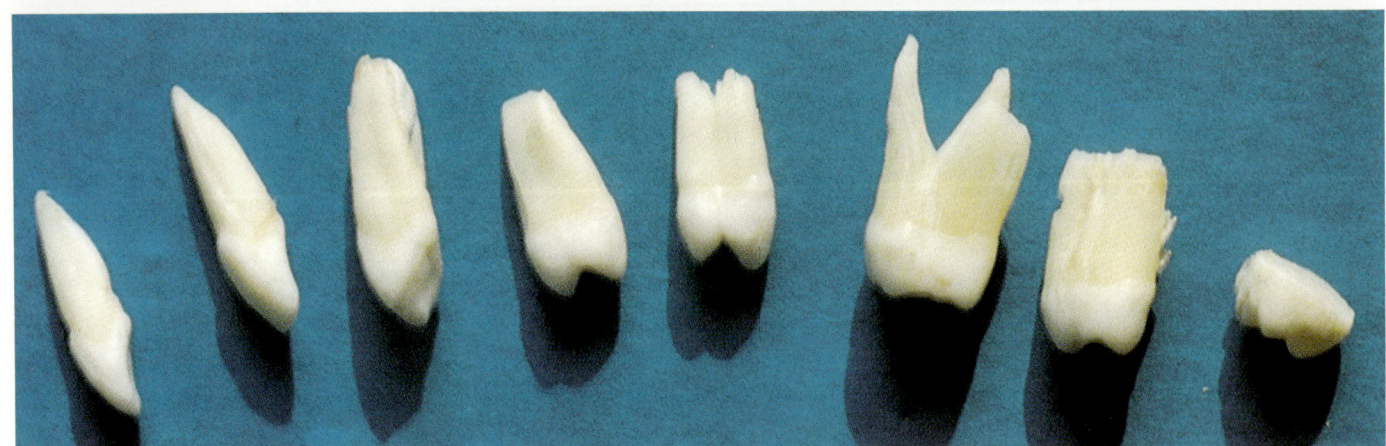

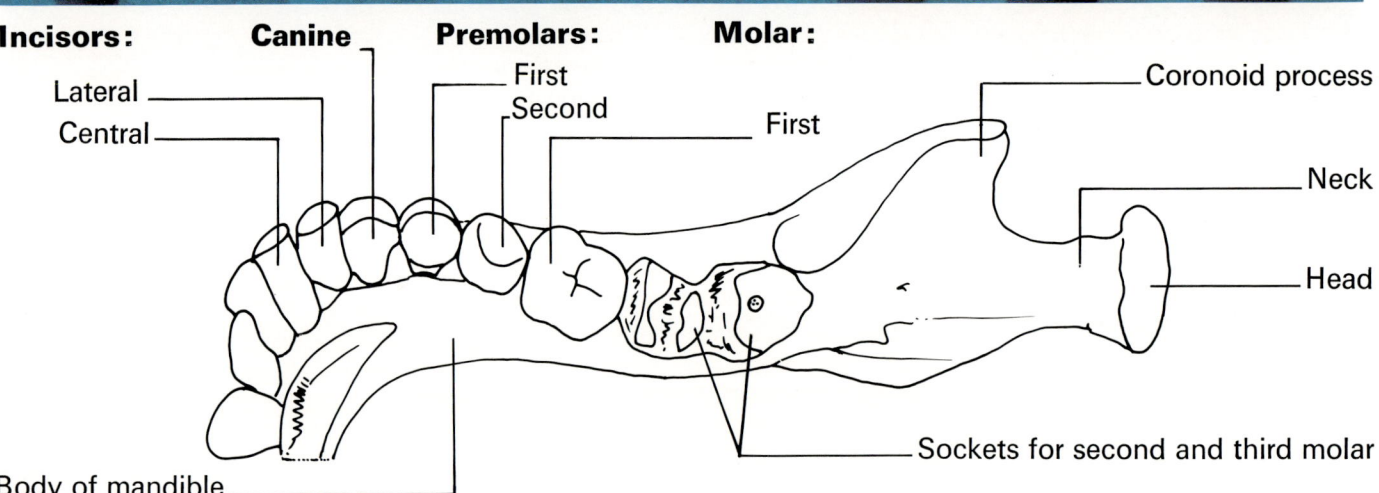

A. Right half of mandible, from above.

B. Representative teeth.

SECTION 2.   Maxilla.

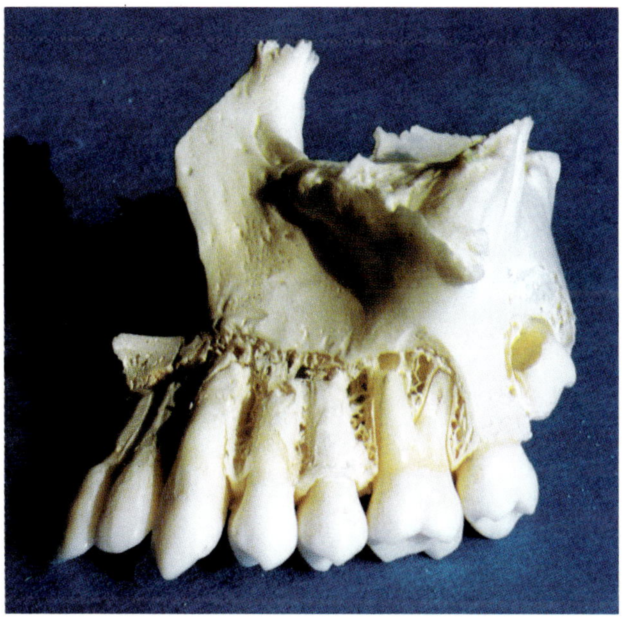

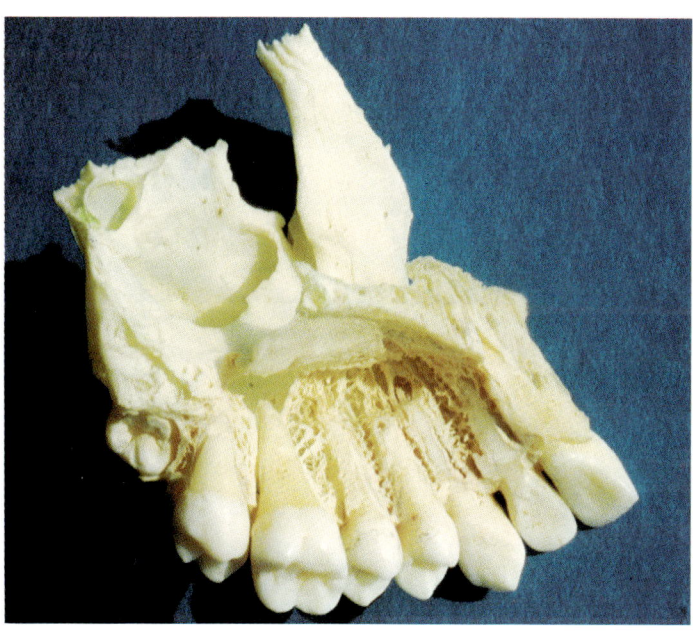

A. Left maxilla. Alveolar process removed to expose roots of teeth. Seen from outer surface.

B. From oral surface.

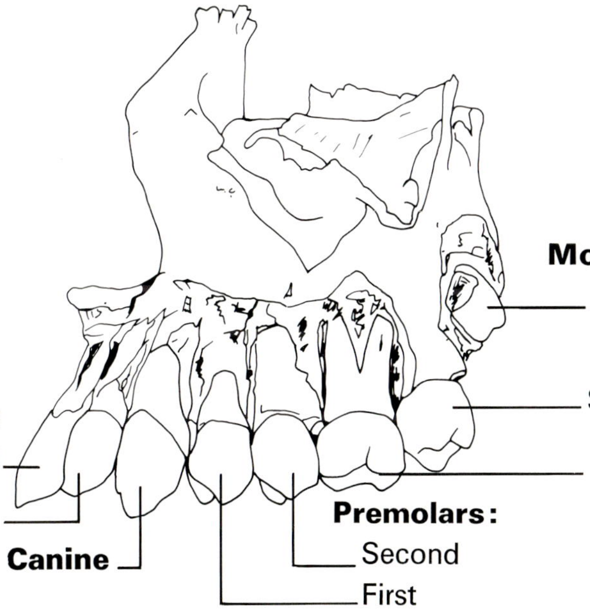

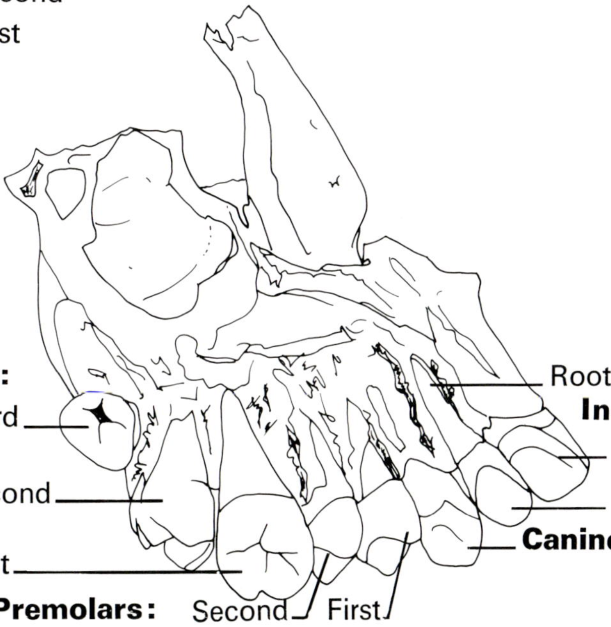

Molars:
- Third
- Second
- First

Incisors:
- Central
- Lateral

Canine

Premolars:
- Second
- First

Molars:
- Third
- Second
- First

Premolars:
- Second
- First

Canine

Incisors:
- Lateral
- Central

Root

Maxilla.

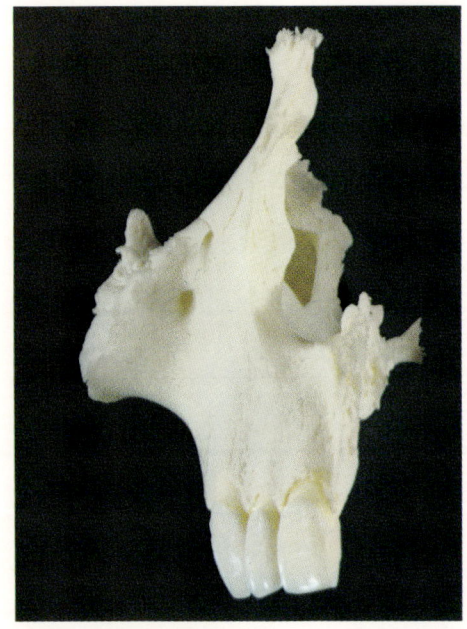

A. Right, from front.

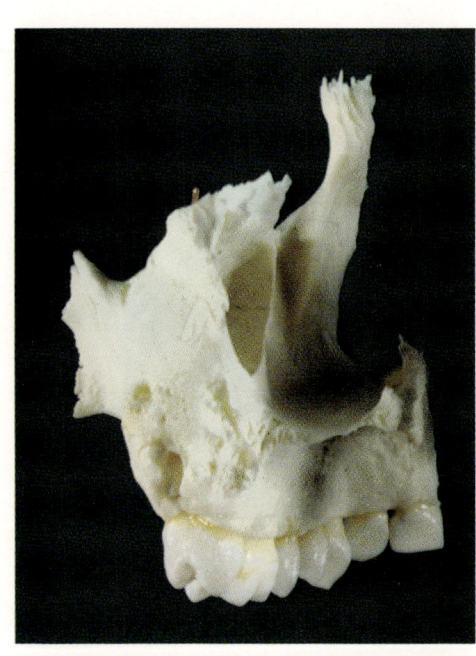

B. Left, from behind, nasal surface.

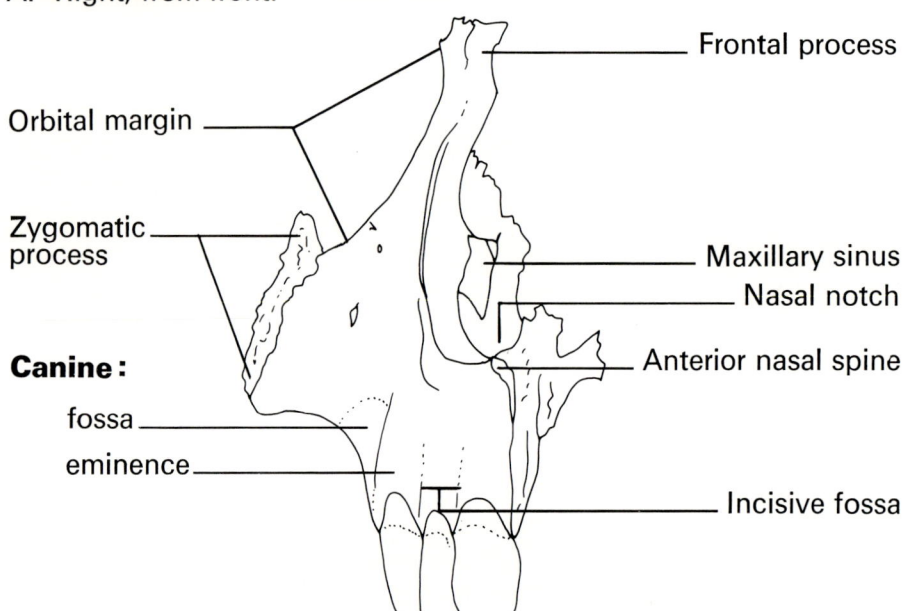

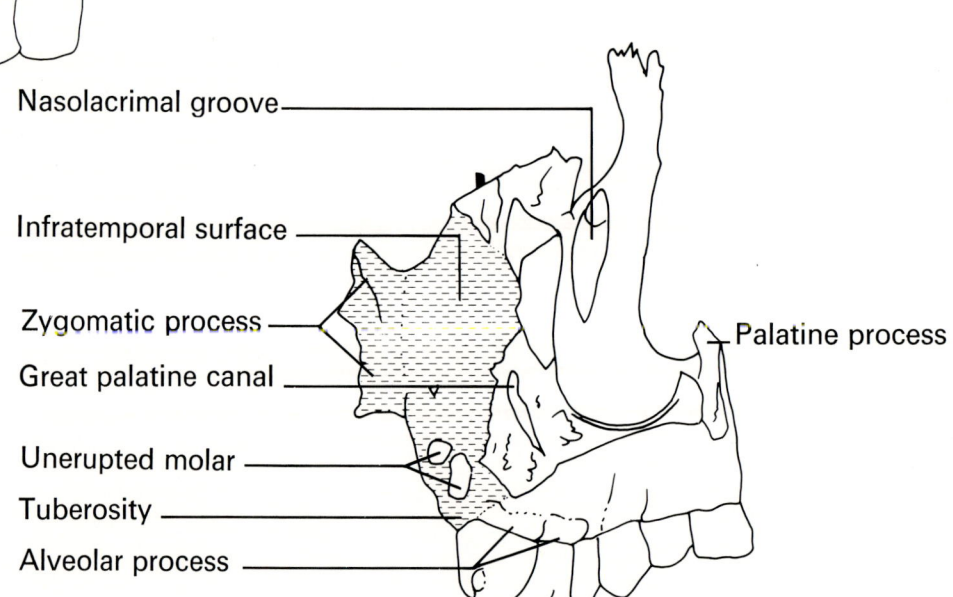

Maxilla.

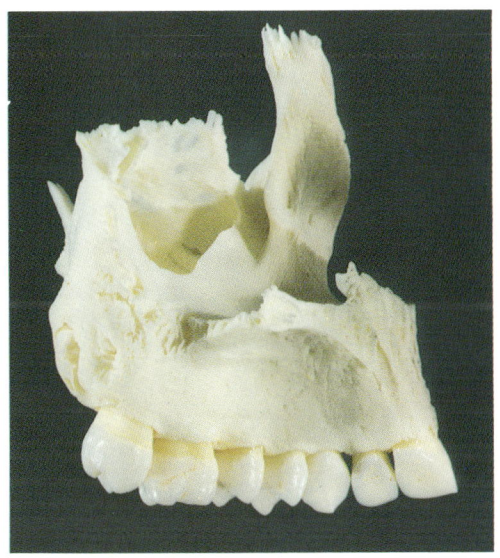

A. Left, from nasal side.

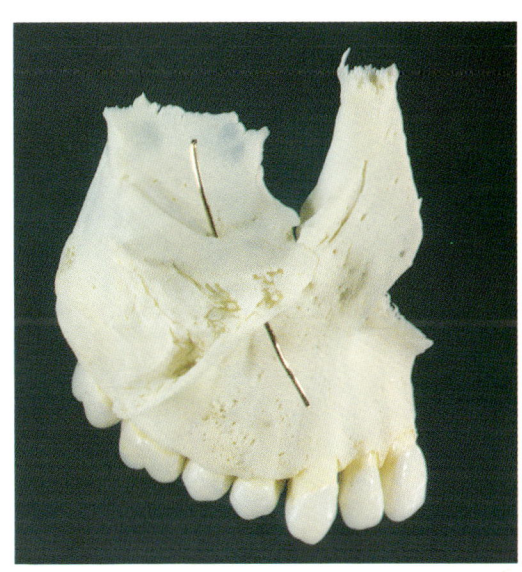

B. Right, from lateral side.

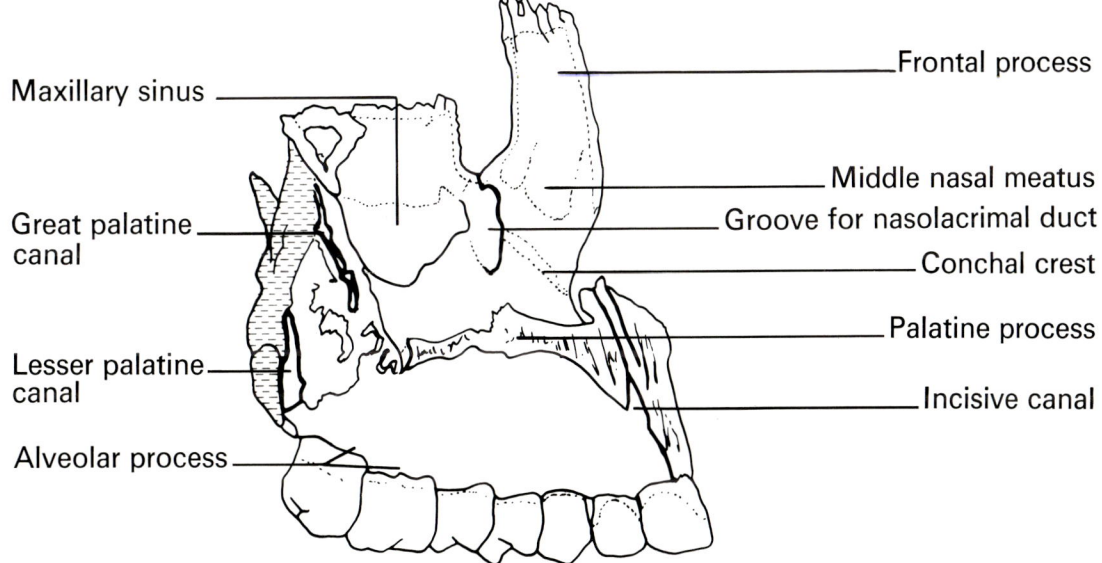

- Maxillary sinus
- Great palatine canal
- Lesser palatine canal
- Alveolar process
- Frontal process
- Middle nasal meatus
- Groove for nasolacrimal duct
- Conchal crest
- Palatine process
- Incisive canal

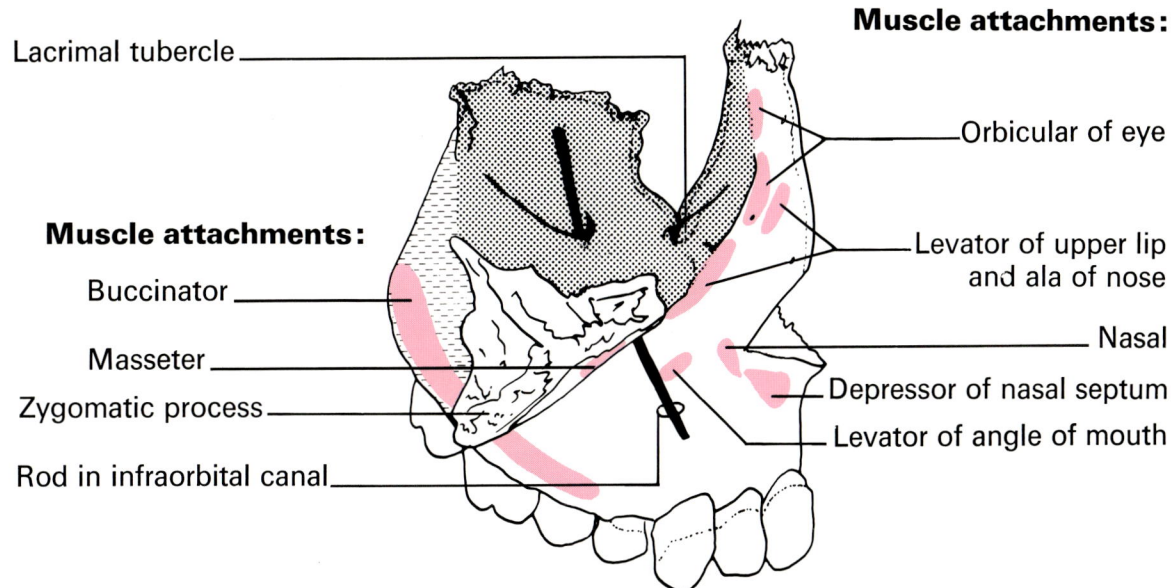

Muscle attachments:
- Lacrimal tubercle
- Buccinator
- Masseter
- Zygomatic process
- Rod in infraorbital canal

Muscle attachments:
- Orbicular of eye
- Levator of upper lip and ala of nose
- Nasal
- Depressor of nasal septum
- Levator of angle of mouth

Maxilla.

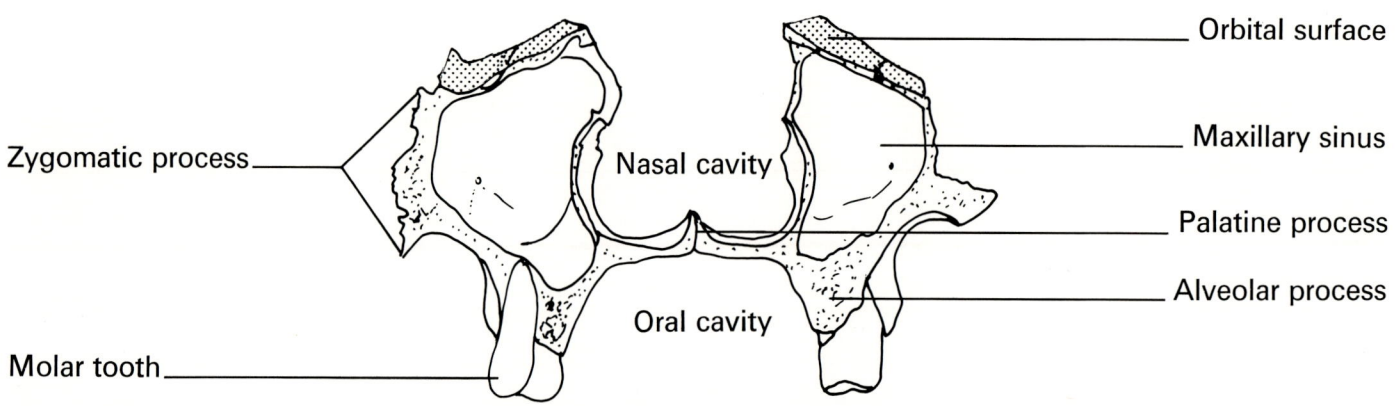

A. Frontal section of both maxillas through sinus cavities.

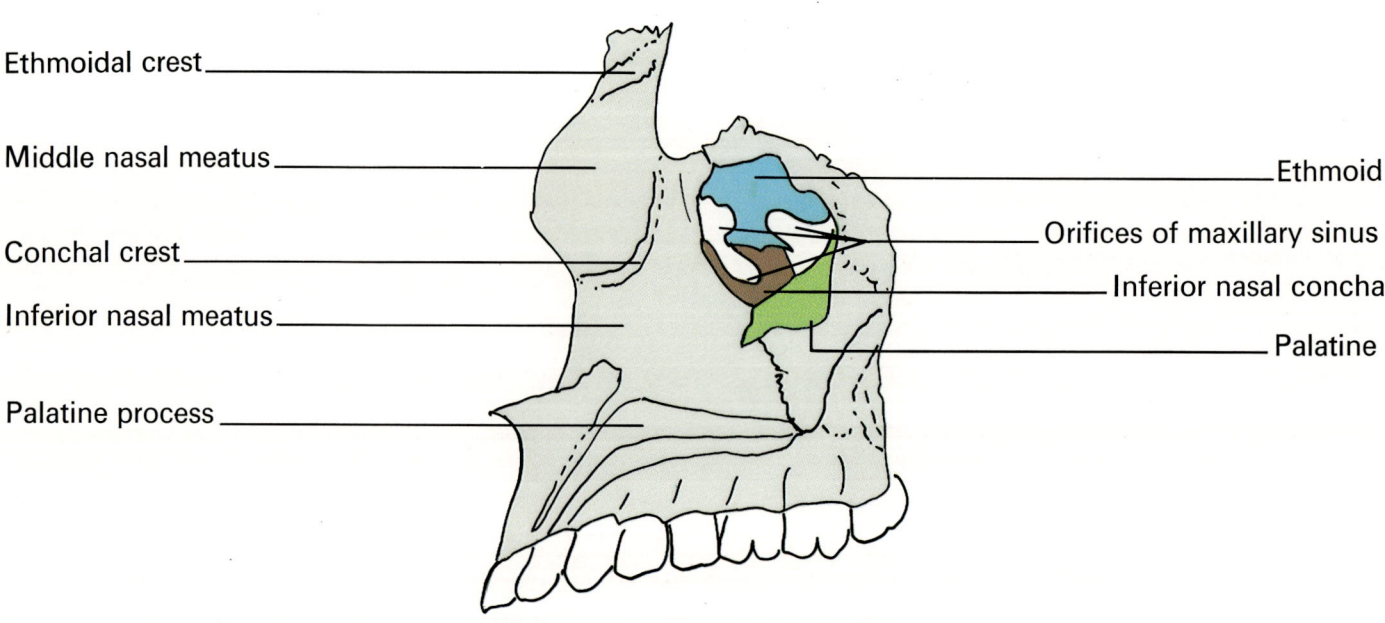

B. From side, orifices of right maxillary sinus and adjacent bones.

Maxilla.

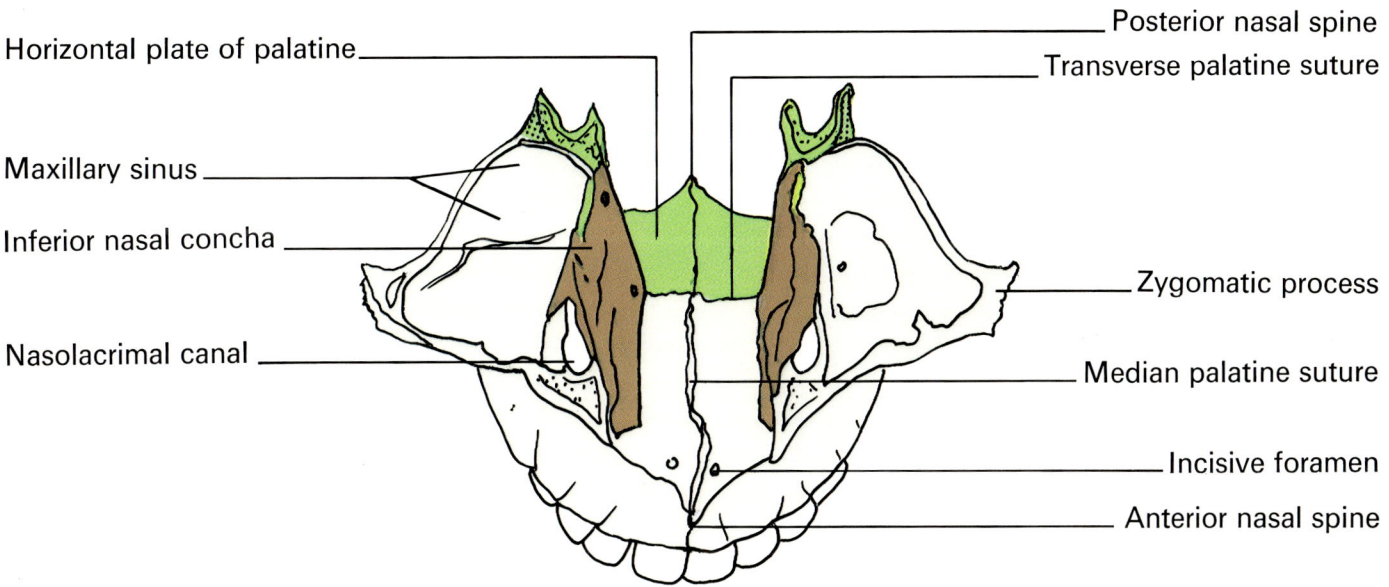

A. Horizontal section through maxillary sinuses. View of floor of nasal cavity.

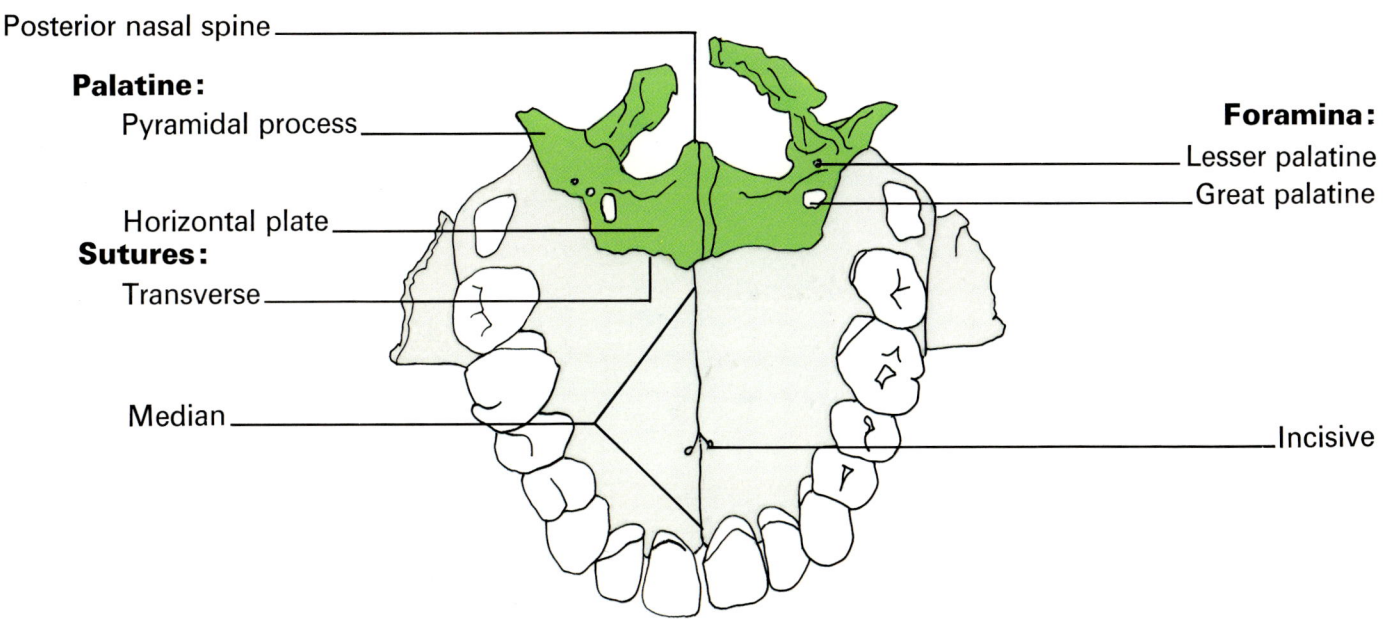

B. Roof of oral cavity.

SECTION 3. Zygomatic.

A. Temporal fossa. Composite of zygomatic and adjacent bones, from above.

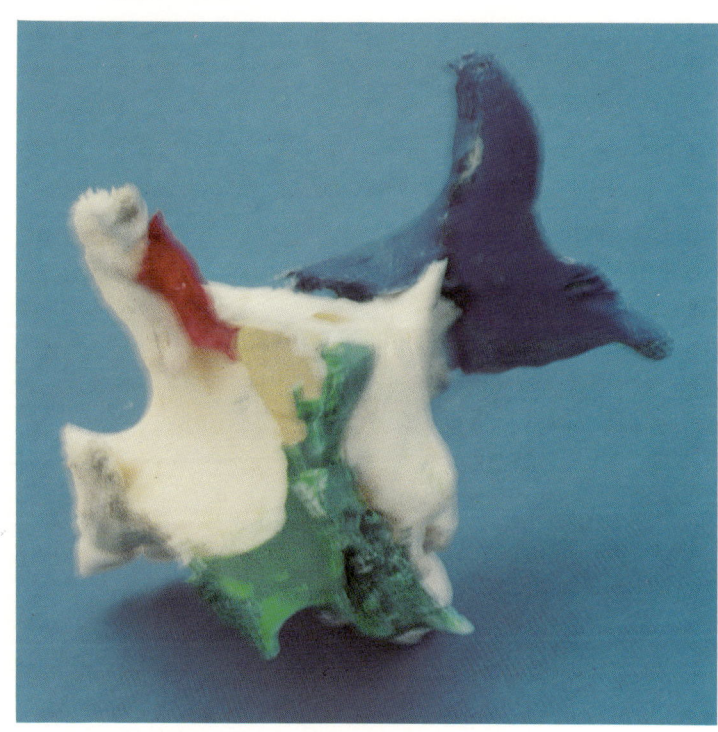

B. From behind (anterior portion of temporal fossa).

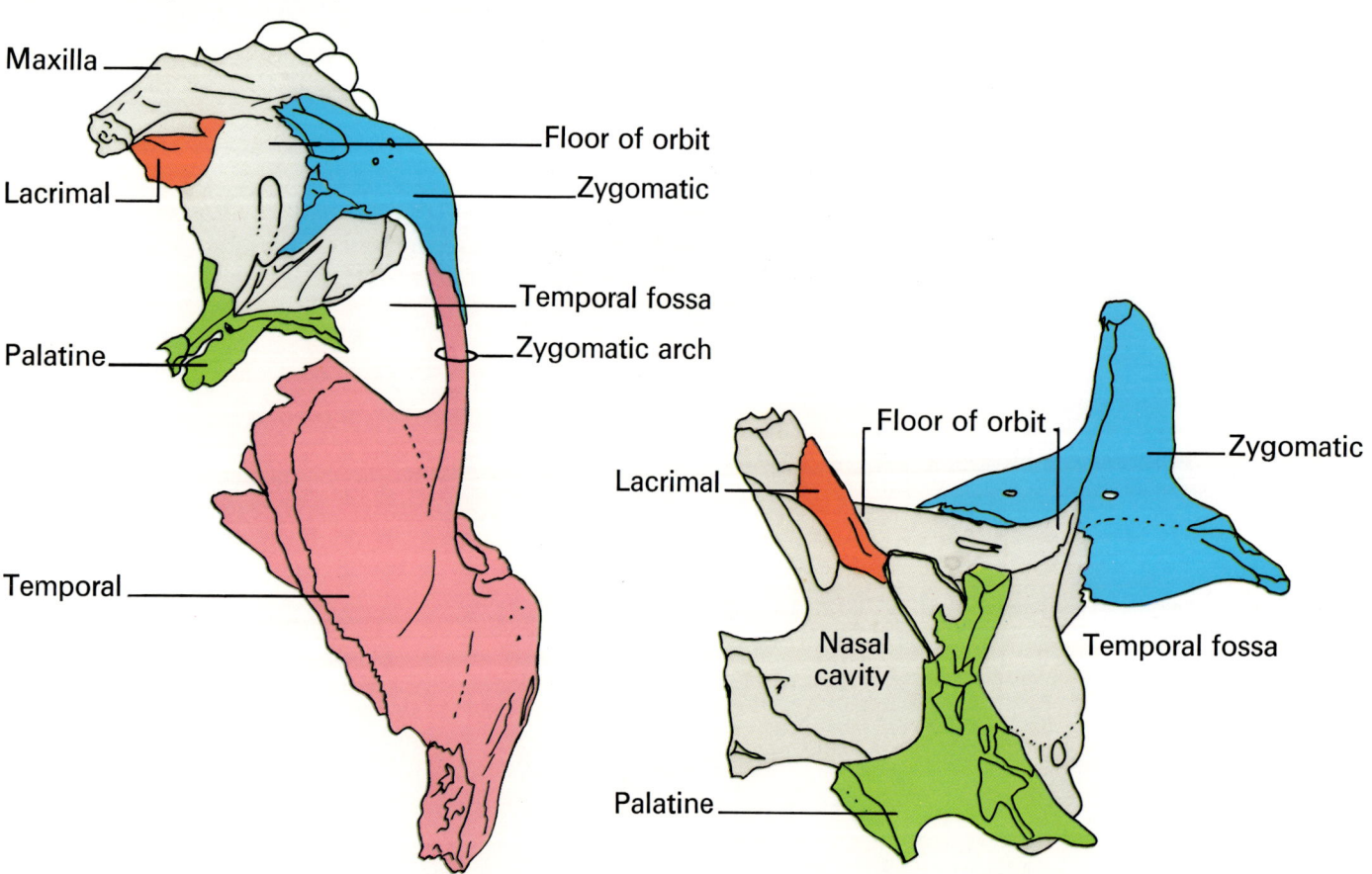

Zygomatic.

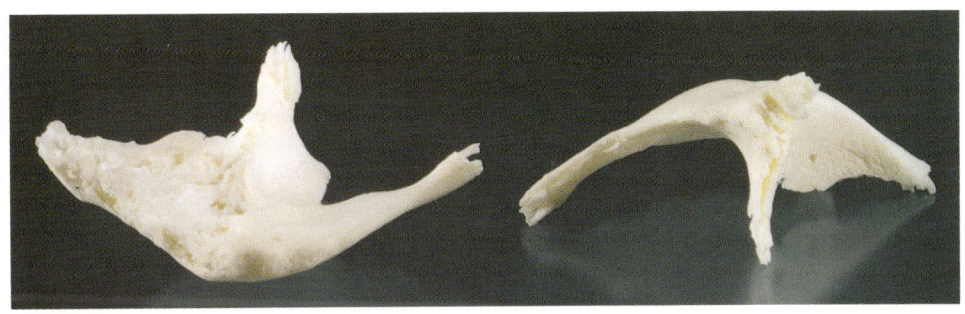

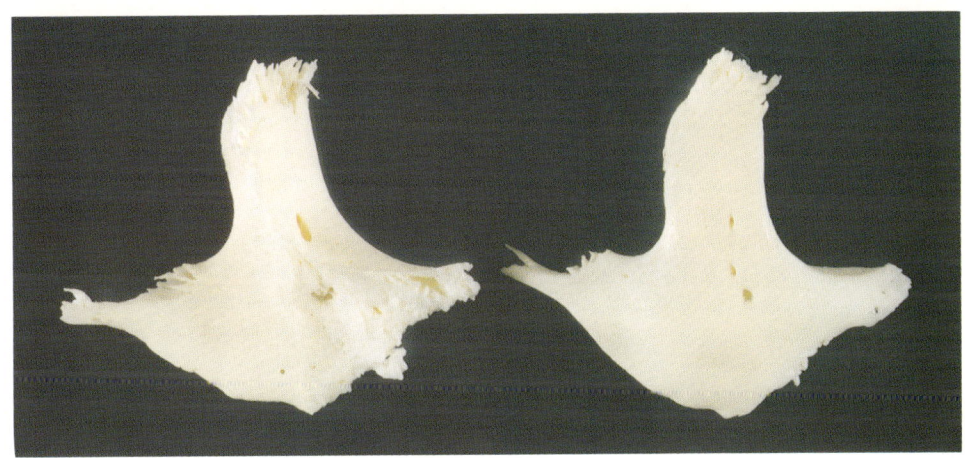

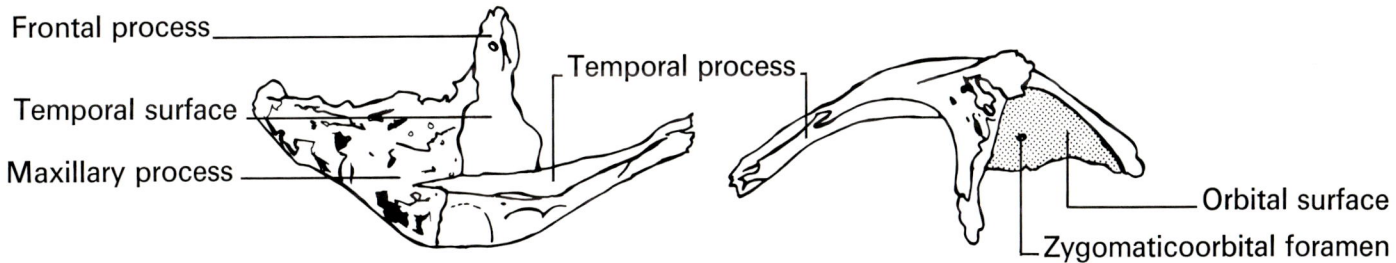

A. Right, zygomatic from above, and left, from below.

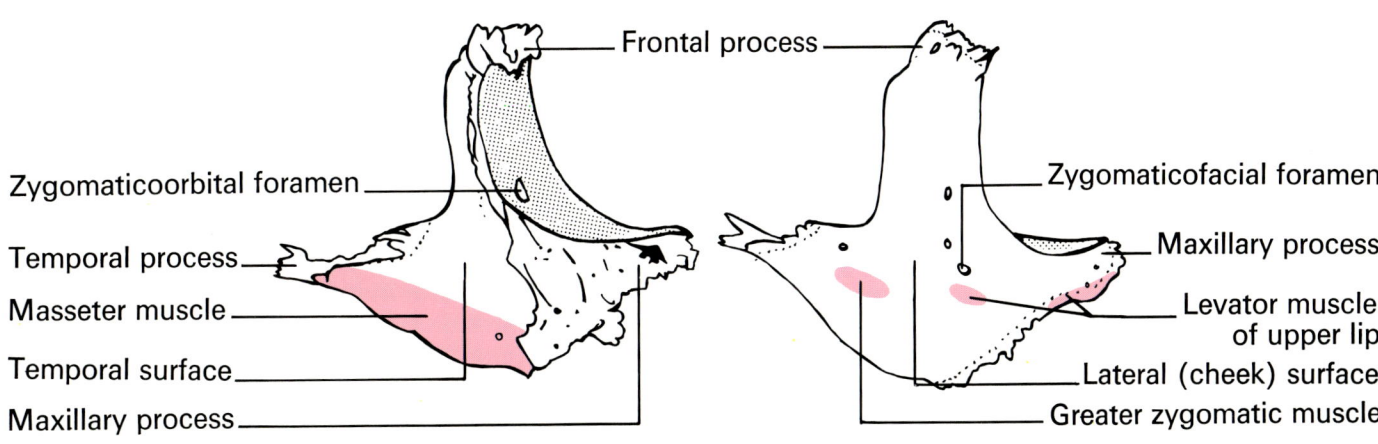

B. Right, zygomatic from front, and left, from behind. Muscle attachments.

SECTION 4. Palatine.

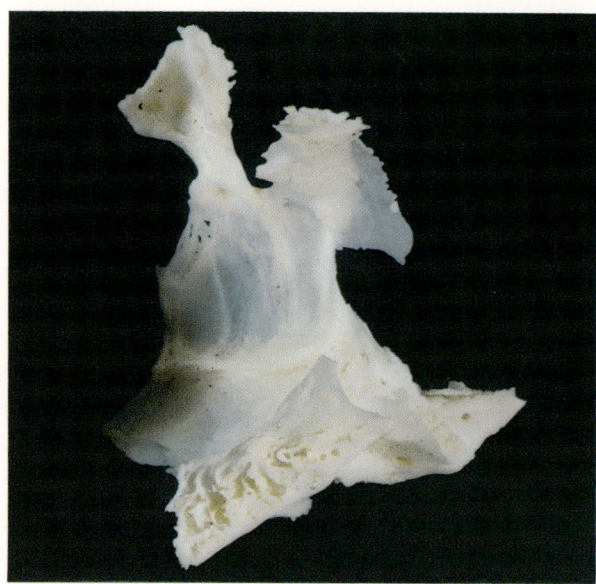

A. Right from medial side.

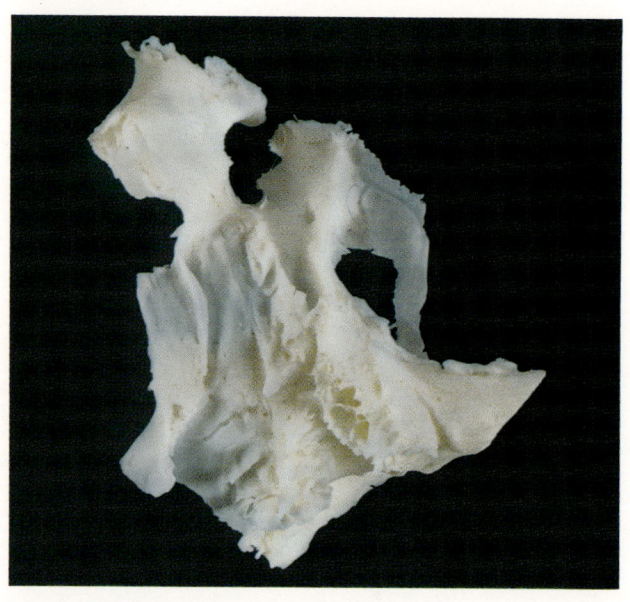

B. Left from lateral side.

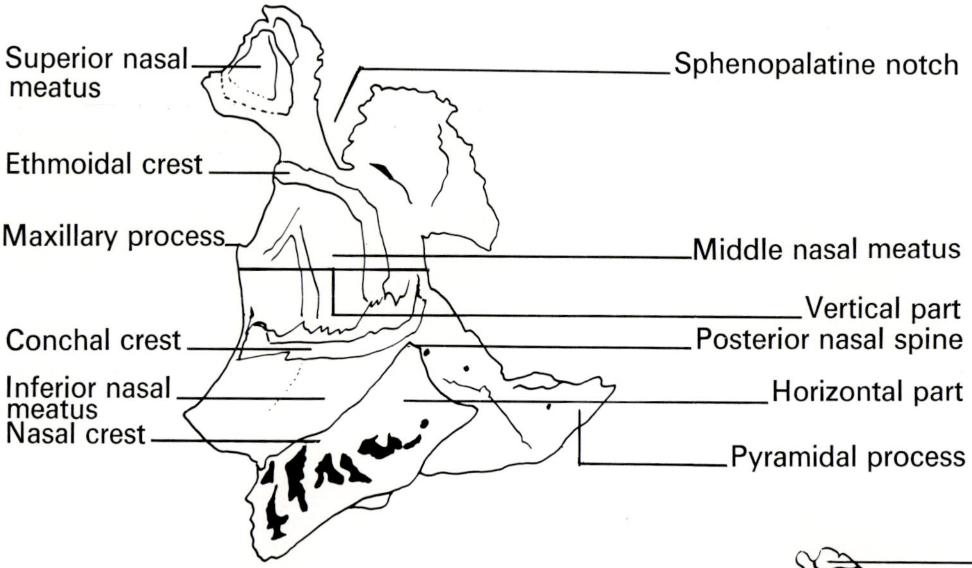

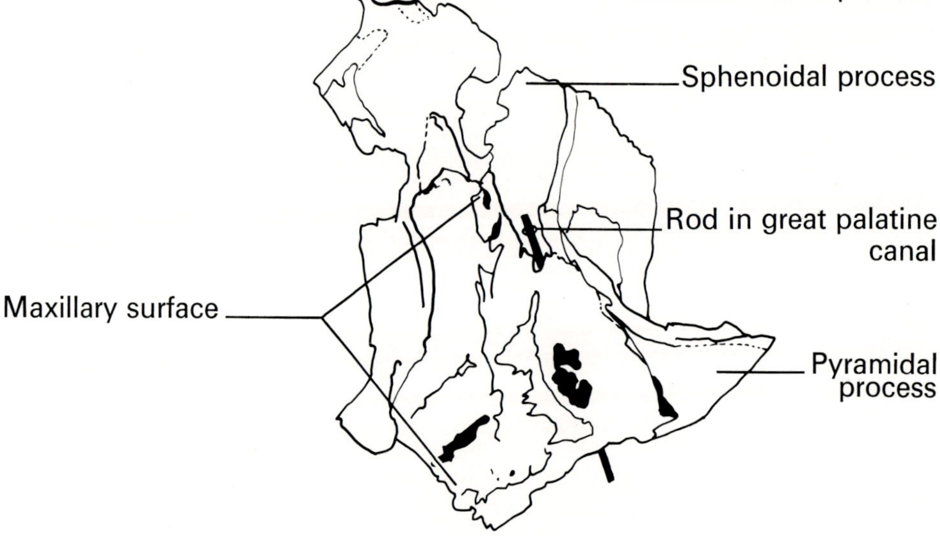

- Superior nasal meatus
- Ethmoidal crest
- Maxillary process
- Conchal crest
- Inferior nasal meatus
- Nasal crest
- Sphenopalatine notch
- Middle nasal meatus
- Vertical part
- Posterior nasal spine
- Horizontal part
- Pyramidal process

- Orbital process
- Sphenoidal process
- Rod in great palatine canal
- Maxillary surface
- Pyramidal process

Palatine.

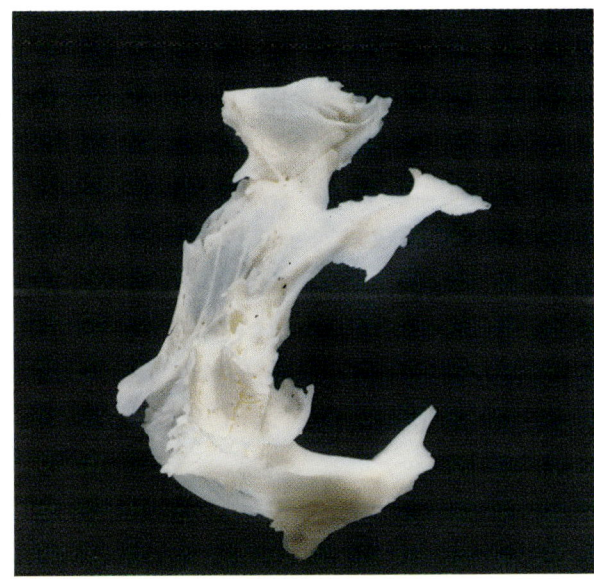

A. Left from behind.

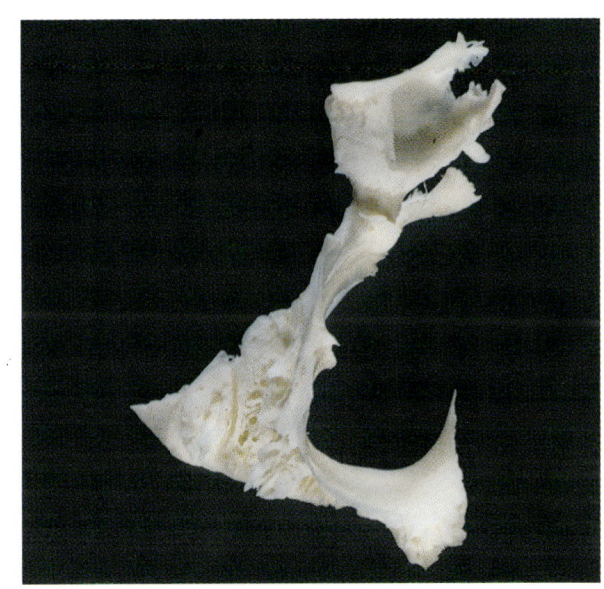

B. Right from front.

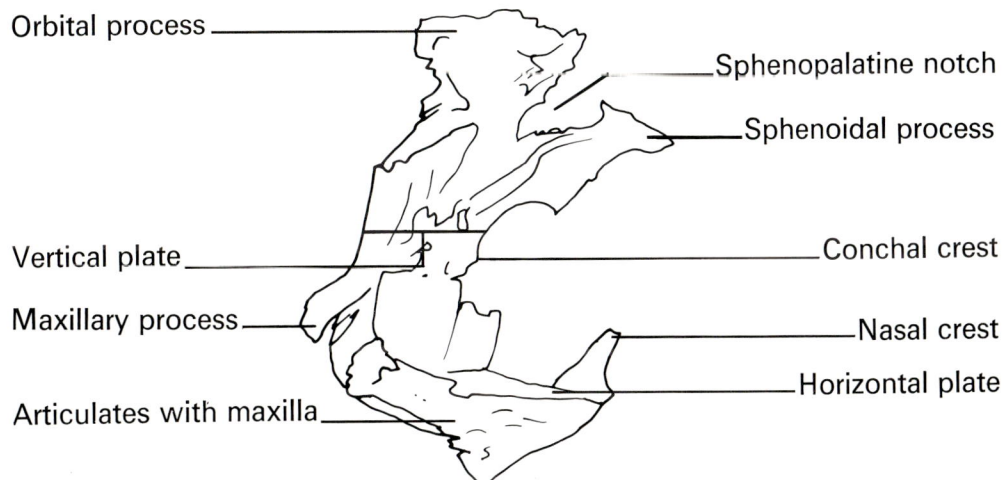

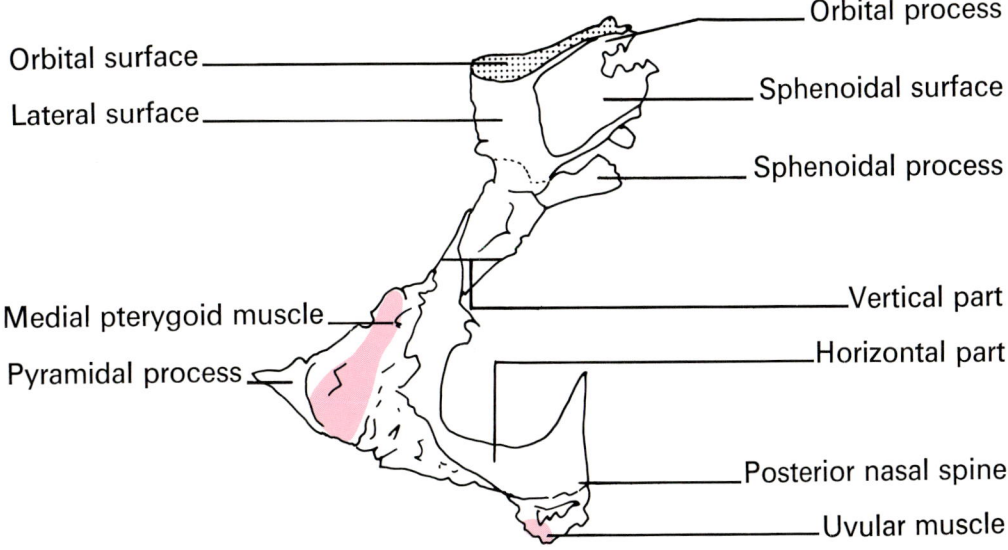

SECTION 5. Small facial bones.

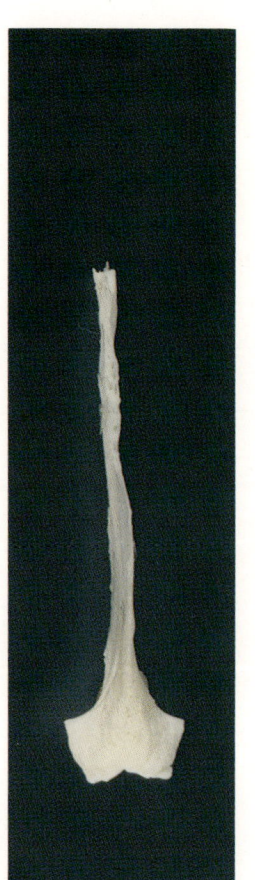
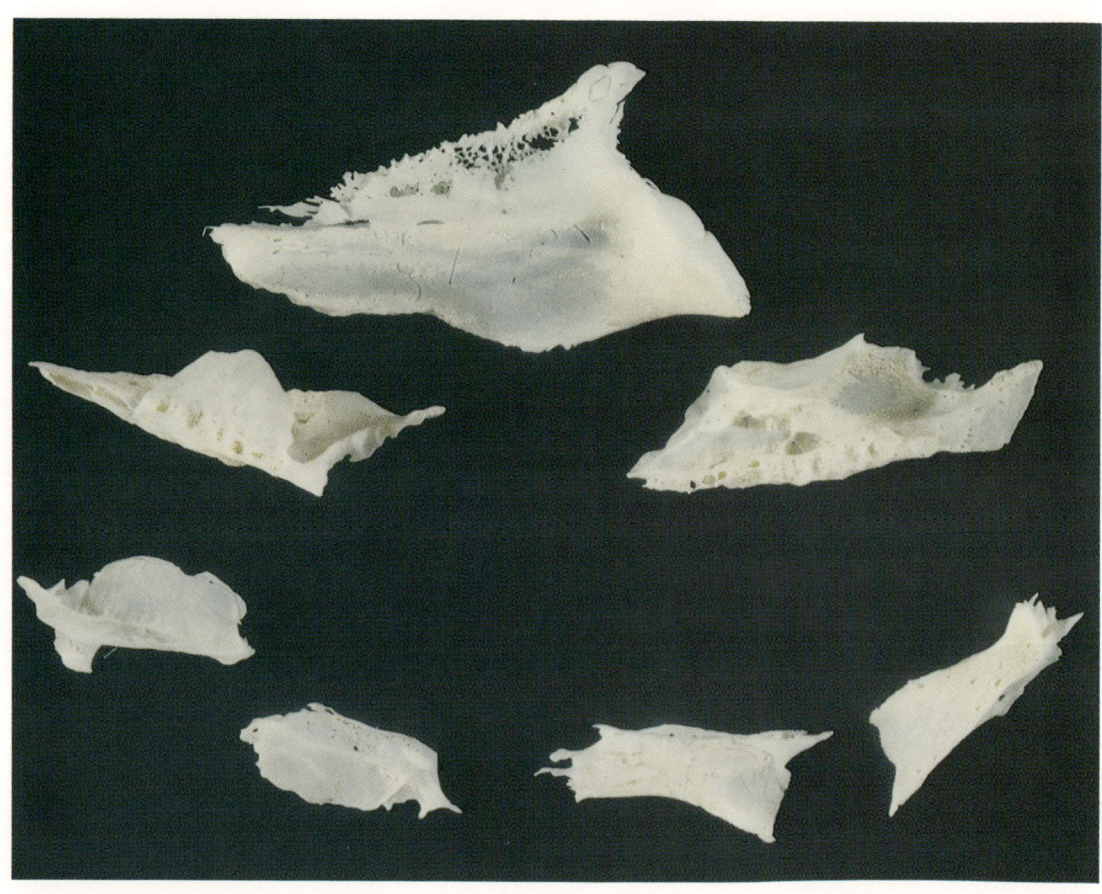

Vomer, inferior conchae, lacrimal and nasal bones.

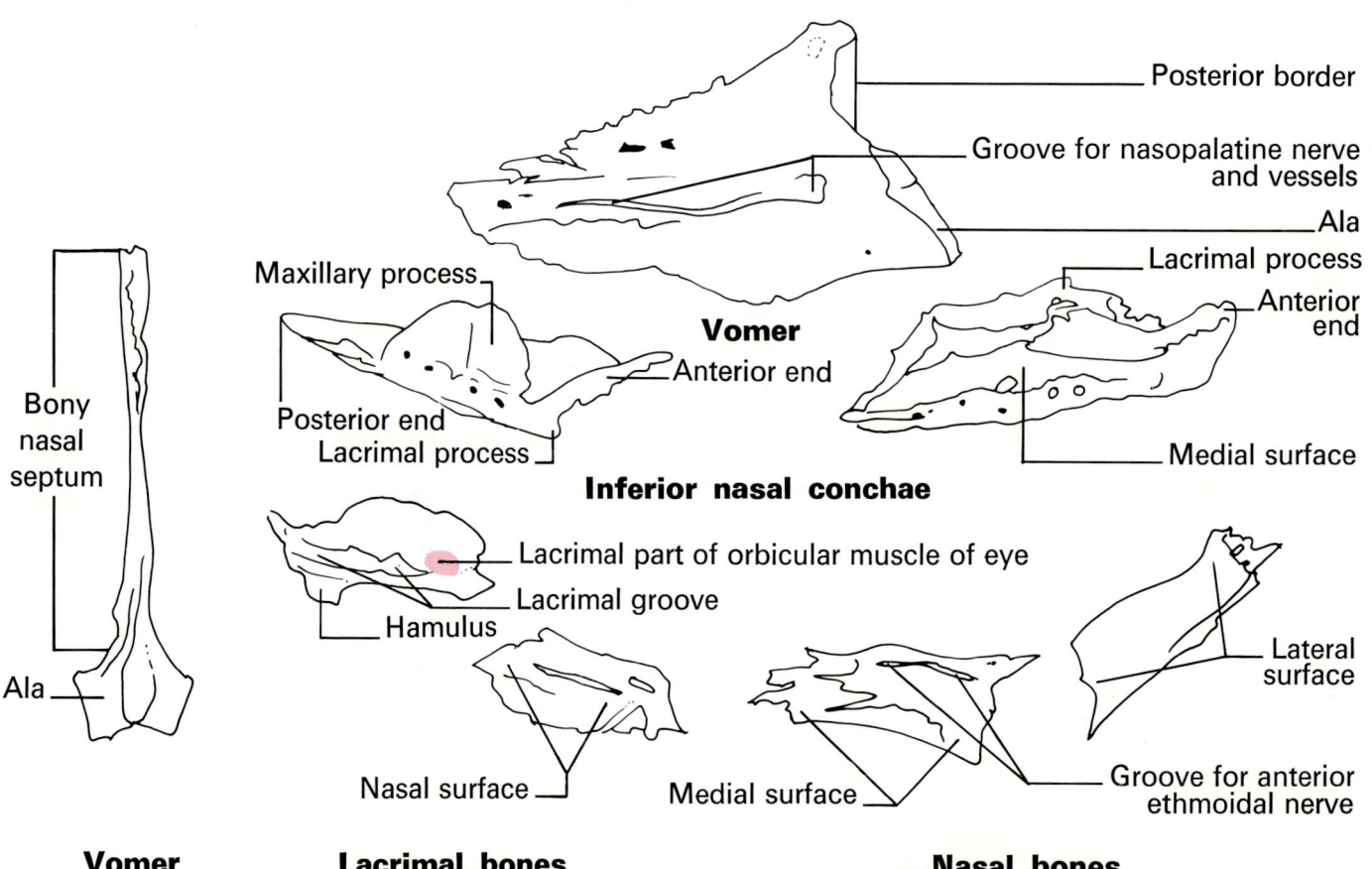

Vomer **Lacrimal bones** **Nasal bones**

PART III. CRANIAL BONES

SECTION 1. Frontal.

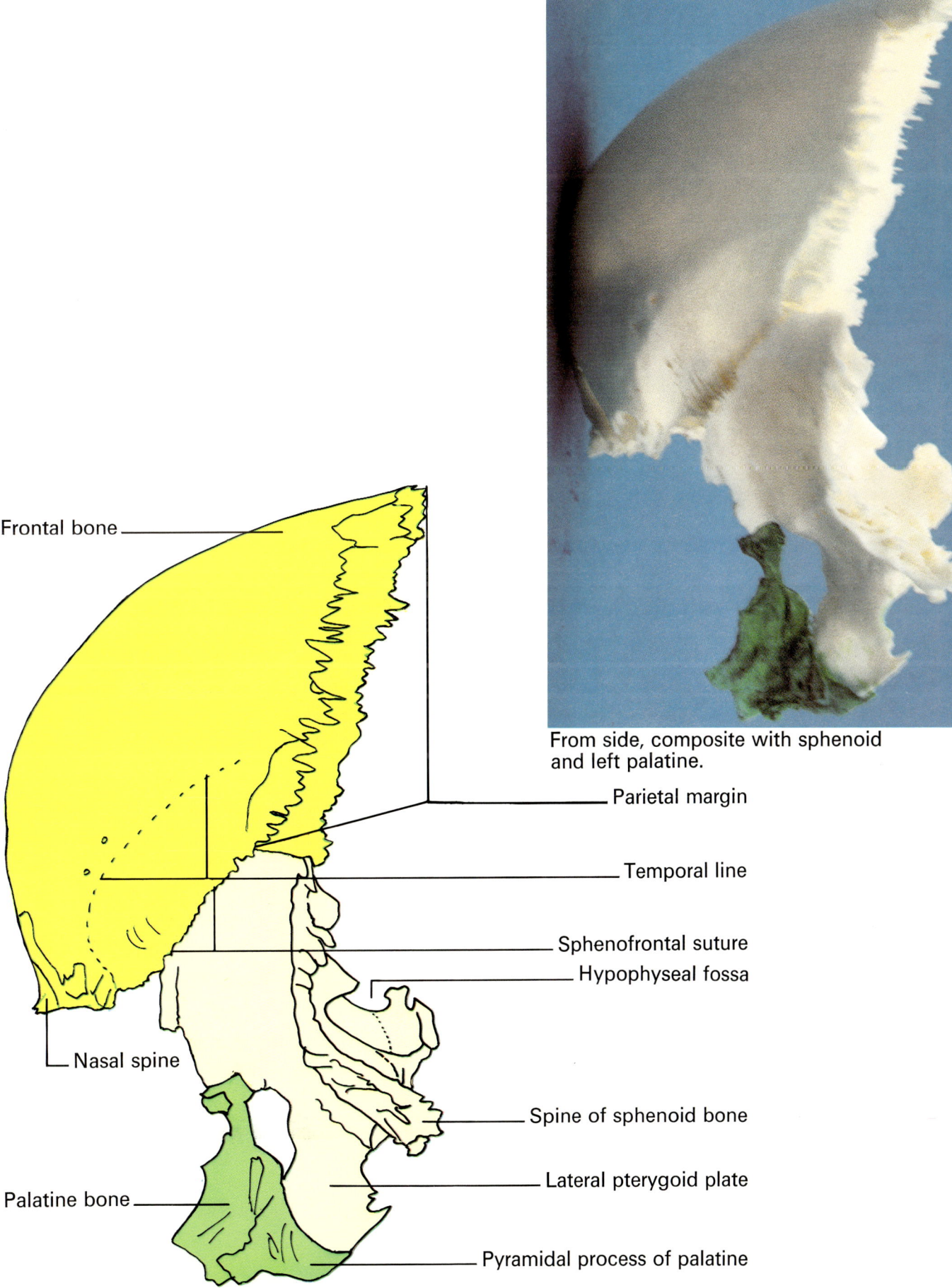

From side, composite with sphenoid and left palatine.

Frontal.

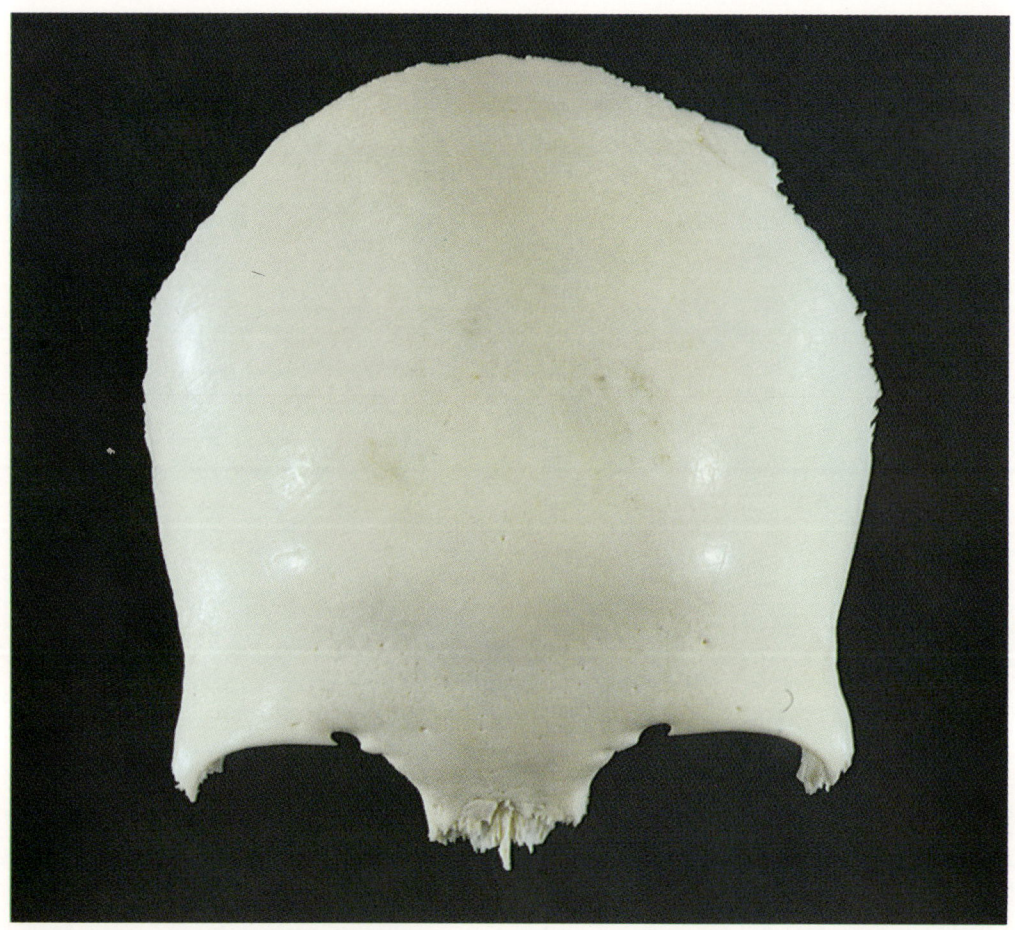

From front. Muscle attachments.

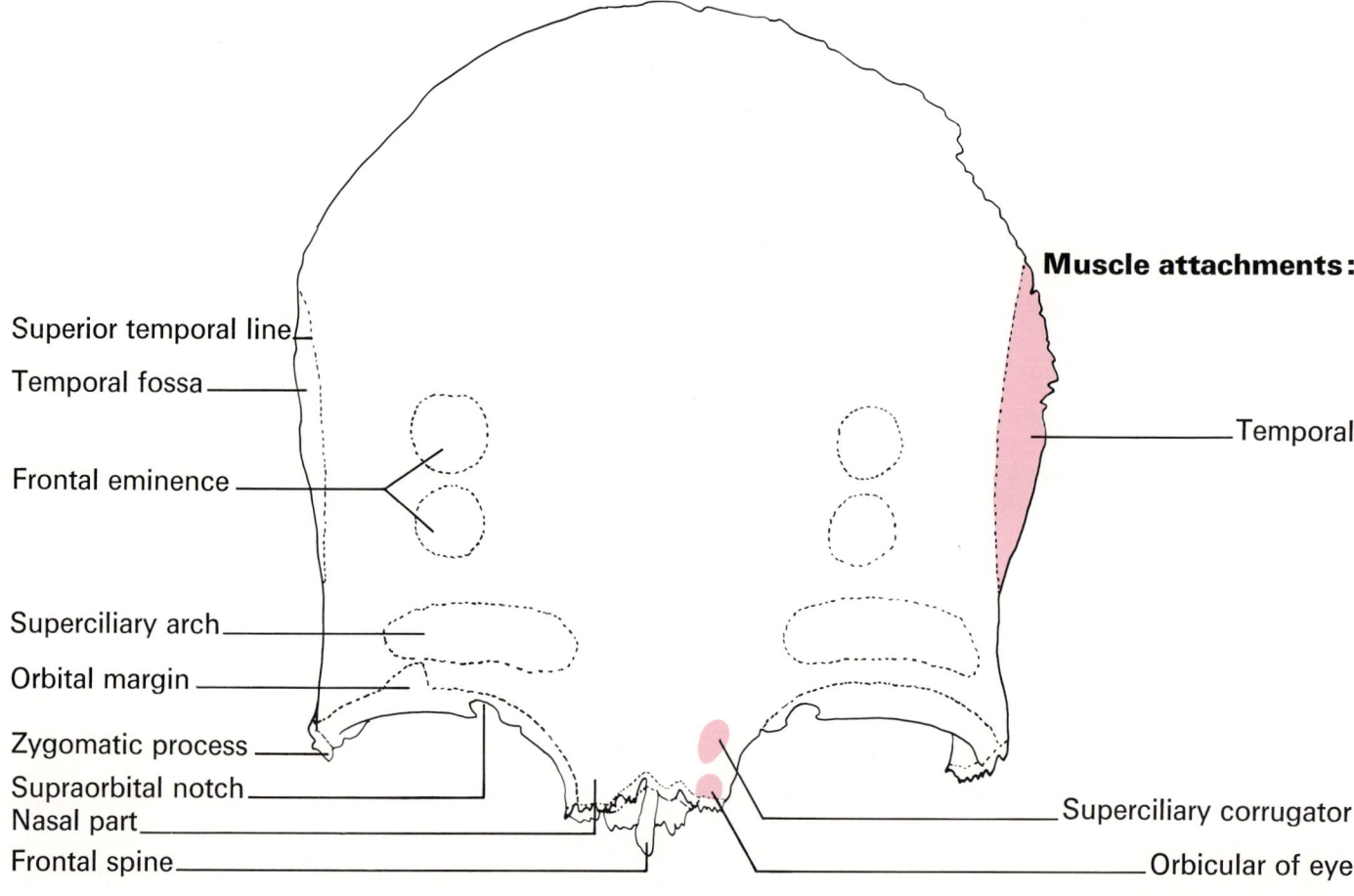

Superior temporal line
Temporal fossa
Frontal eminence
Superciliary arch
Orbital margin
Zygomatic process
Supraorbital notch
Nasal part
Frontal spine

Muscle attachments:
Temporal
Superciliary corrugator
Orbicular of eye

Frontal.

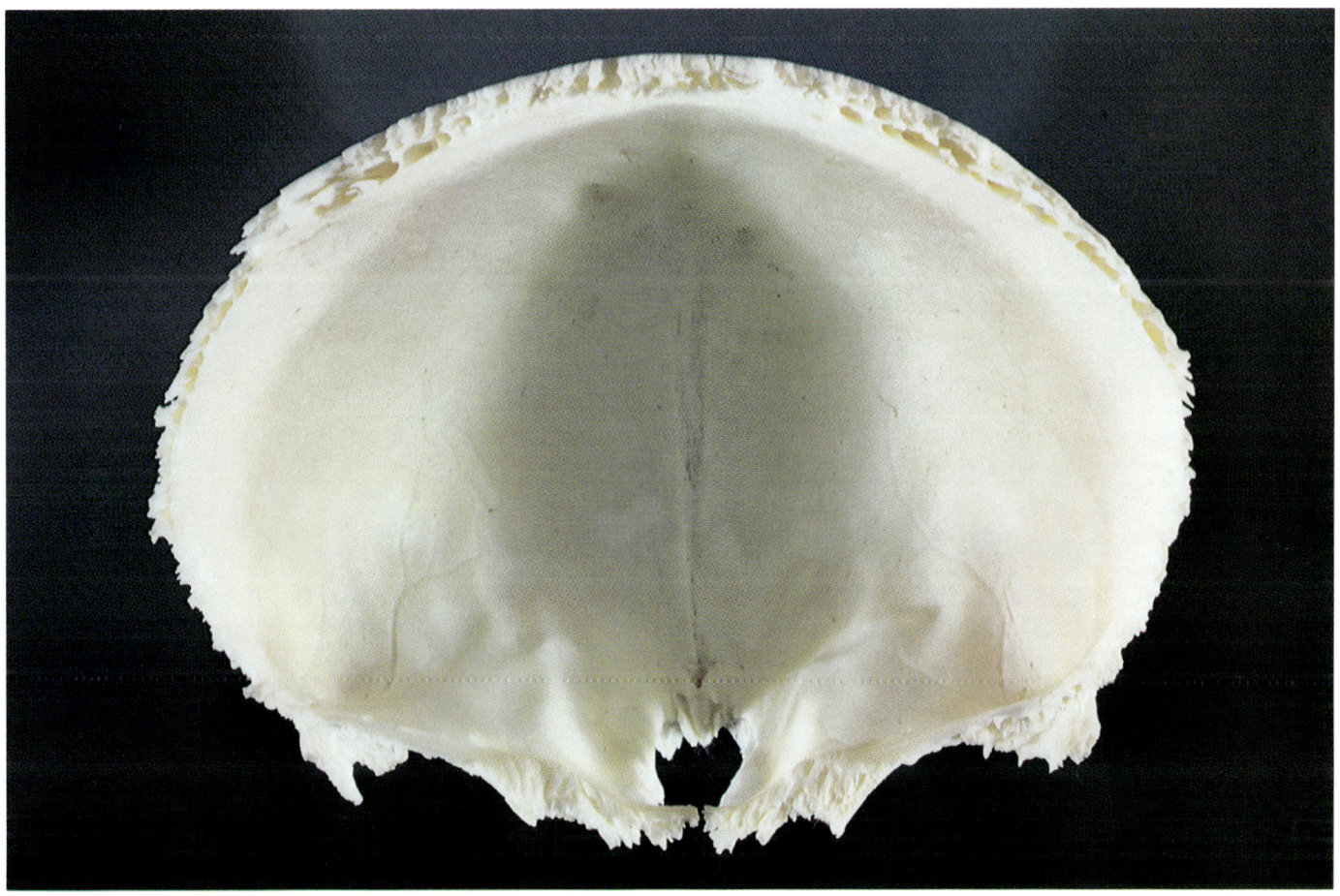

From behind. Cranial surface. (For view from below, see page 48.)

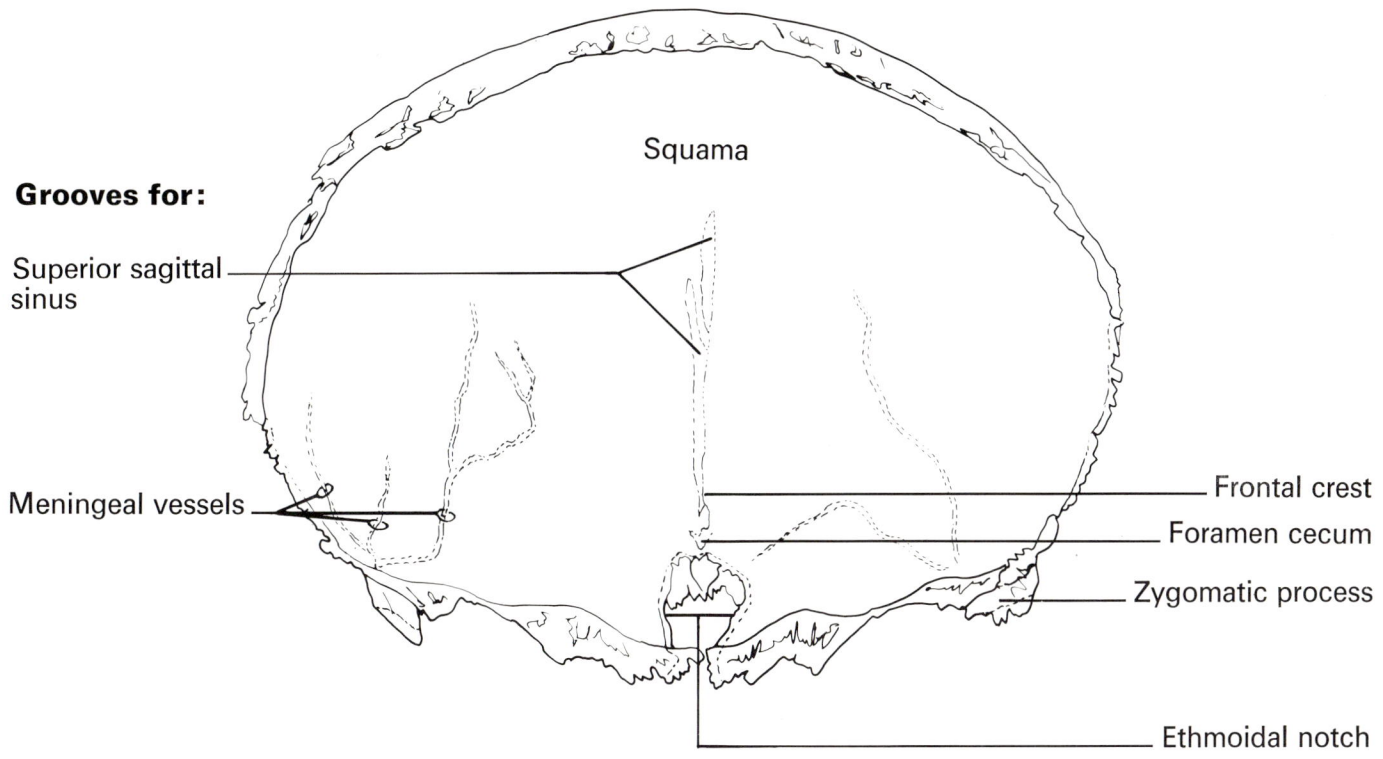

Grooves for:
Superior sagittal sinus
Meningeal vessels
Squama
Frontal crest
Foramen cecum
Zygomatic process
Ethmoidal notch

SECTION 2.   Ethmoid.

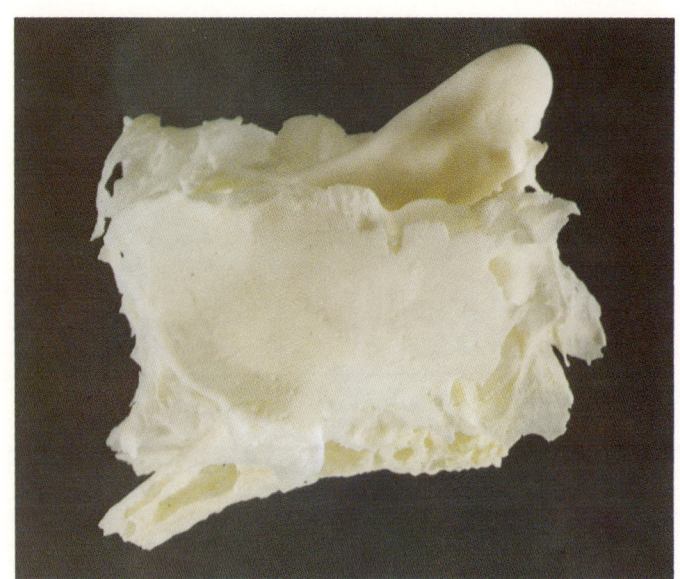

A. From side.

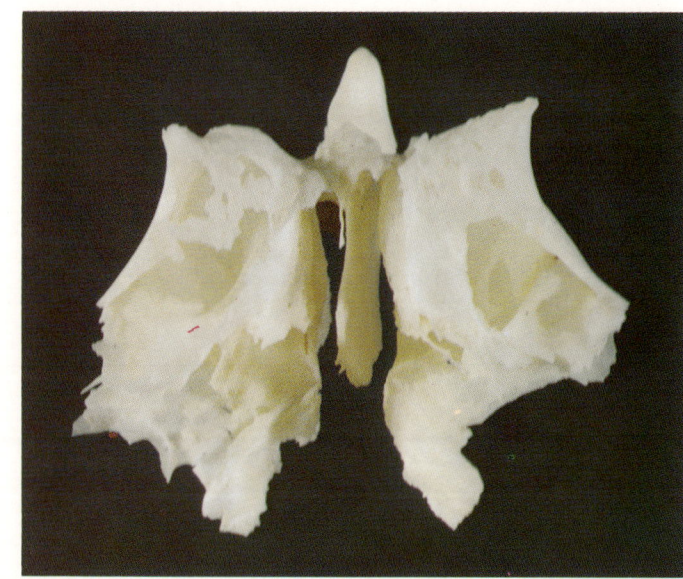

B. From behind magnification 2X.

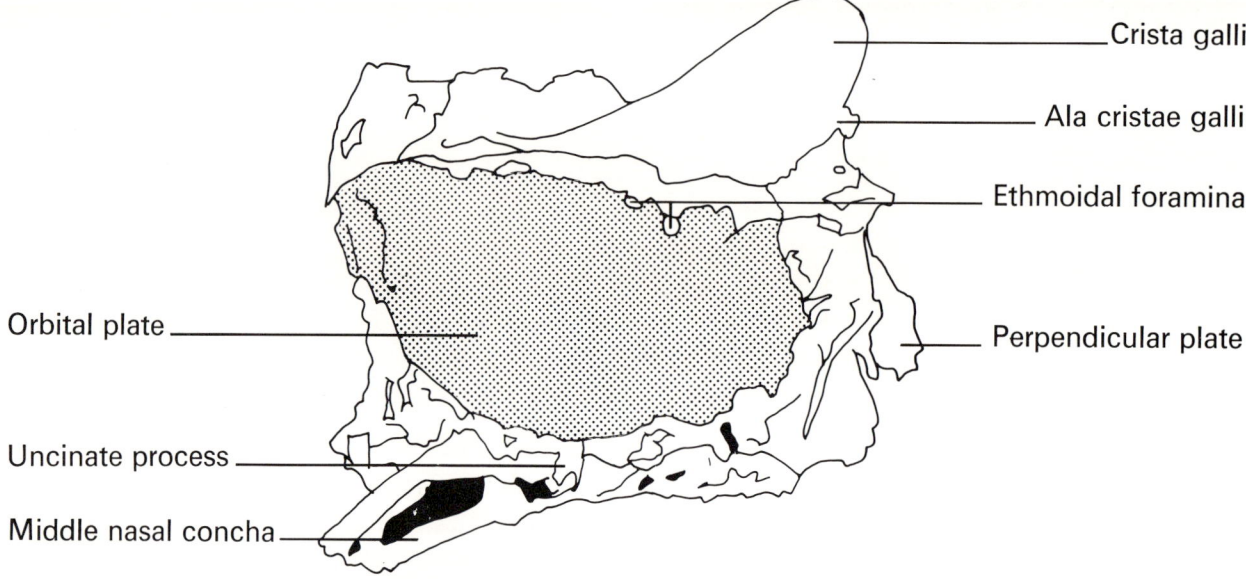

- Crista galli
- Ala cristae galli
- Ethmoidal foramina
- Perpendicular plate
- Orbital plate
- Uncinate process
- Middle nasal concha

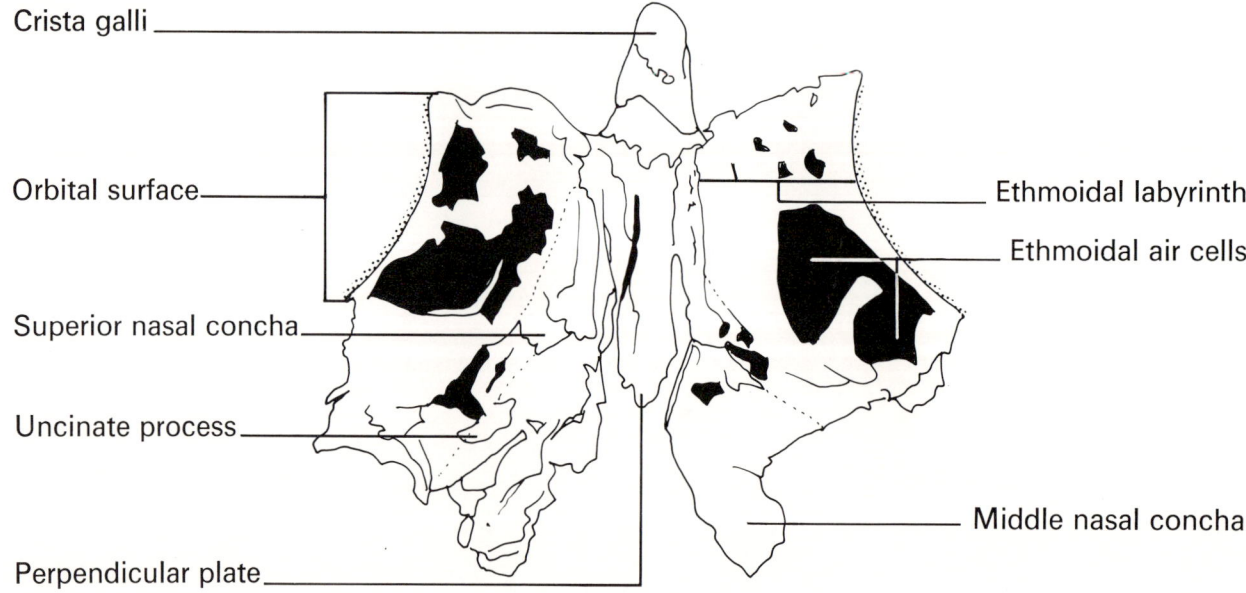

- Crista galli
- Orbital surface
- Ethmoidal labyrinth
- Ethmoidal air cells
- Superior nasal concha
- Uncinate process
- Middle nasal concha
- Perpendicular plate

Ethmoid.

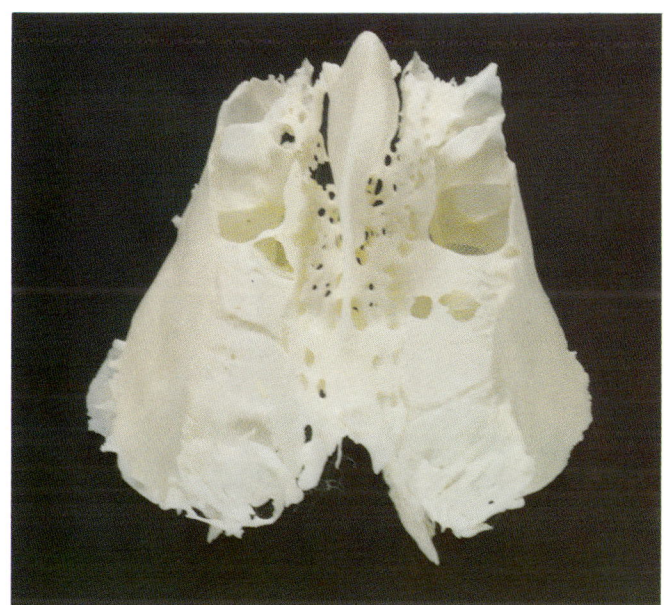

A. From above.
(For view from front, see page 60 and 61)

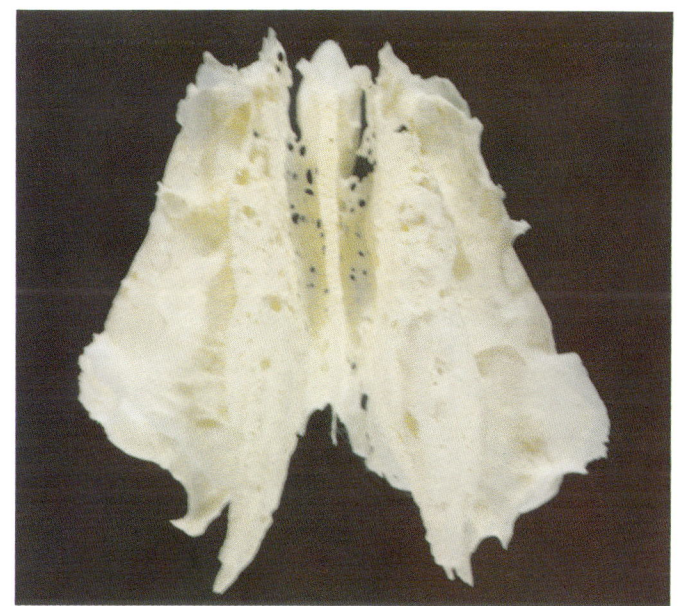

B. From below.

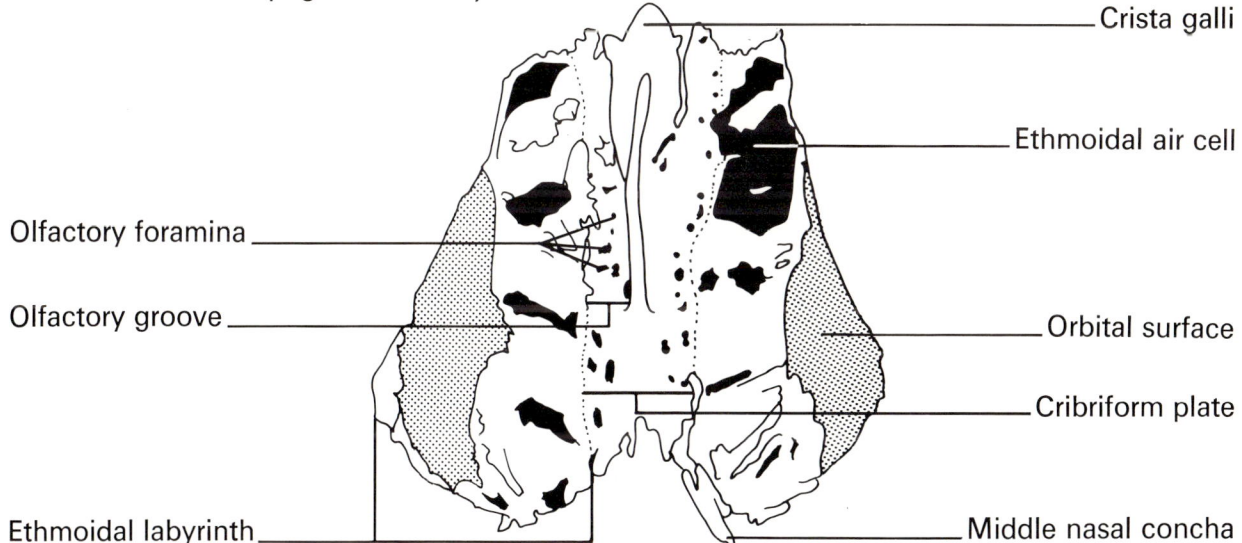

- Crista galli
- Ethmoidal air cell
- Olfactory foramina
- Olfactory groove
- Orbital surface
- Cribriform plate
- Ethmoidal labyrinth
- Middle nasal concha

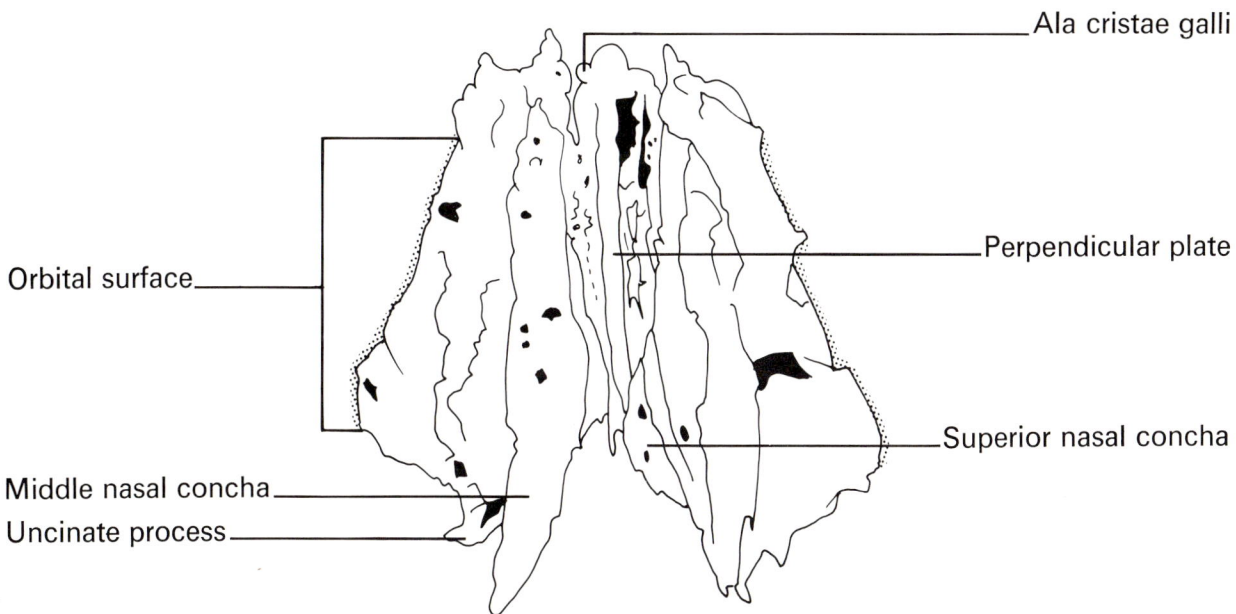

- Ala cristae galli
- Orbital surface
- Perpendicular plate
- Superior nasal concha
- Middle nasal concha
- Uncinate process

SECTION 3.   Sphenoid.

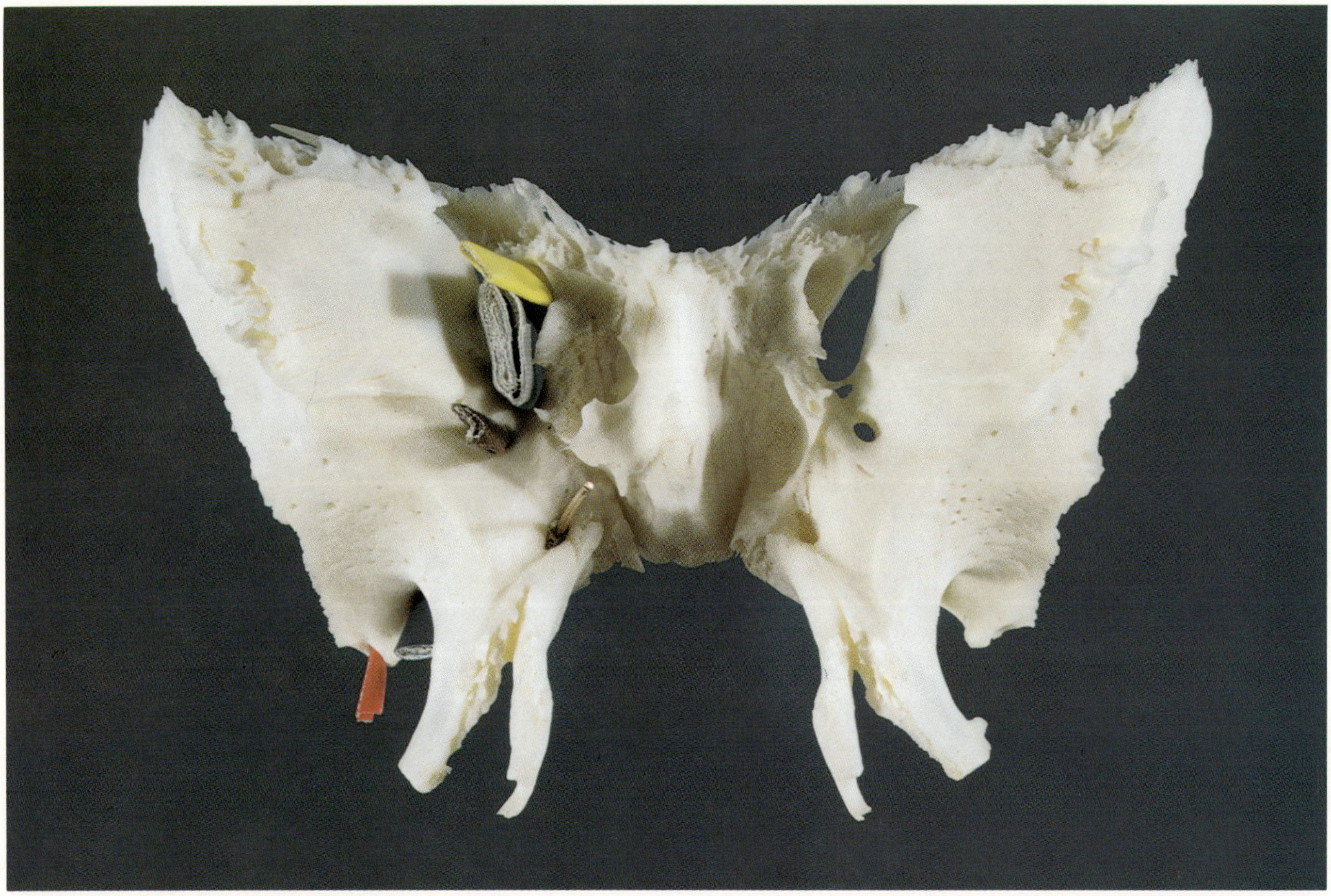

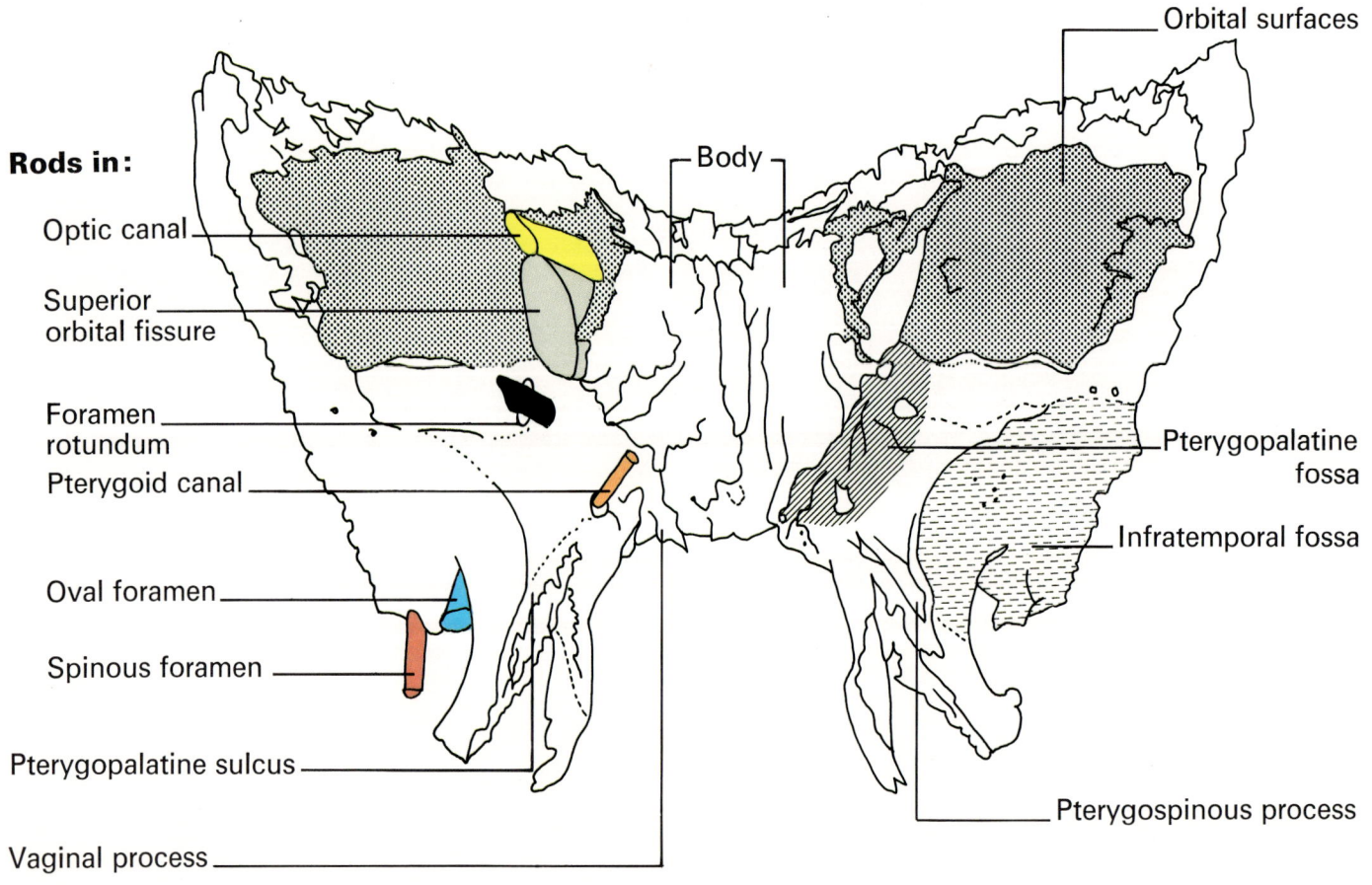

From front. Magnification 1.5X.

Rods in:
- Optic canal
- Superior orbital fissure
- Foramen rotundum
- Pterygoid canal
- Oval foramen
- Spinous foramen
- Pterygopalatine sulcus
- Vaginal process

- Orbital surfaces
- Body
- Pterygopalatine fossa
- Infratemporal fossa
- Pterygospinous process

Sphenoid.

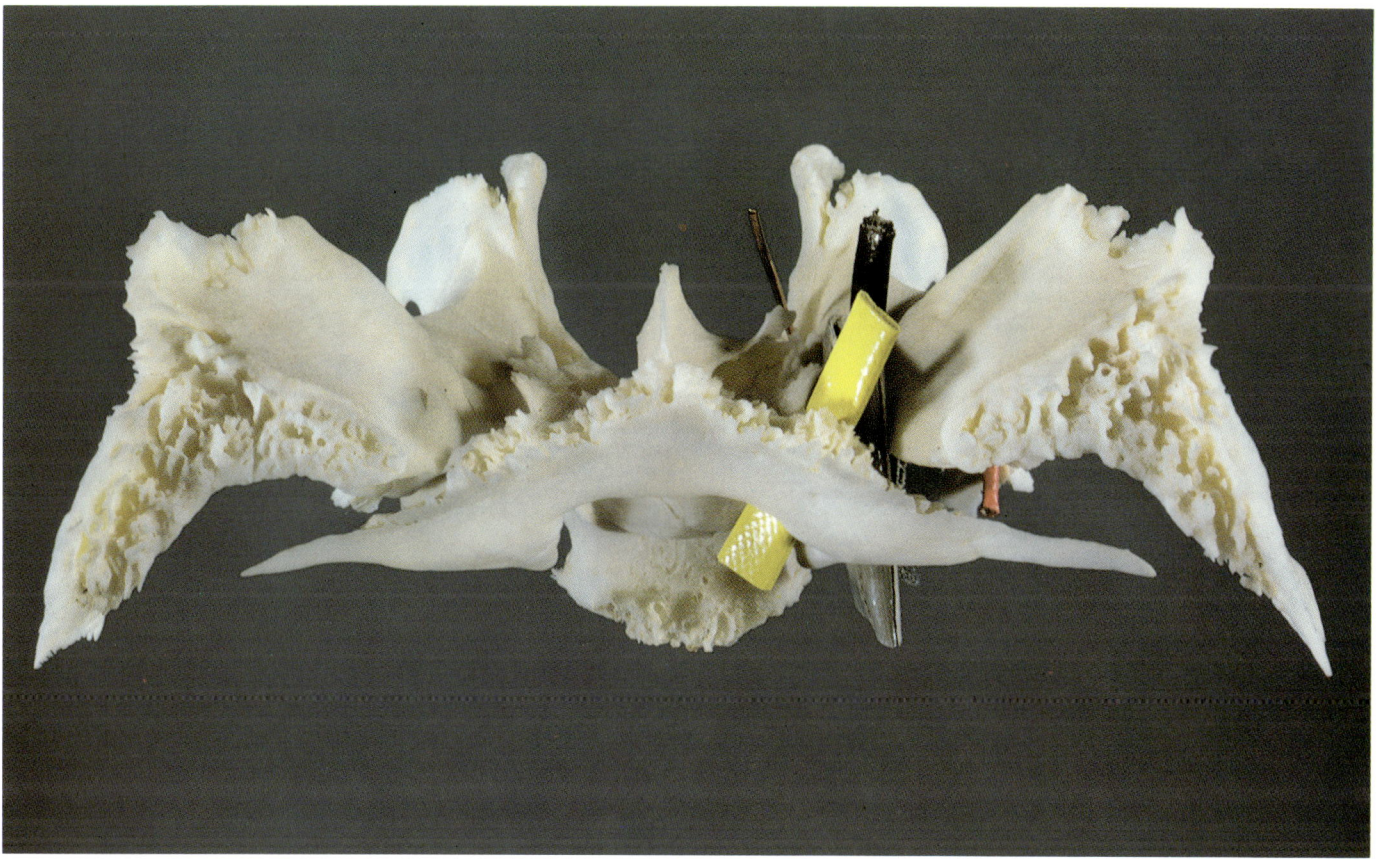

From above, onto orbital surface.

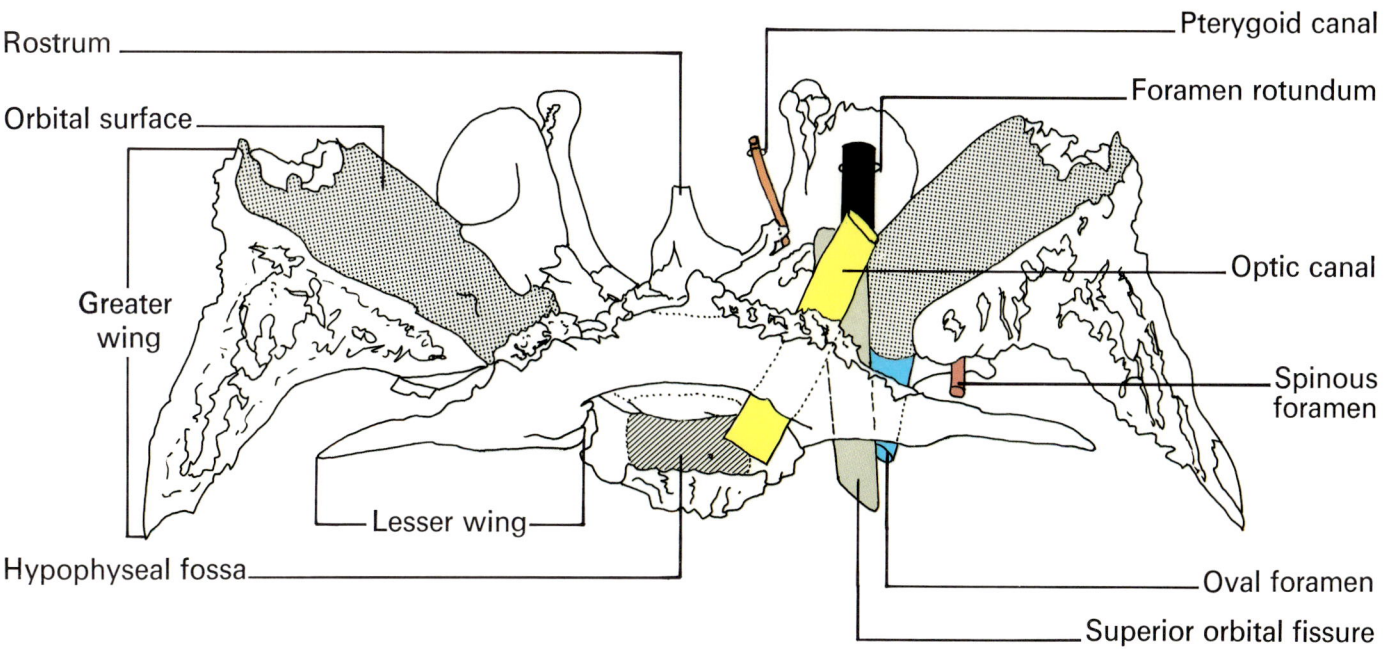

Rostrum
Orbital surface
Greater wing
Lesser wing
Hypophyseal fossa

Rods in:
Pterygoid canal
Foramen rotundum
Optic canal
Spinous foramen
Oval foramen
Superior orbital fissure

Sphenoid.

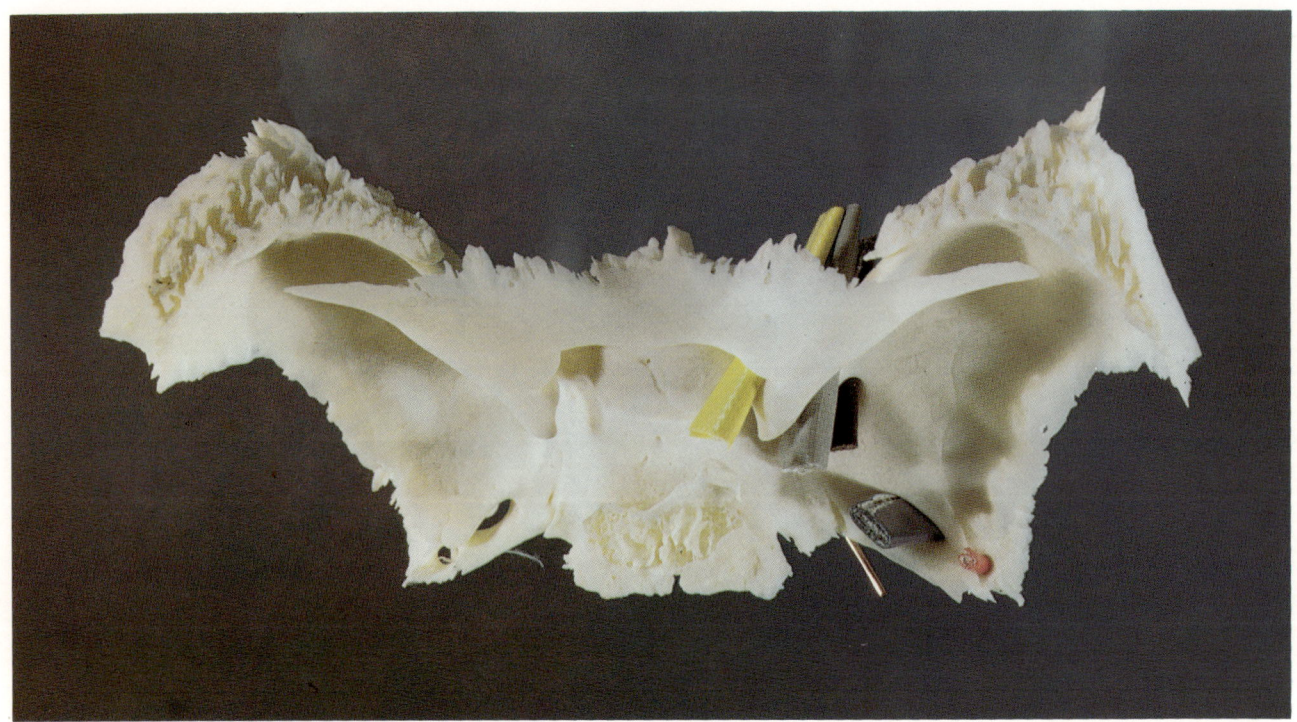

From above, onto cranial surface.

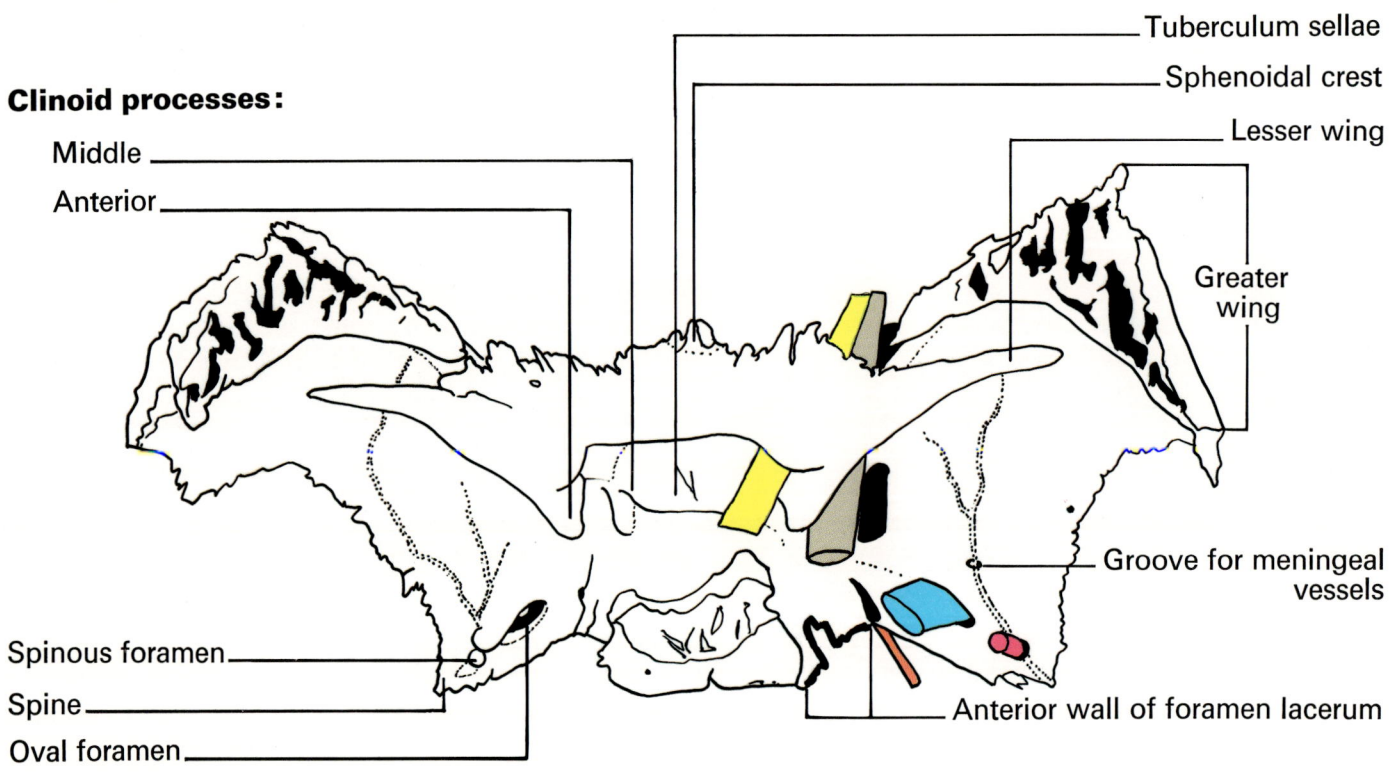

Clinoid processes:
Middle
Anterior

Spinous foramen
Spine
Oval foramen

Tuberculum sellae
Sphenoidal crest
Lesser wing
Greater wing
Groove for meningeal vessels
Anterior wall of foramen lacerum

Sphenoid.

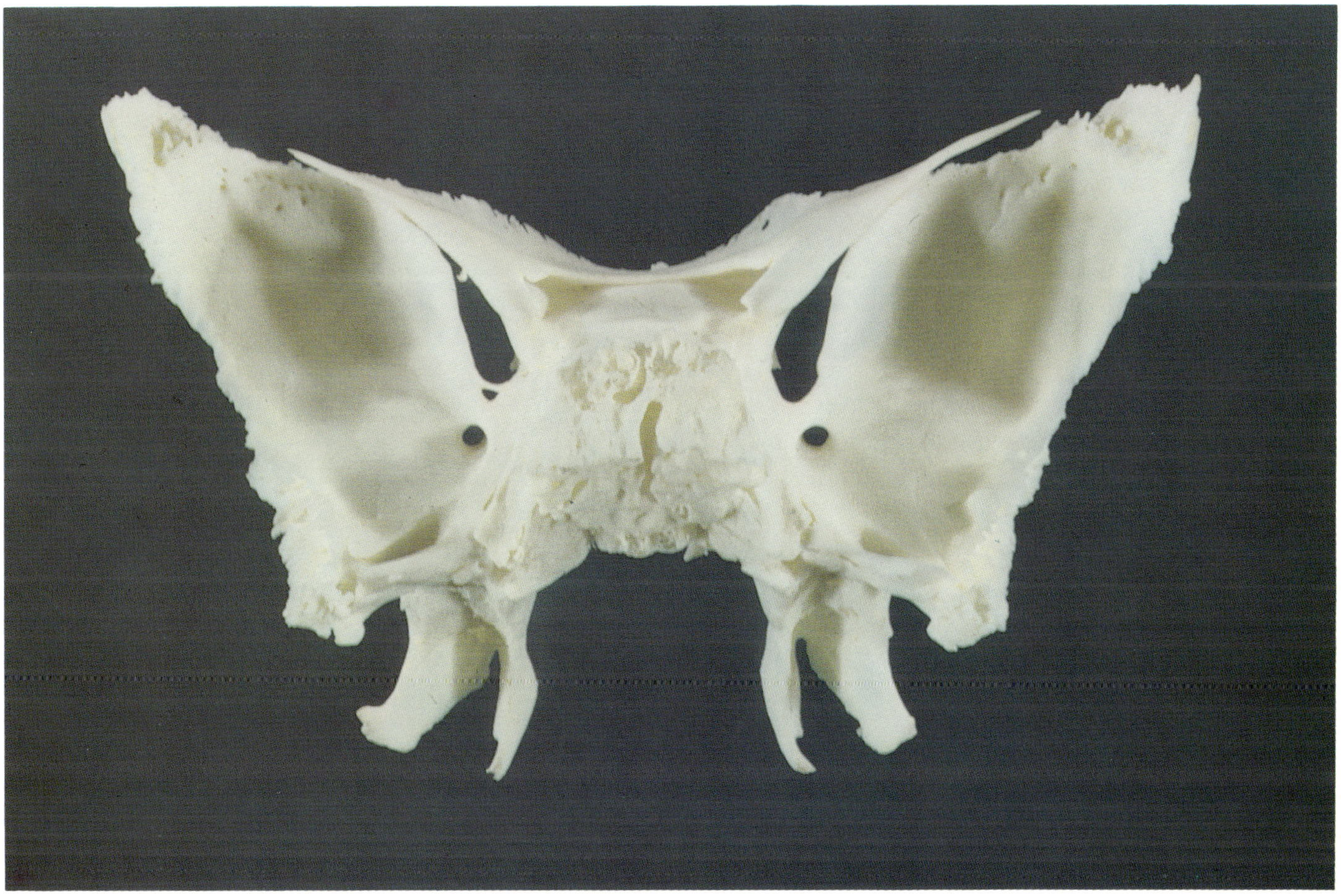

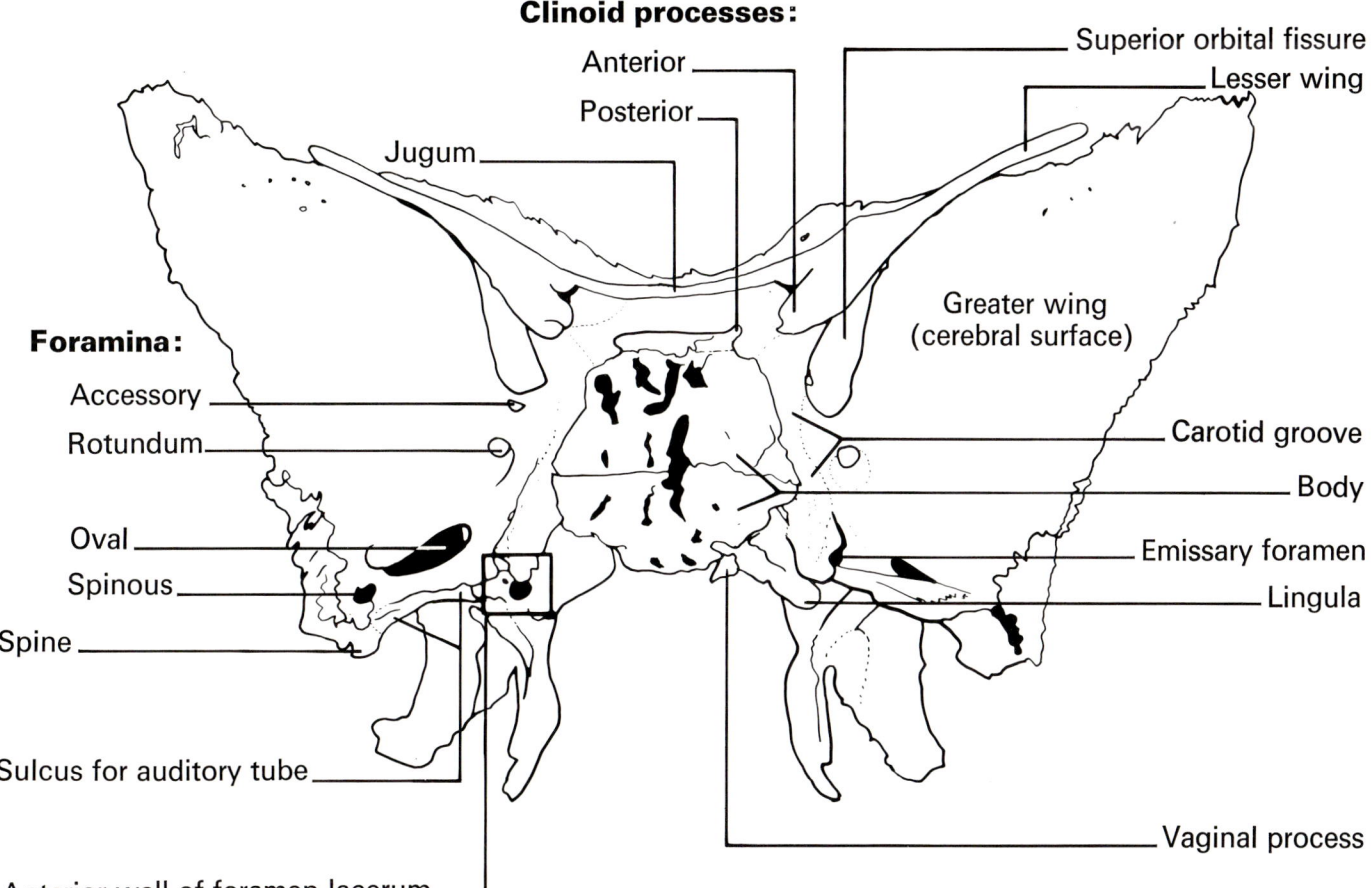

From behind, onto cranial surface of great wings.

Clinoid processes: Anterior, Posterior
Jugum
Superior orbital fissure
Lesser wing
Greater wing (cerebral surface)
Carotid groove
Body
Emissary foramen
Lingula
Vaginal process

Foramina: Accessory, Rotundum, Oval, Spinous
Spine
Sulcus for auditory tube
Anterior wall of foramen lacerum

Sphenoid.

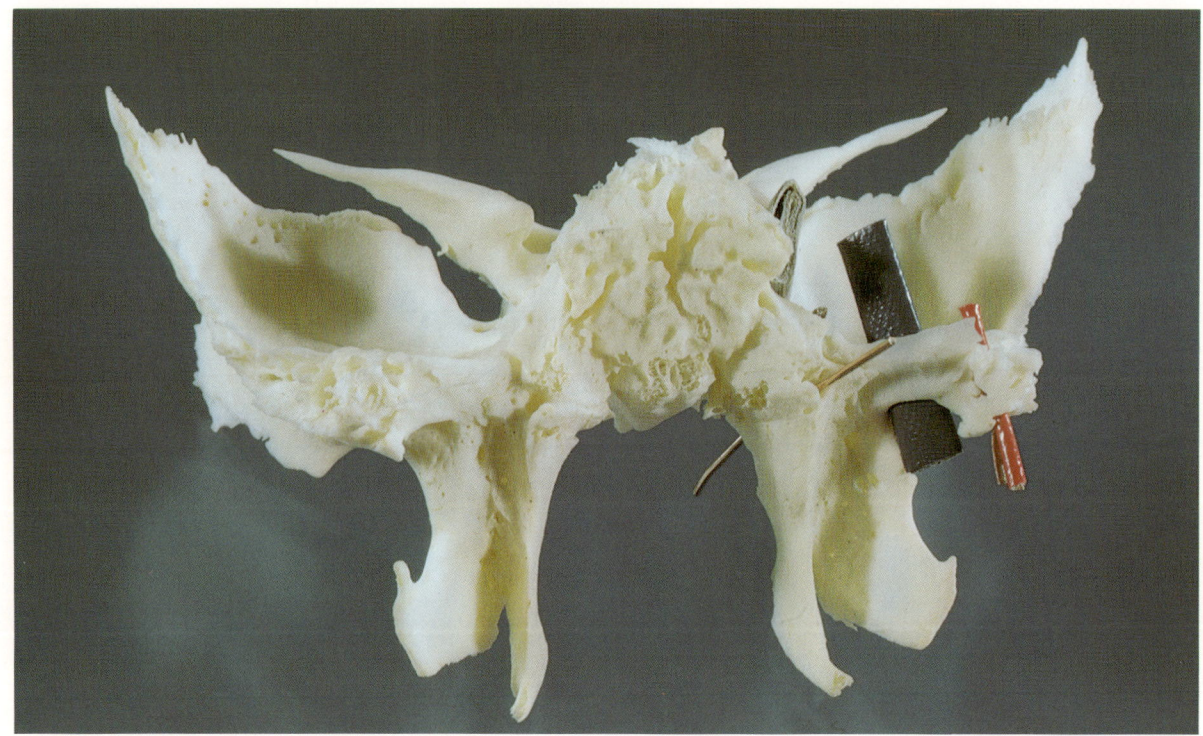

From behind, view of pterygoid plates.

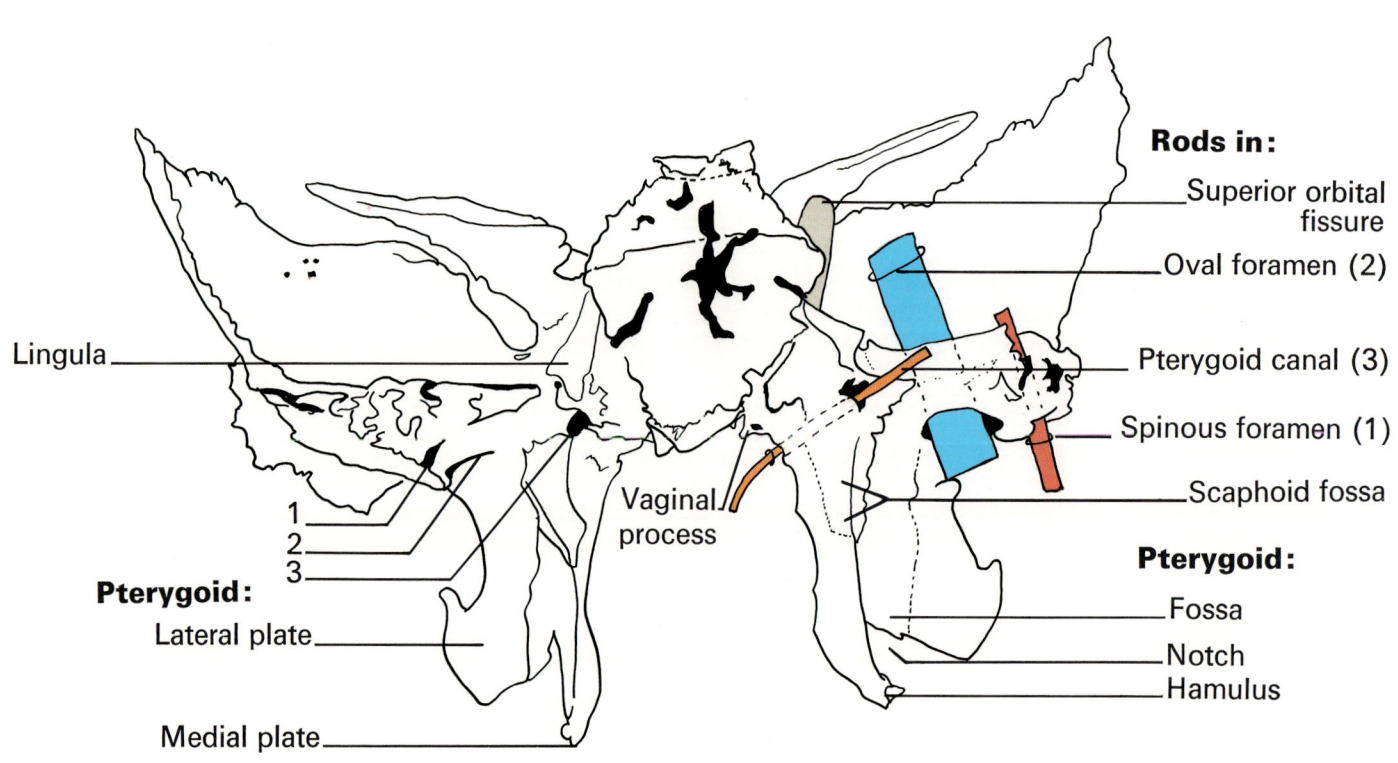

Numbers correspond to identical structures

111

Sphenoid.

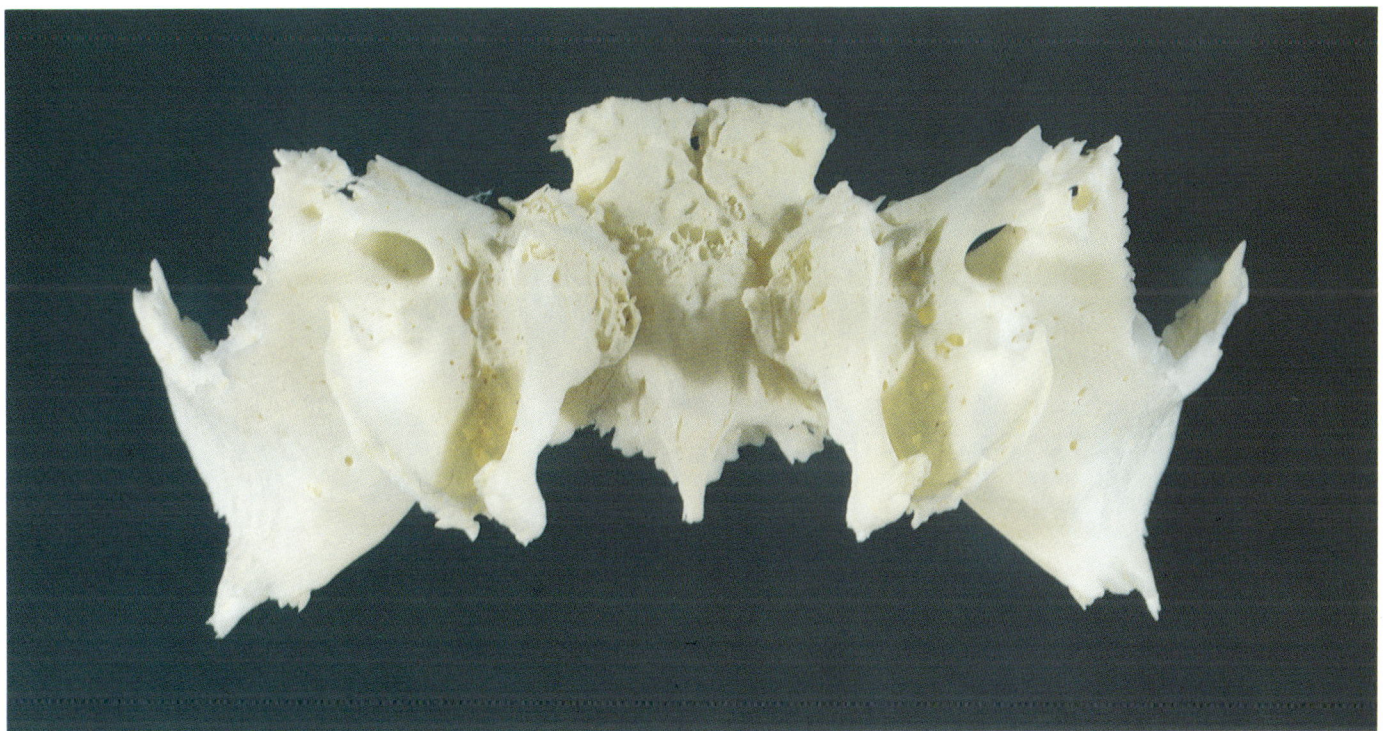

From behind and below.

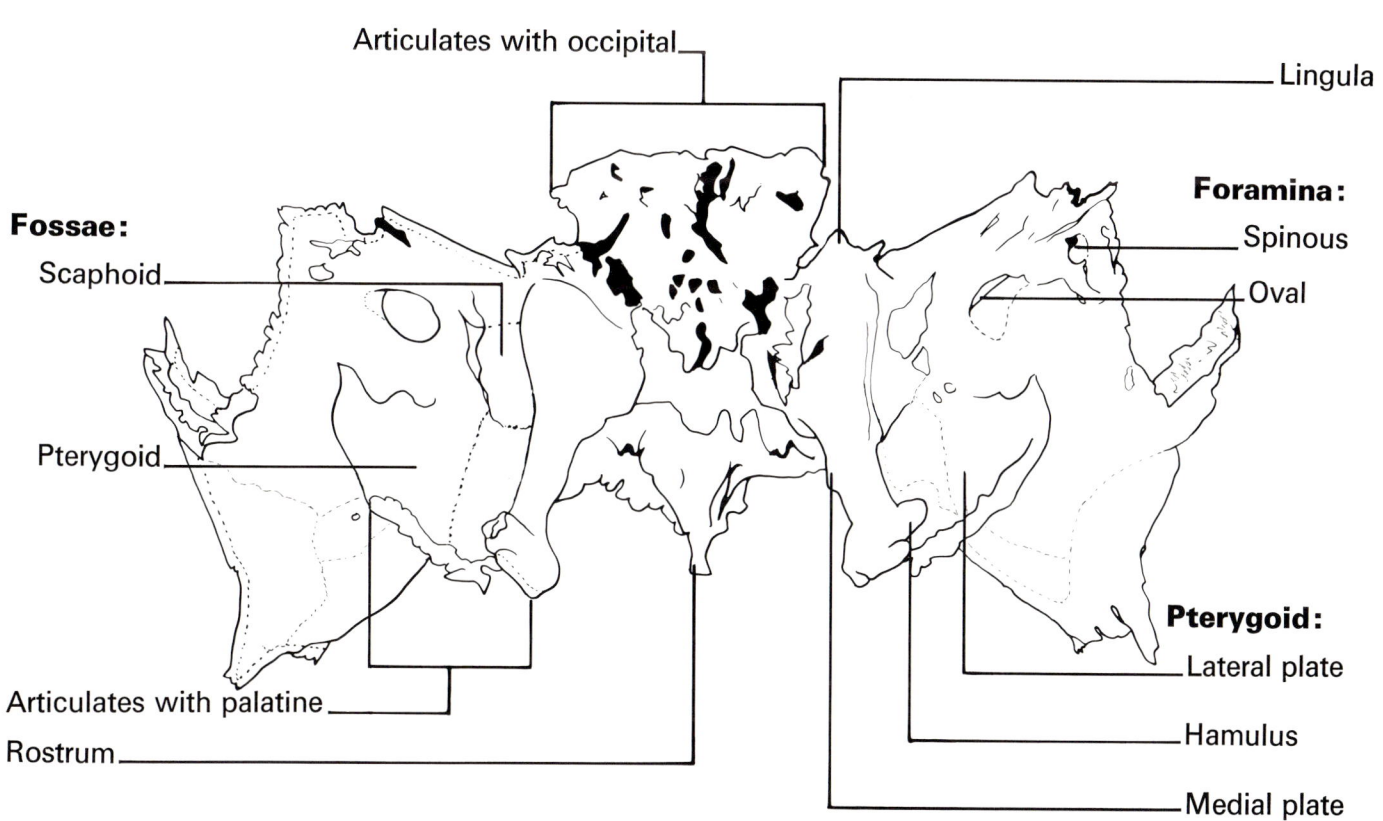

Sphenoid.

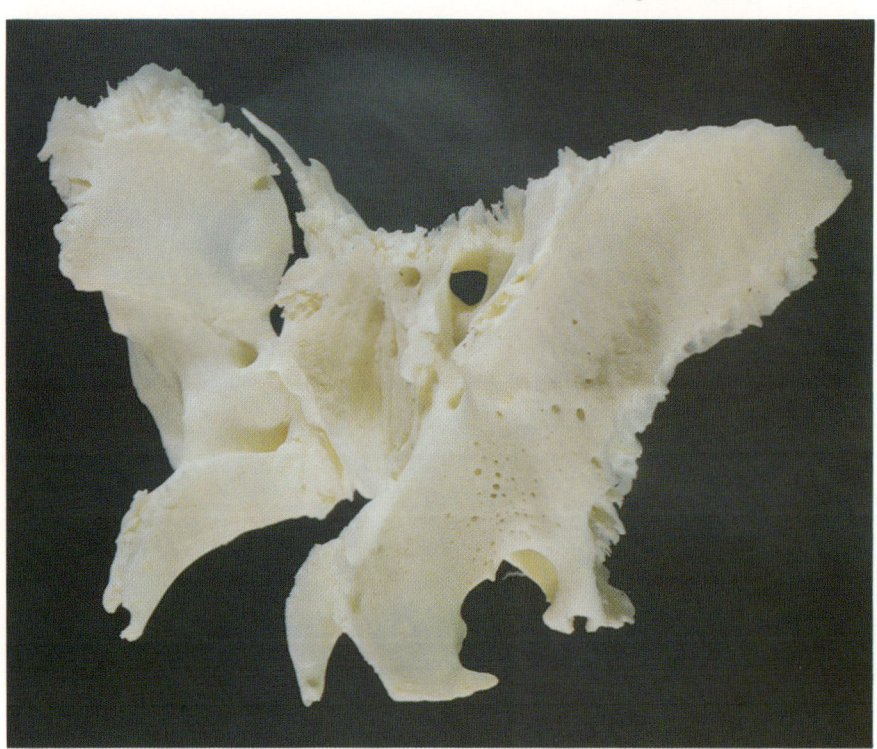

From left side and slightly in front.

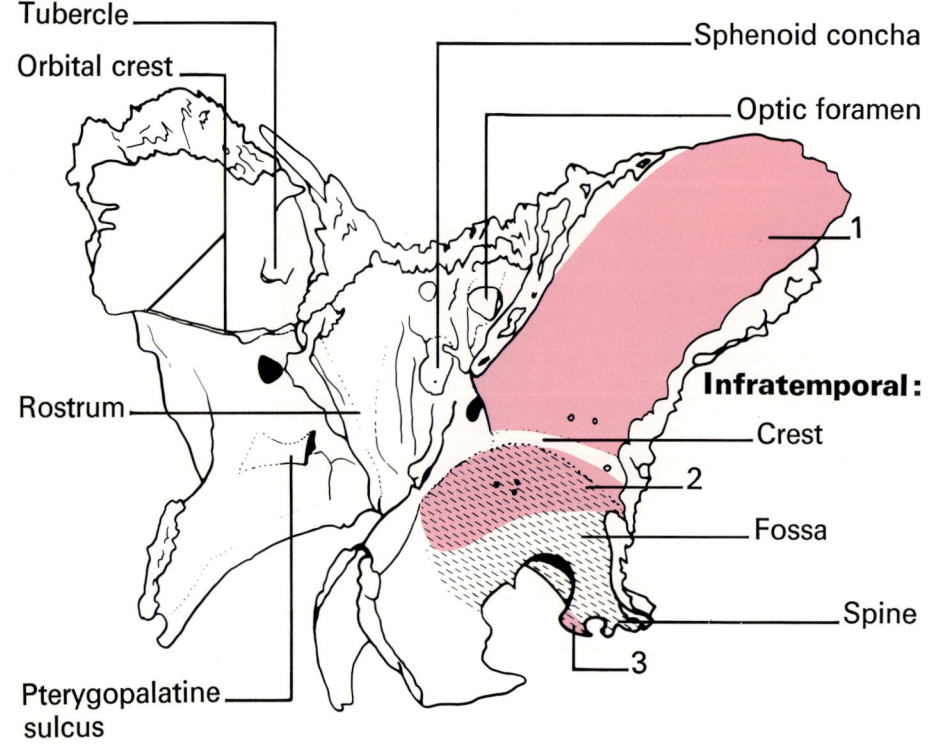

Muscle attachments:
1. Temporal
2. Lateral pterygoid
3. Tensor of velum palatini
4. Medial pterygoid

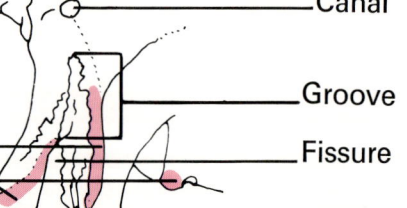

Sphenoid.

Lateral view, from right side.

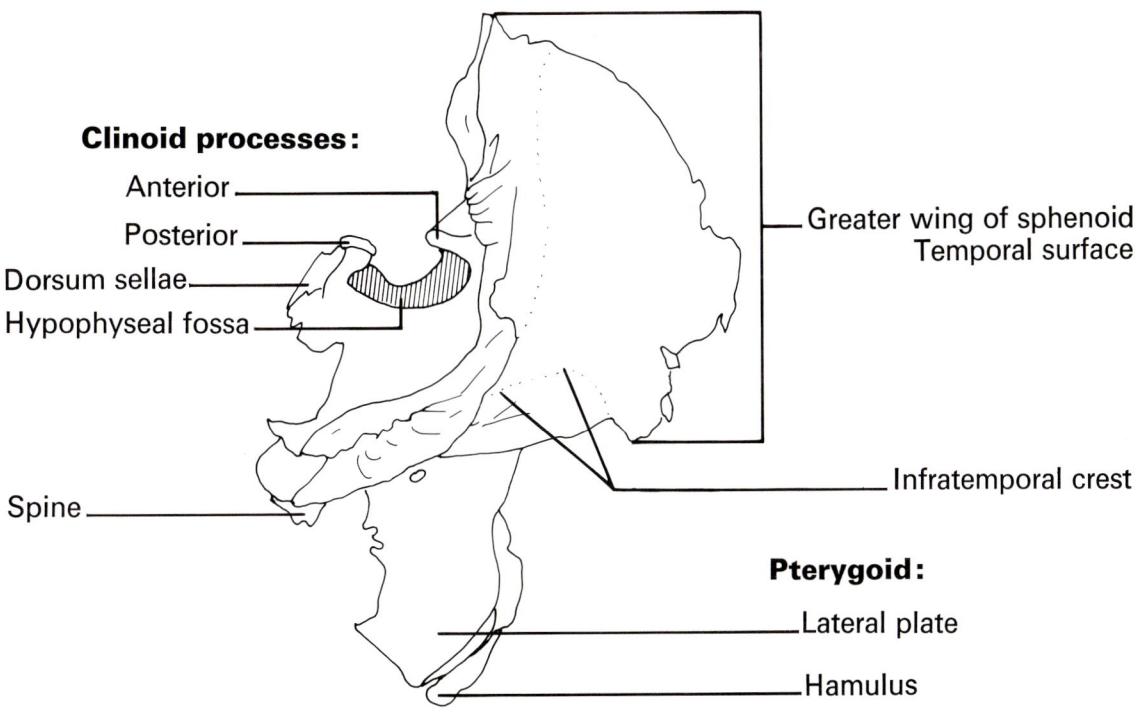

SECTION 4.   Occipital.

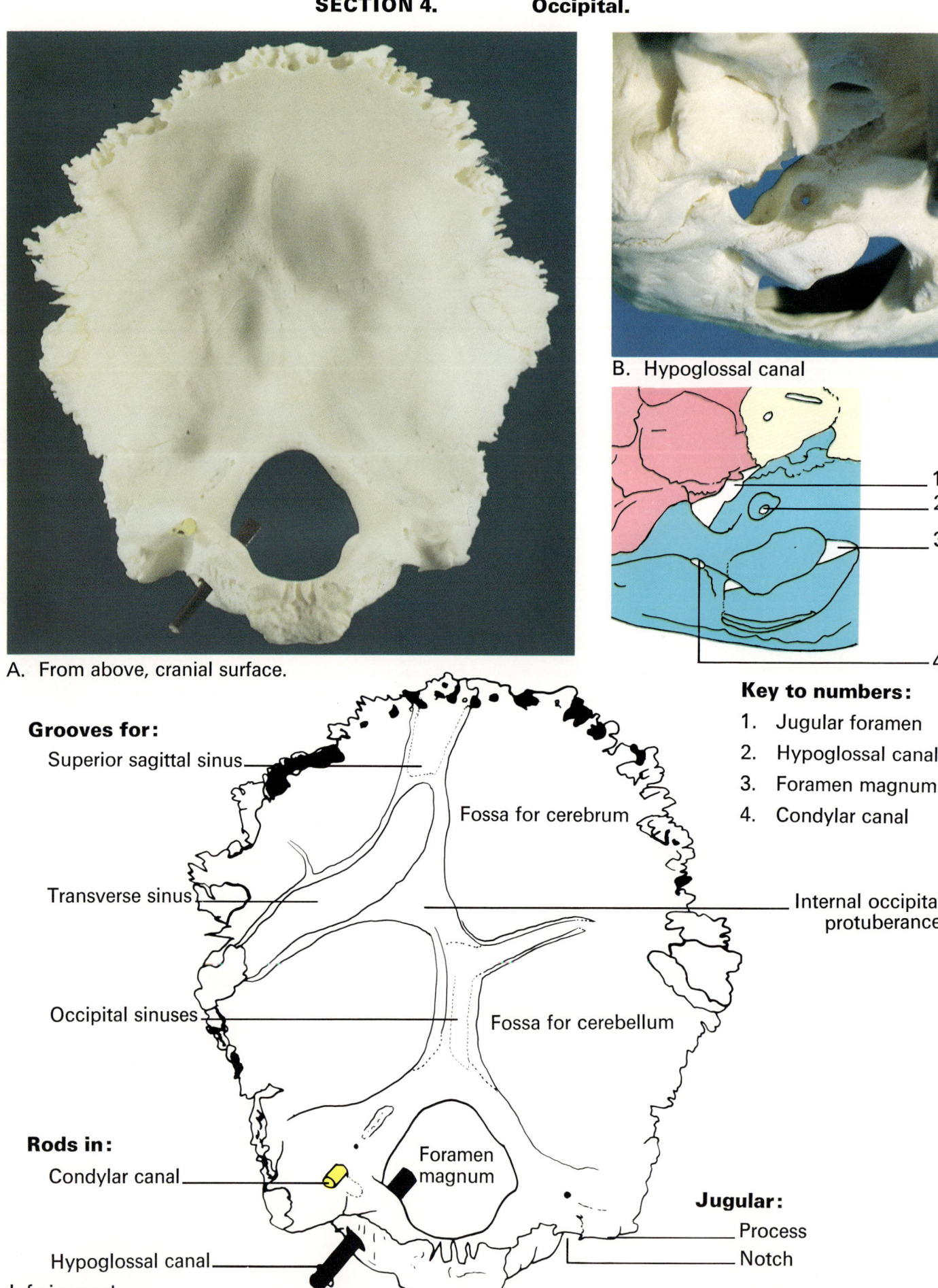

A. From above, cranial surface.

B. Hypoglossal canal

Key to numbers:
1. Jugular foramen
2. Hypoglossal canal
3. Foramen magnum
4. Condylar canal

Grooves for:
- Superior sagittal sinus
- Transverse sinus
- Occipital sinuses

- Fossa for cerebrum
- Internal occipital protuberance
- Fossa for cerebellum

Rods in:
- Condylar canal
- Hypoglossal canal
- Inferior angle

- Foramen magnum

Jugular:
- Process
- Notch

Occipital.

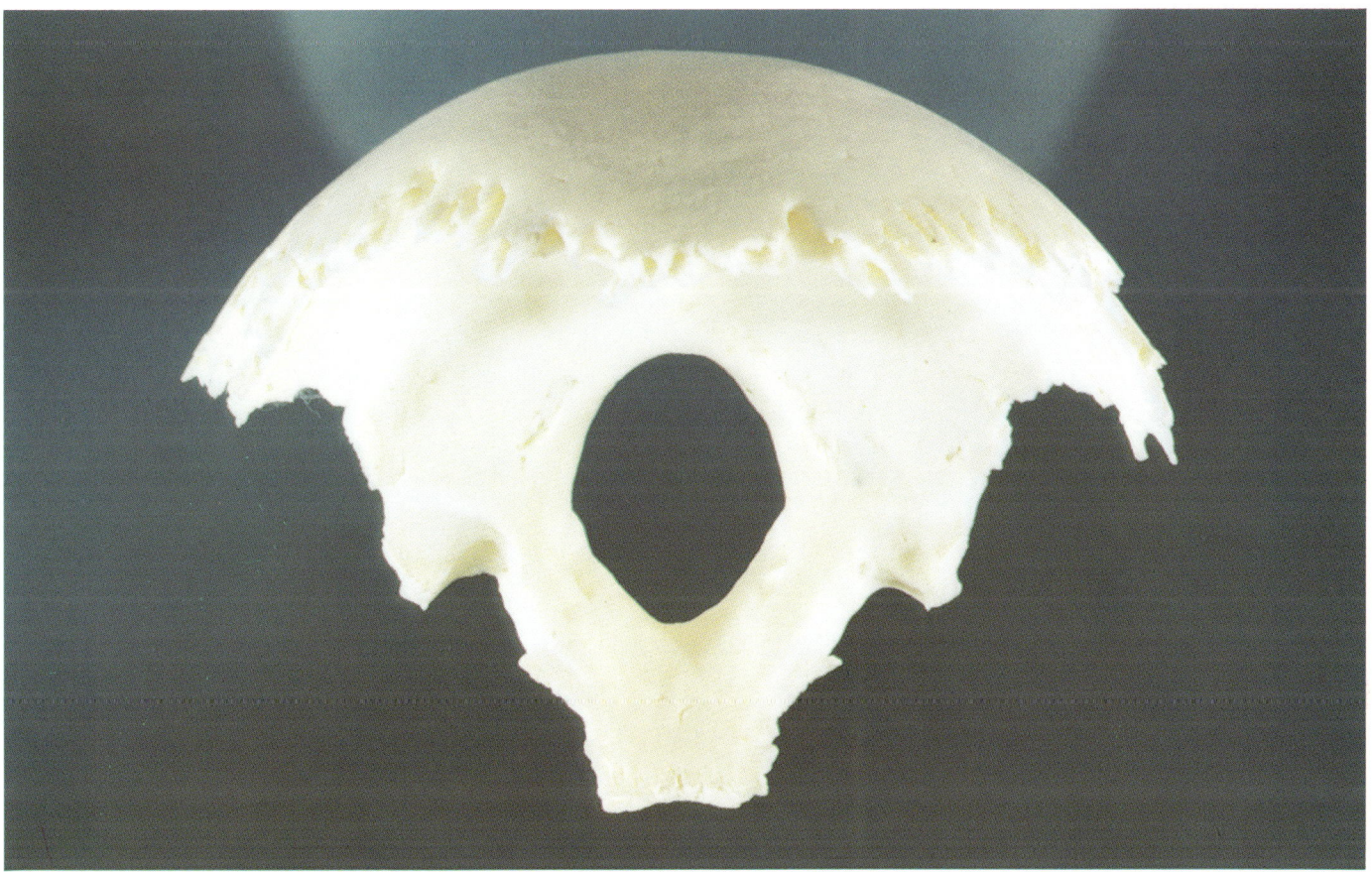

From above, onto basilar portion.

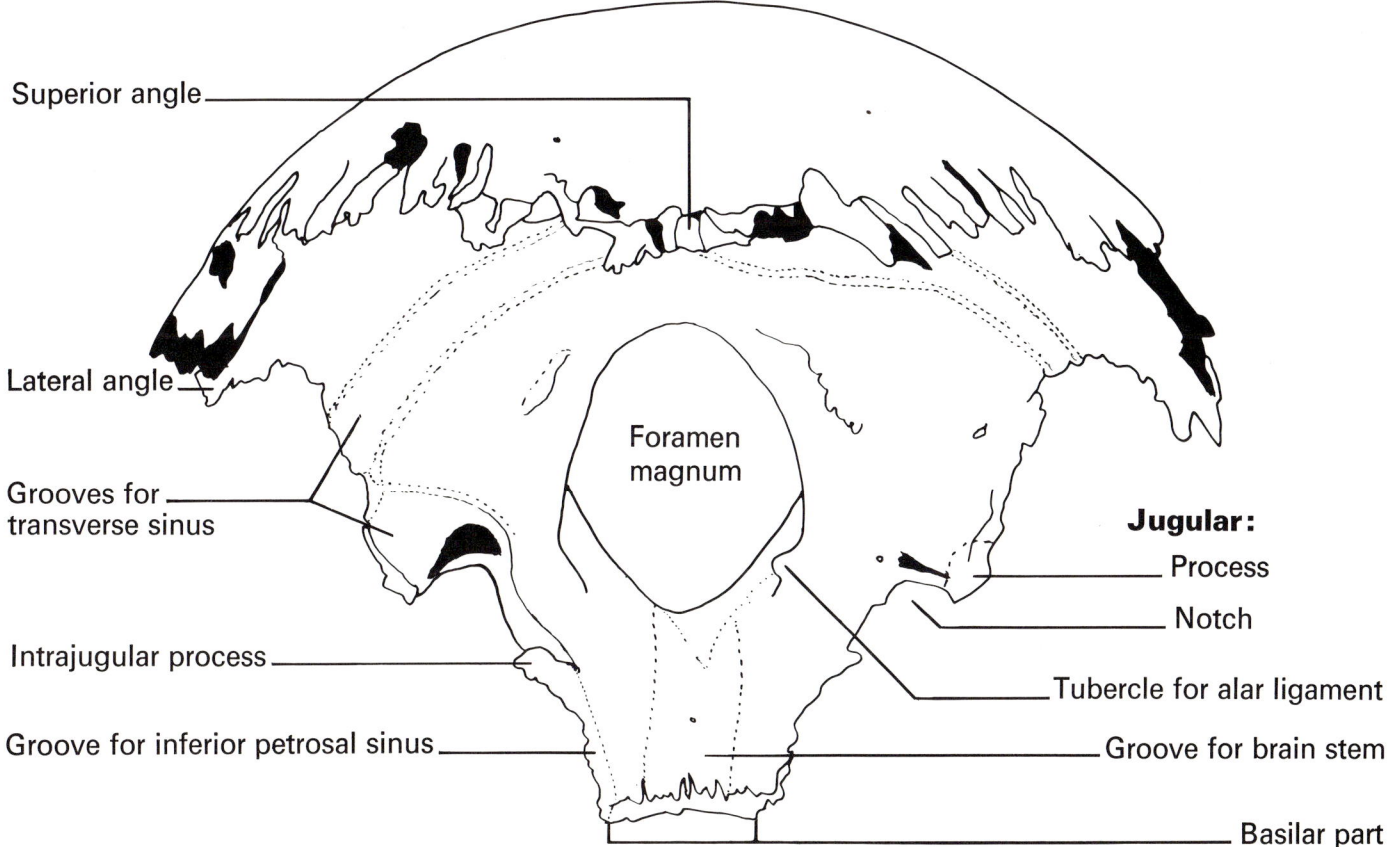

Occipital.

From behind.

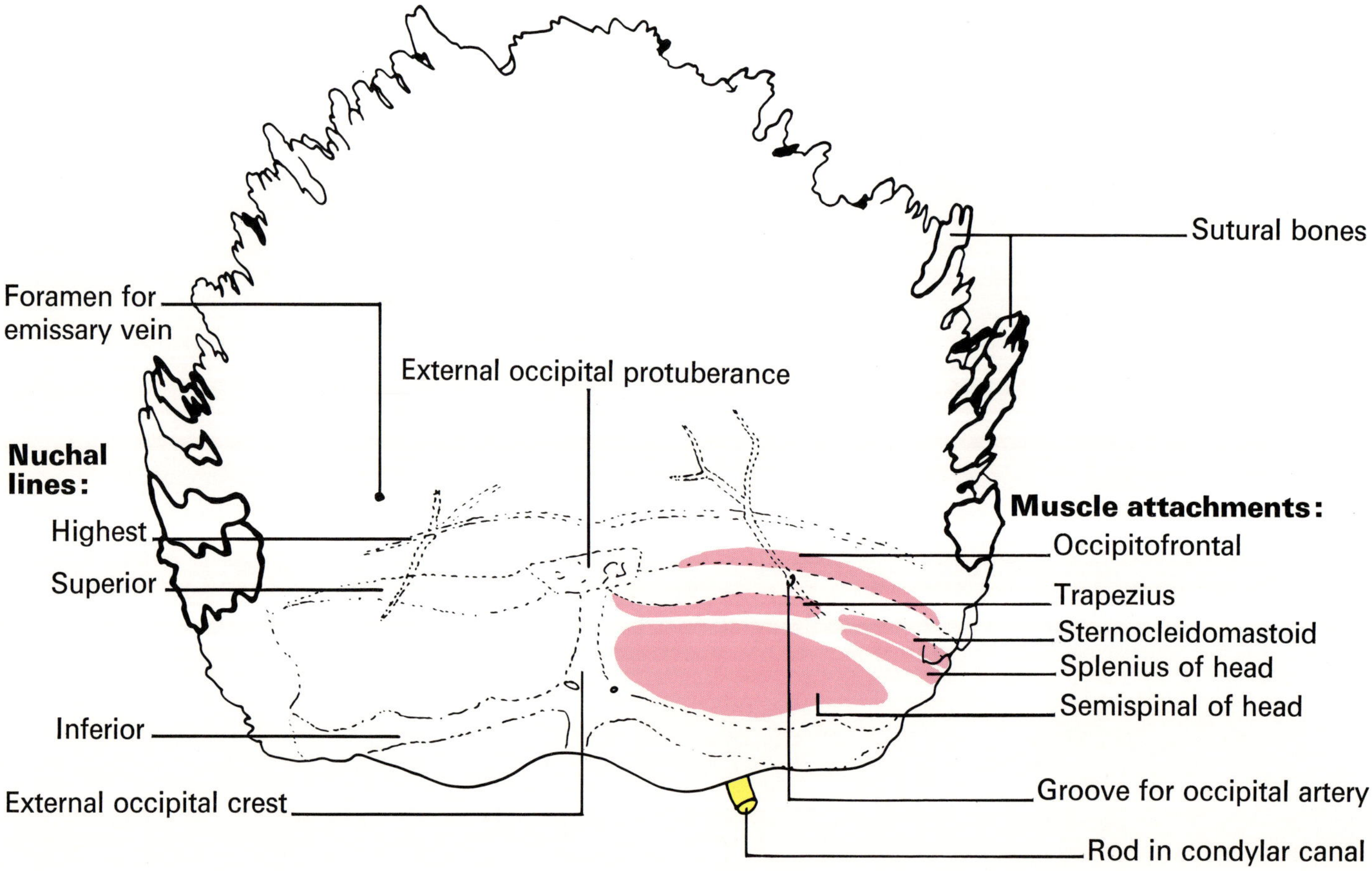

- Sutural bones
- Foramen for emissary vein
- External occipital protuberance
- **Nuchal lines:**
  - Highest
  - Superior
  - Inferior
- External occipital crest
- **Muscle attachments:**
  - Occipitofrontal
  - Trapezius
  - Sternocleidomastoid
  - Splenius of head
  - Semispinal of head
- Groove for occipital artery
- Rod in condylar canal

Occipital.

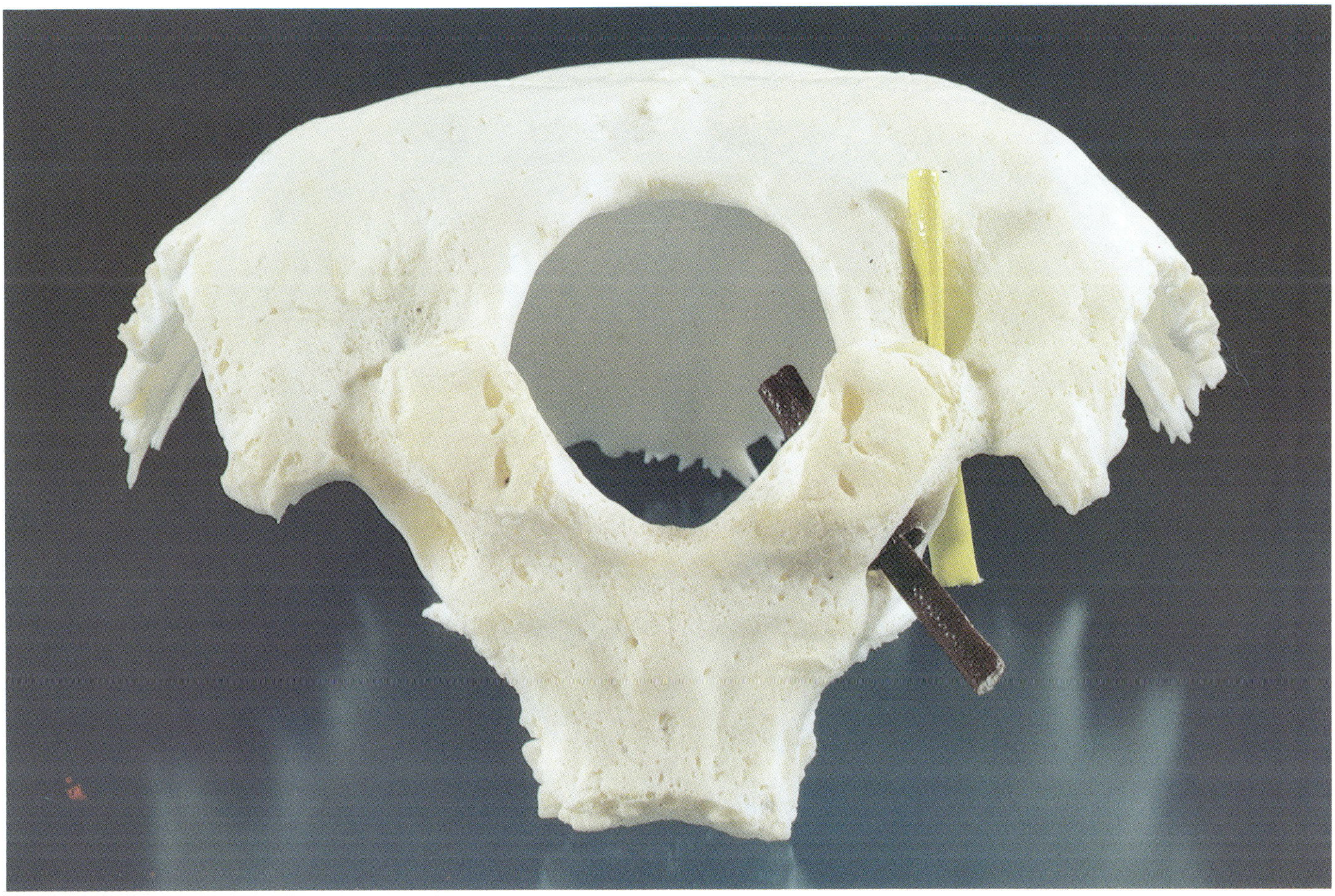

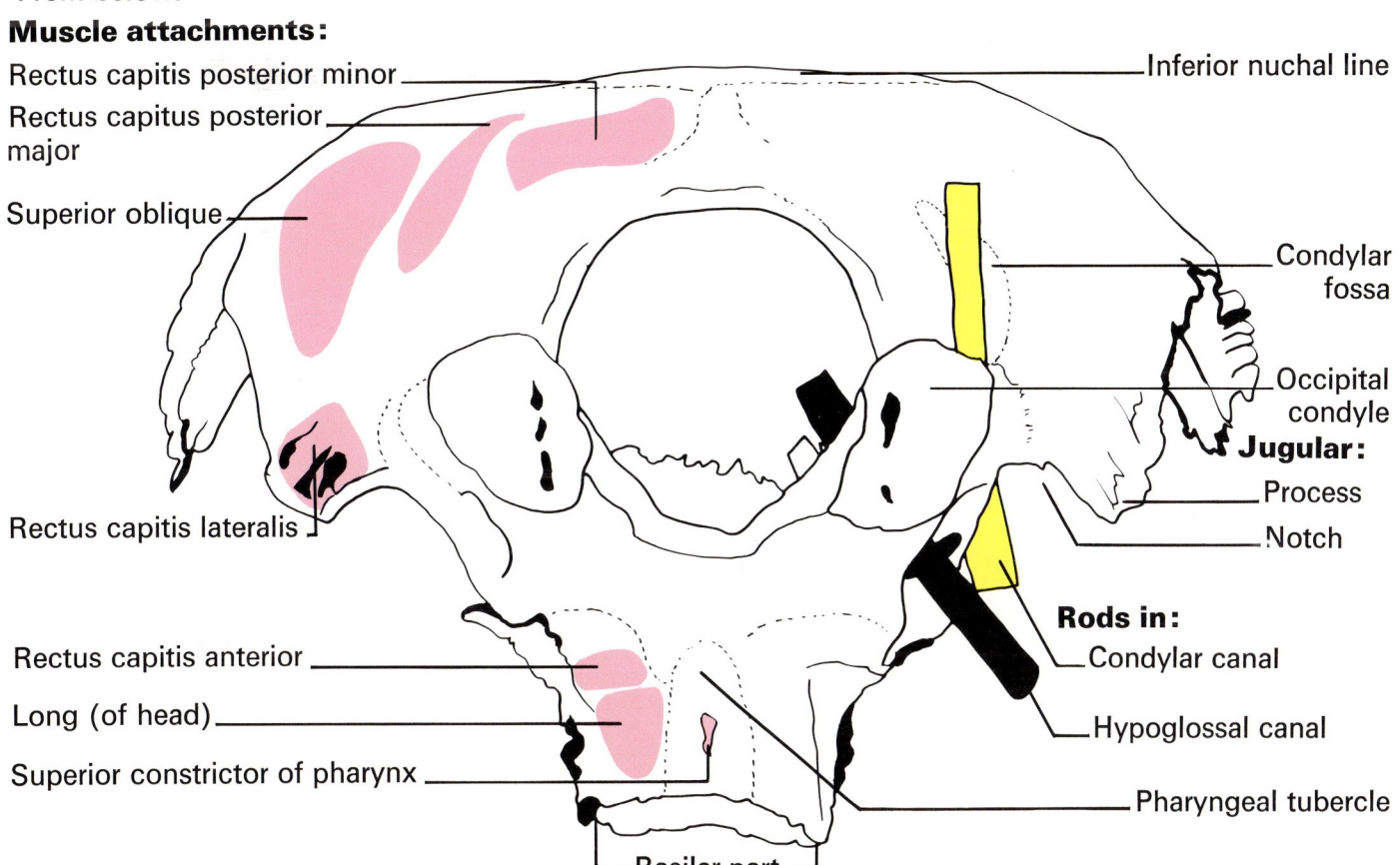

From below.

Muscle attachments:

- Rectus capitis posterior minor
- Rectus capitus posterior major
- Superior oblique
- Rectus capitis lateralis
- Rectus capitis anterior
- Long (of head)
- Superior constrictor of pharynx
- Basilar part
- Inferior nuchal line
- Condylar fossa
- Occipital condyle

Jugular:
- Process
- Notch

Rods in:
- Condylar canal
- Hypoglossal canal
- Pharyngeal tubercle

Occipital.

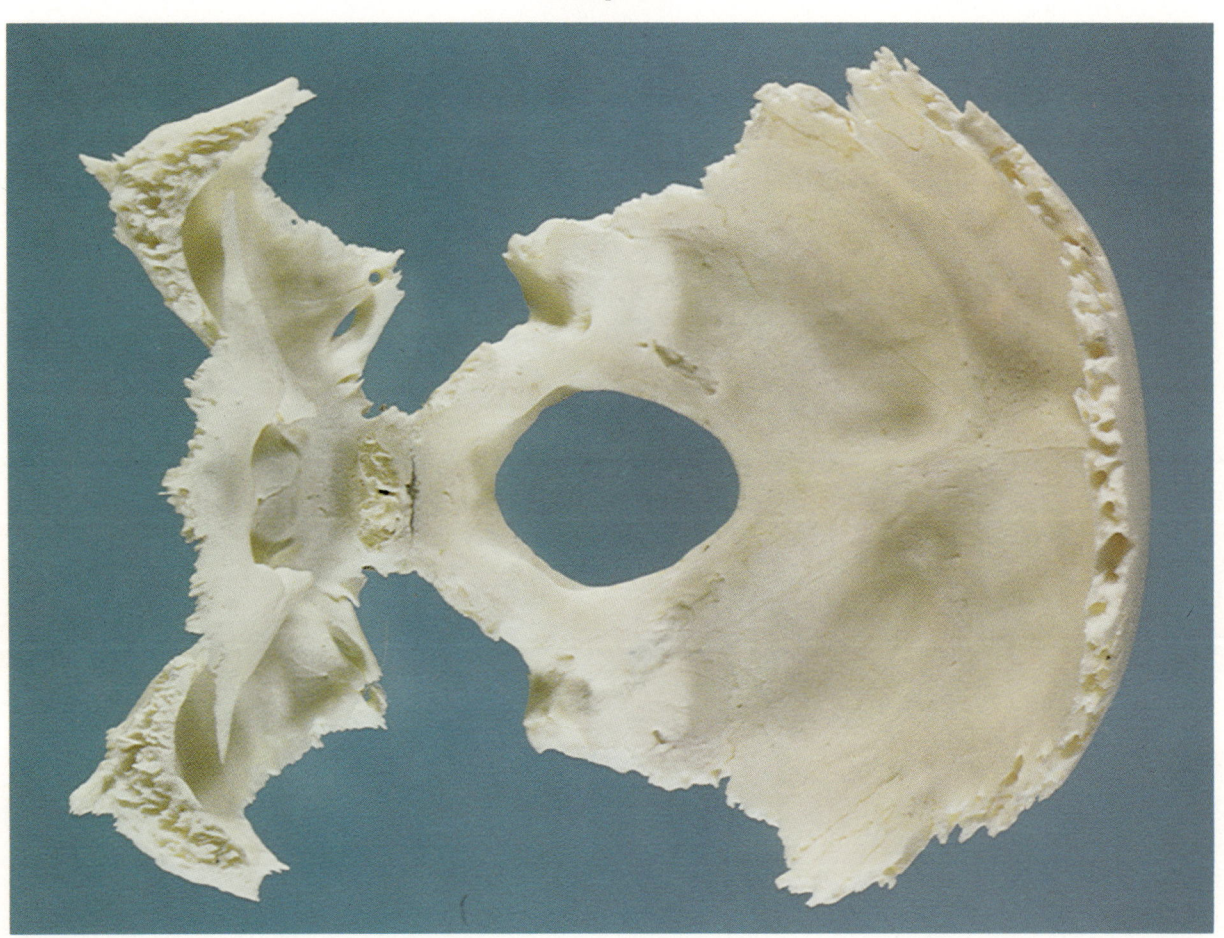

With sphenoid, seen from above.

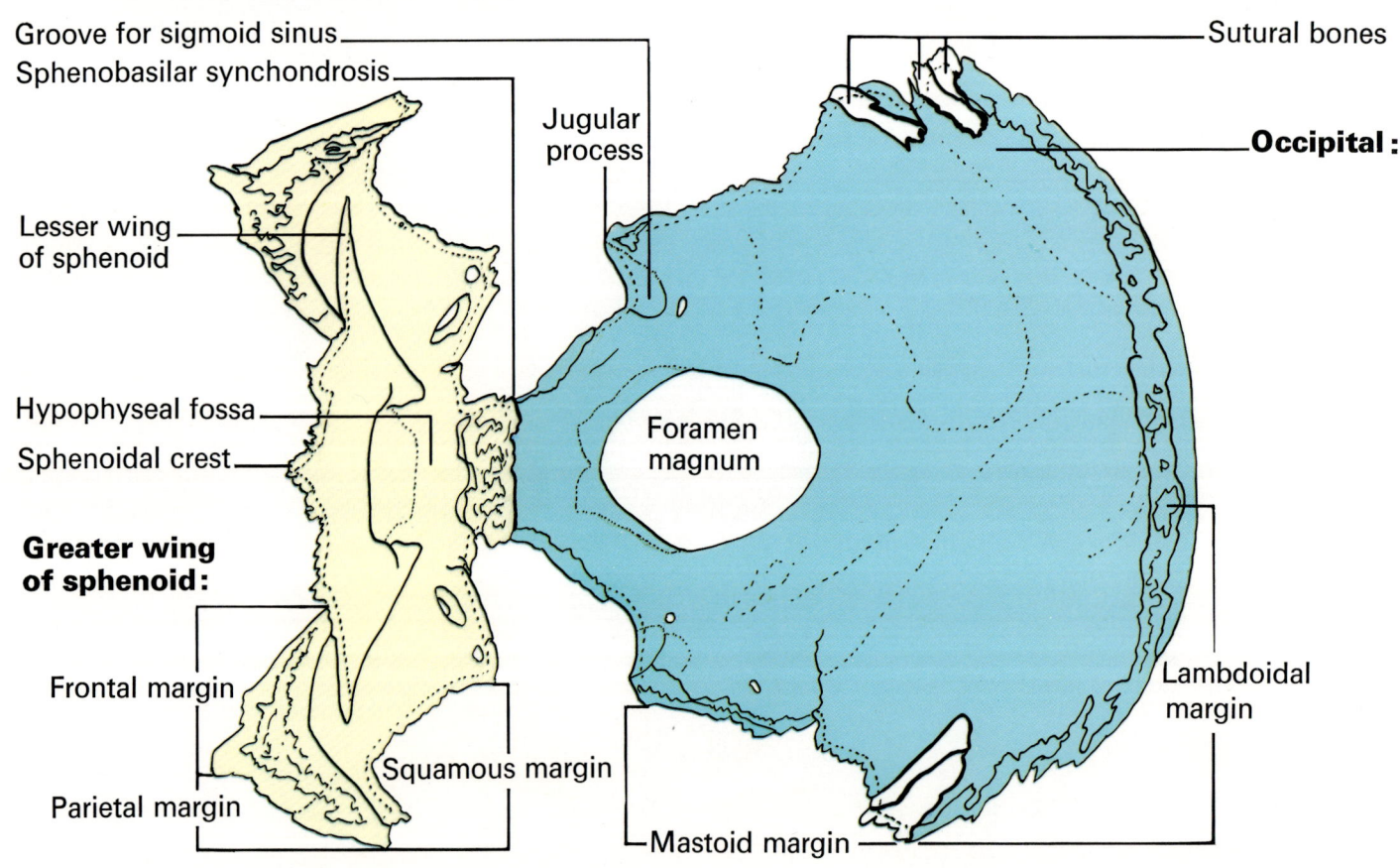

Occipital.

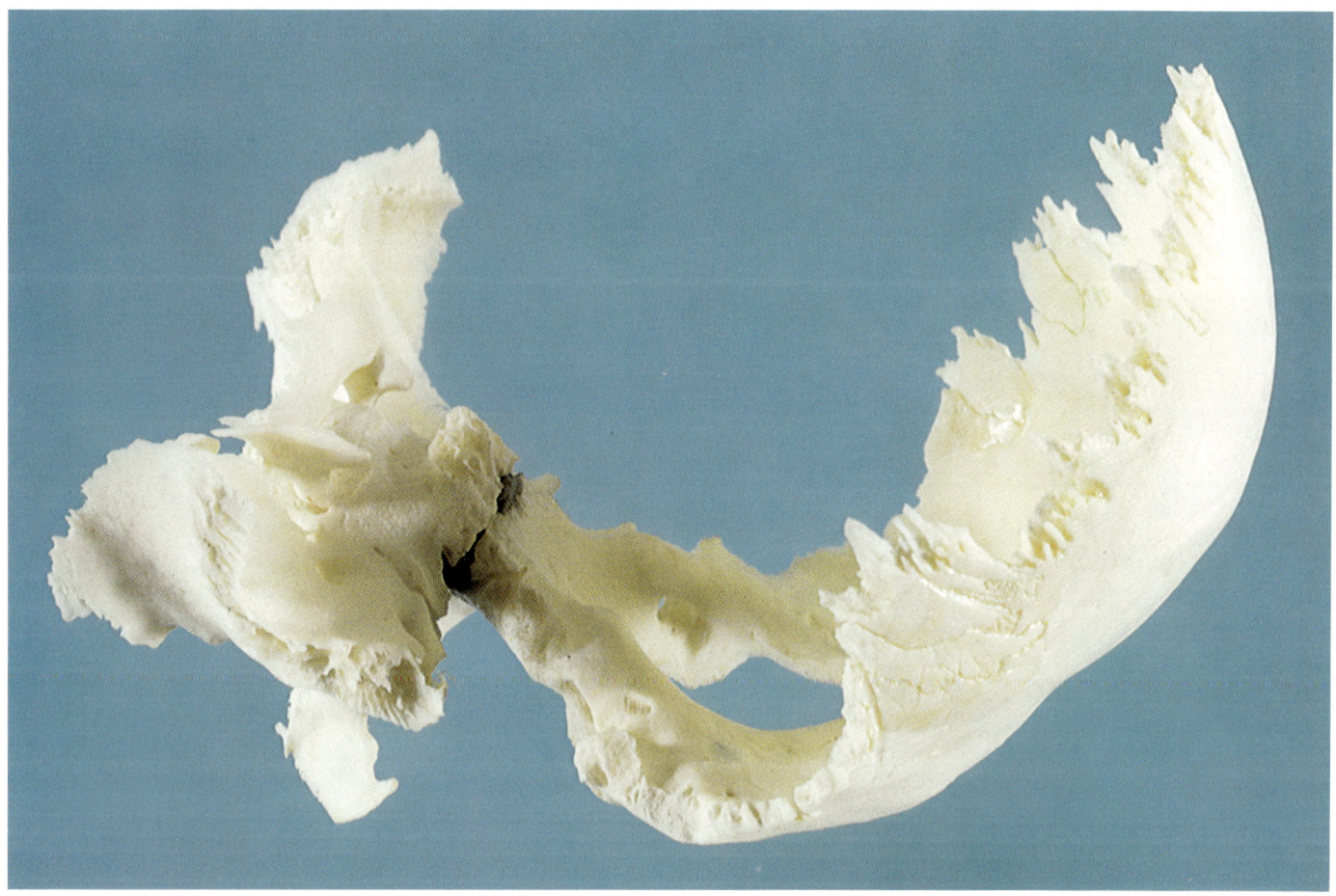

With sphenoid, from left side.

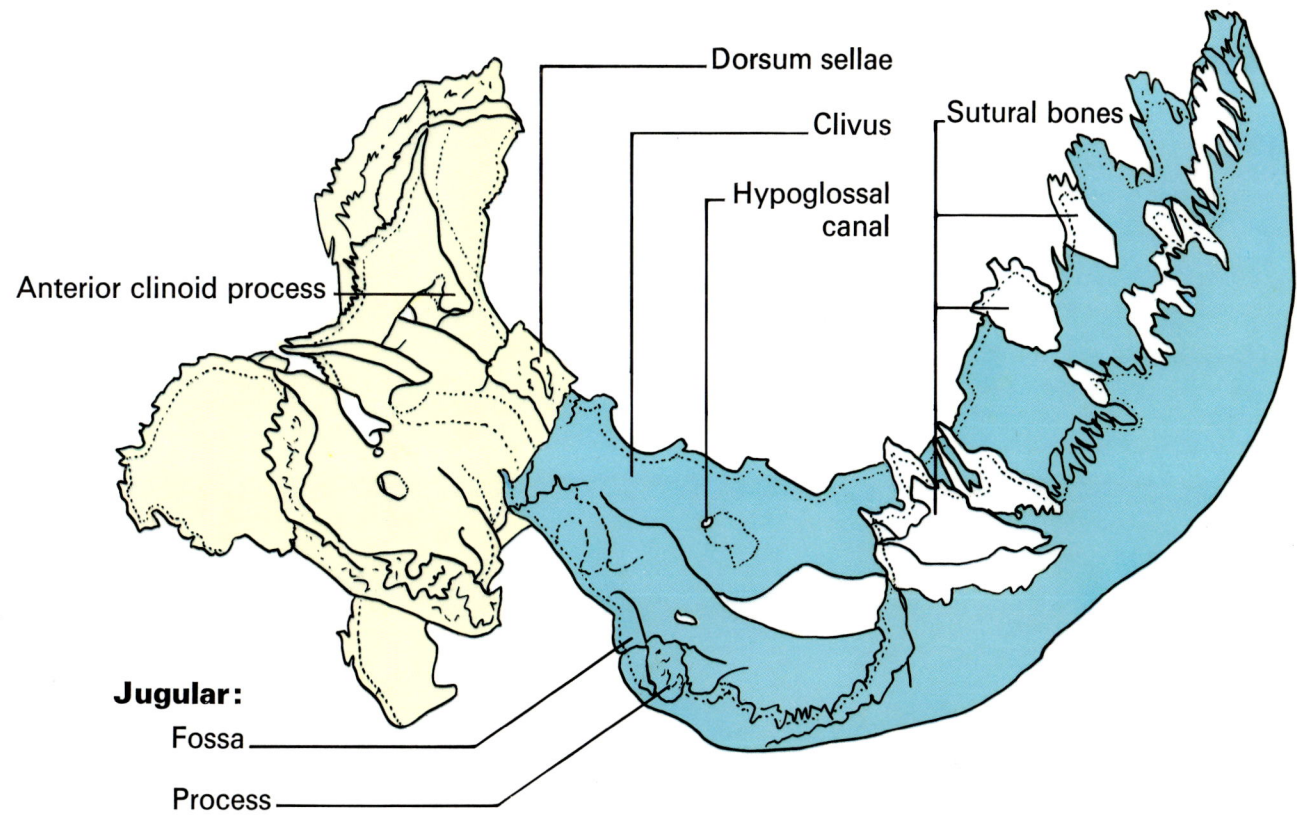

SECTION 5. Parietal.

From right side, external surface.

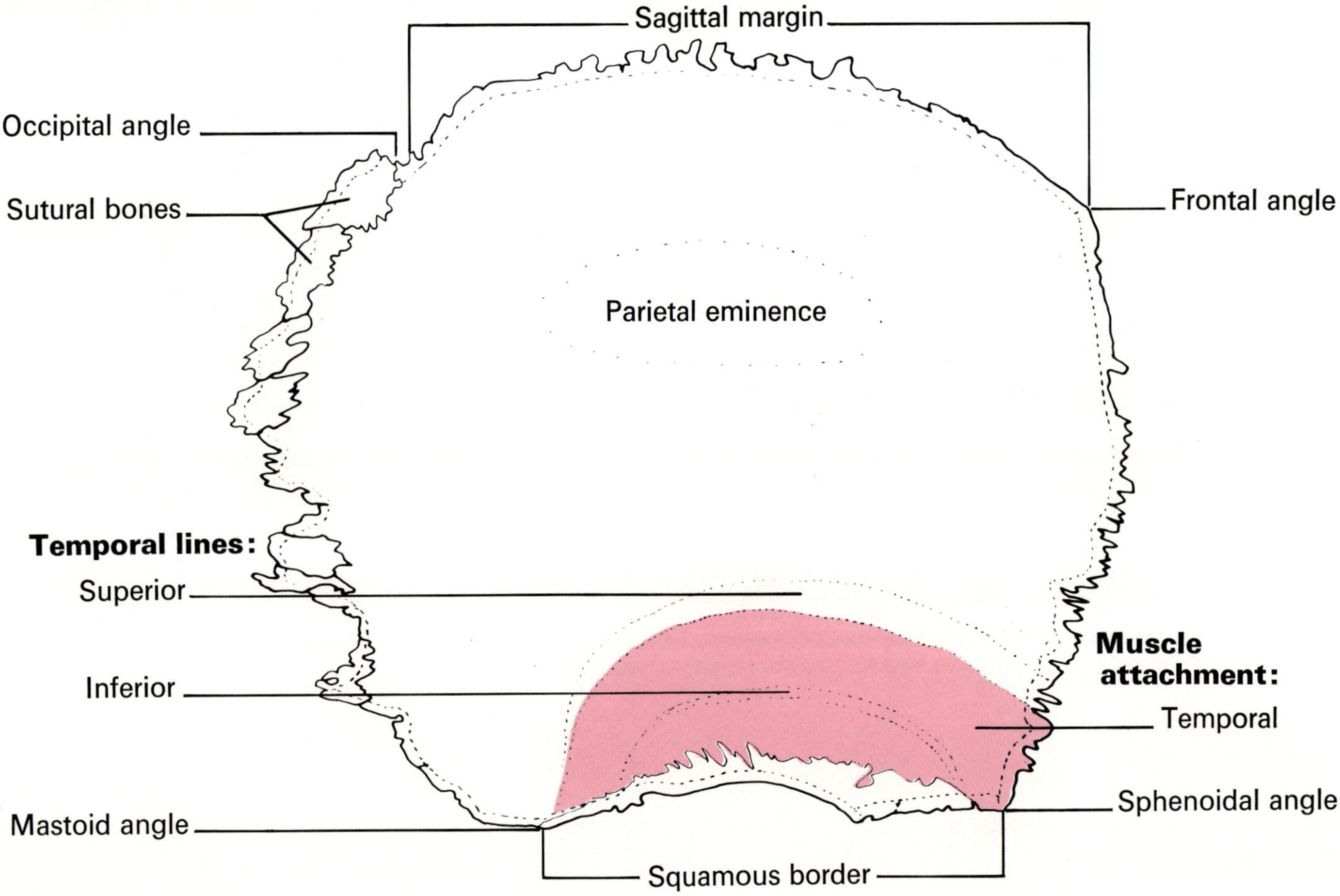

- Sagittal margin
- Occipital angle
- Sutural bones
- Frontal angle
- Parietal eminence
- Temporal lines:
  - Superior
  - Inferior
- Muscle attachment:
  - Temporal
- Mastoid angle
- Sphenoidal angle
- Squamous border

Parietal.

From right side, internal surface.

- Parietal foramen
- Groove for superior sagittal sinus
- Granular foveolae
- Frontal margin
- Occipital margin
- **Grooves for:**
- Middle meningeal vessels
- Transverse sinus
- Grooves for middle meningeal vessels

Parietal.

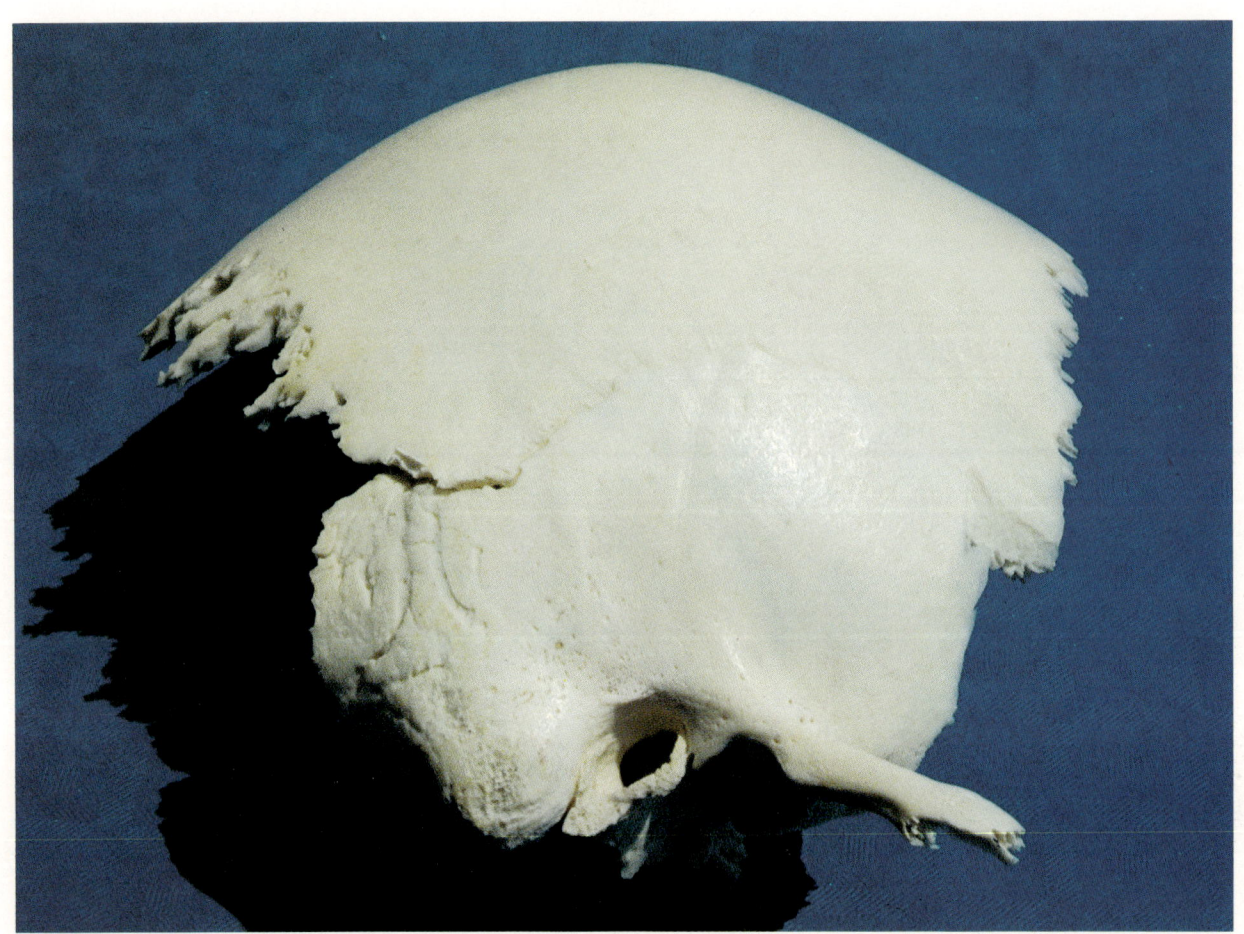

From right side, with temporal.

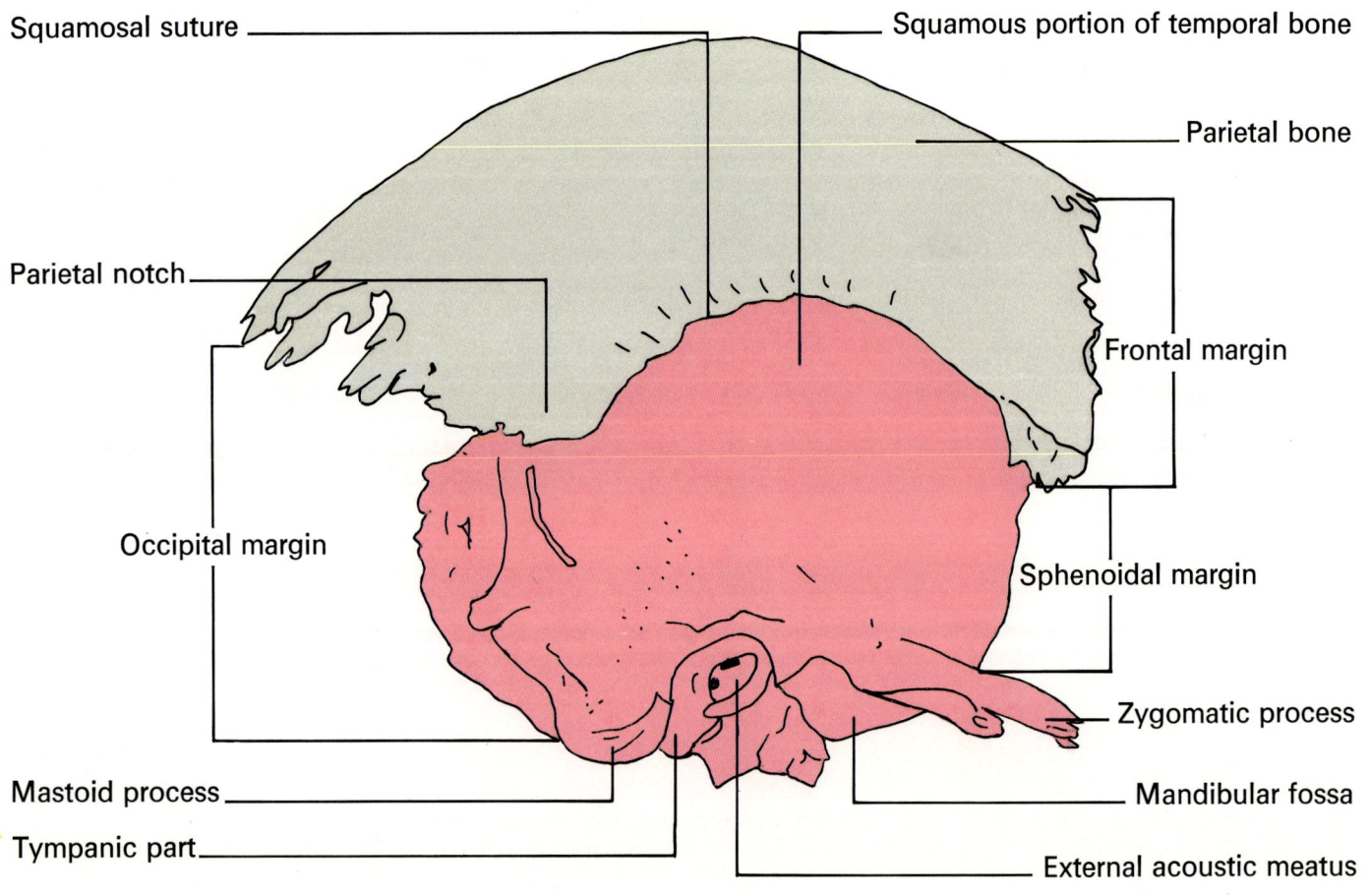

123

SECTION 6.   Temporal.

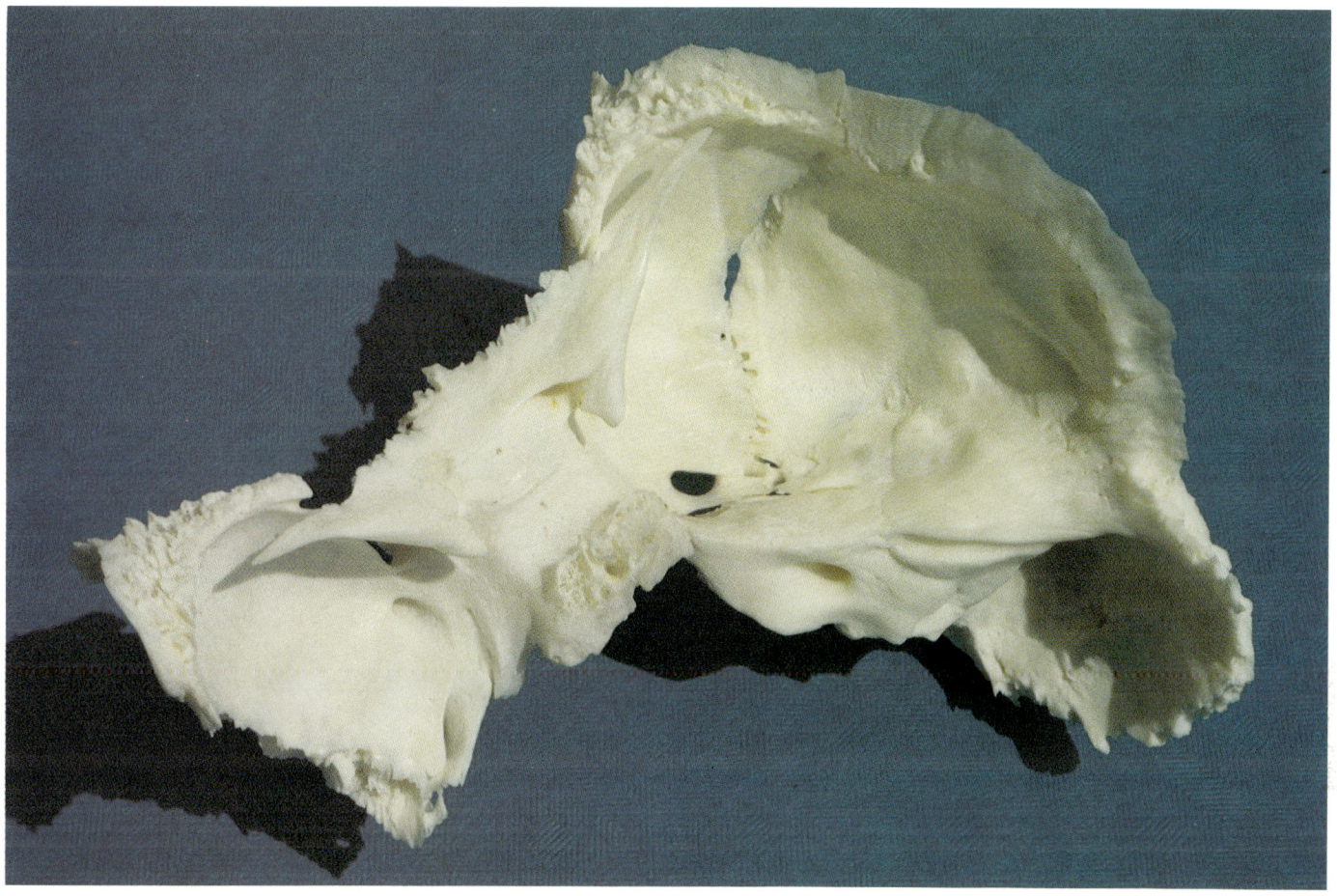

With sphenoid, view of middle cranial fossa.

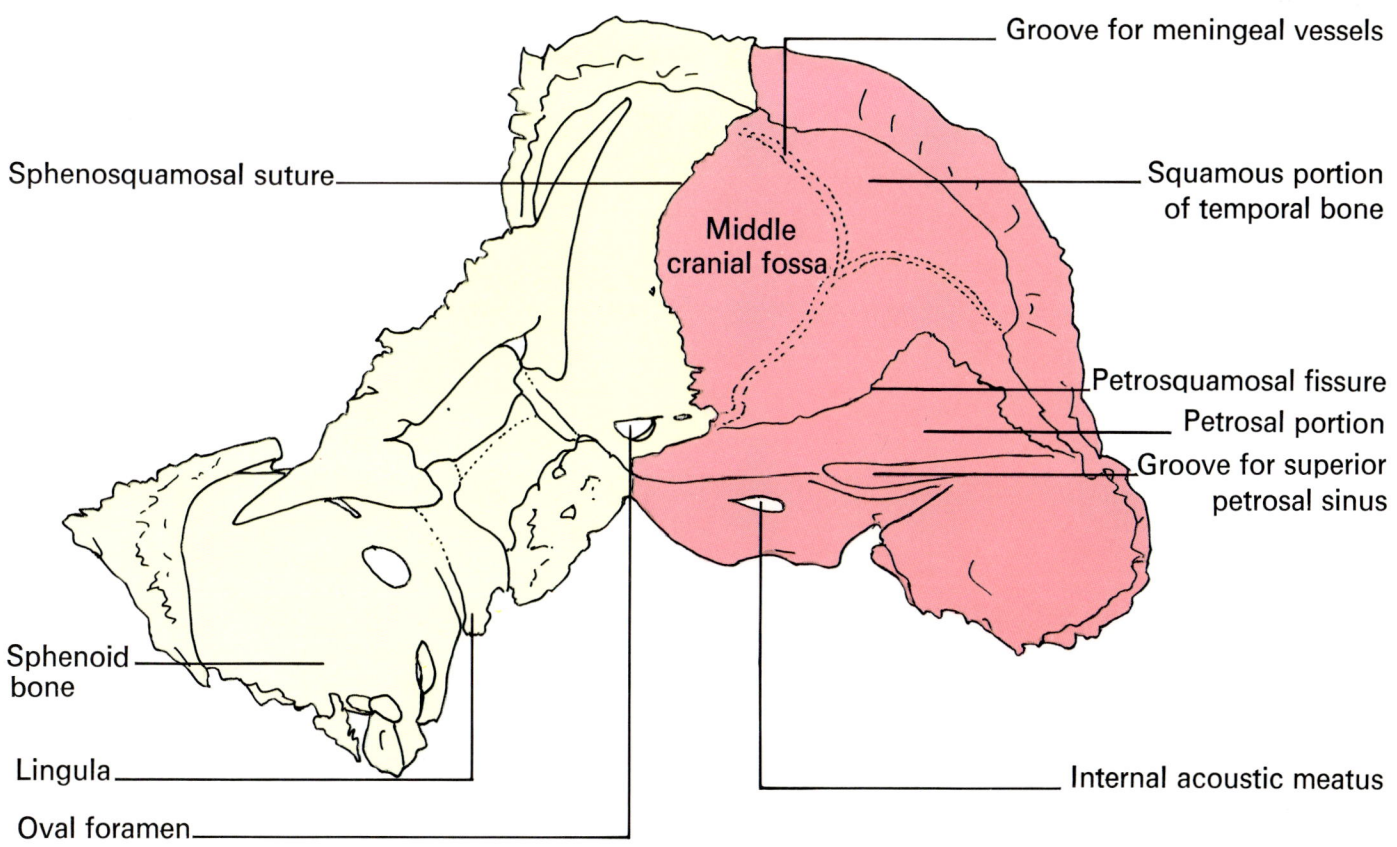

Temporal.

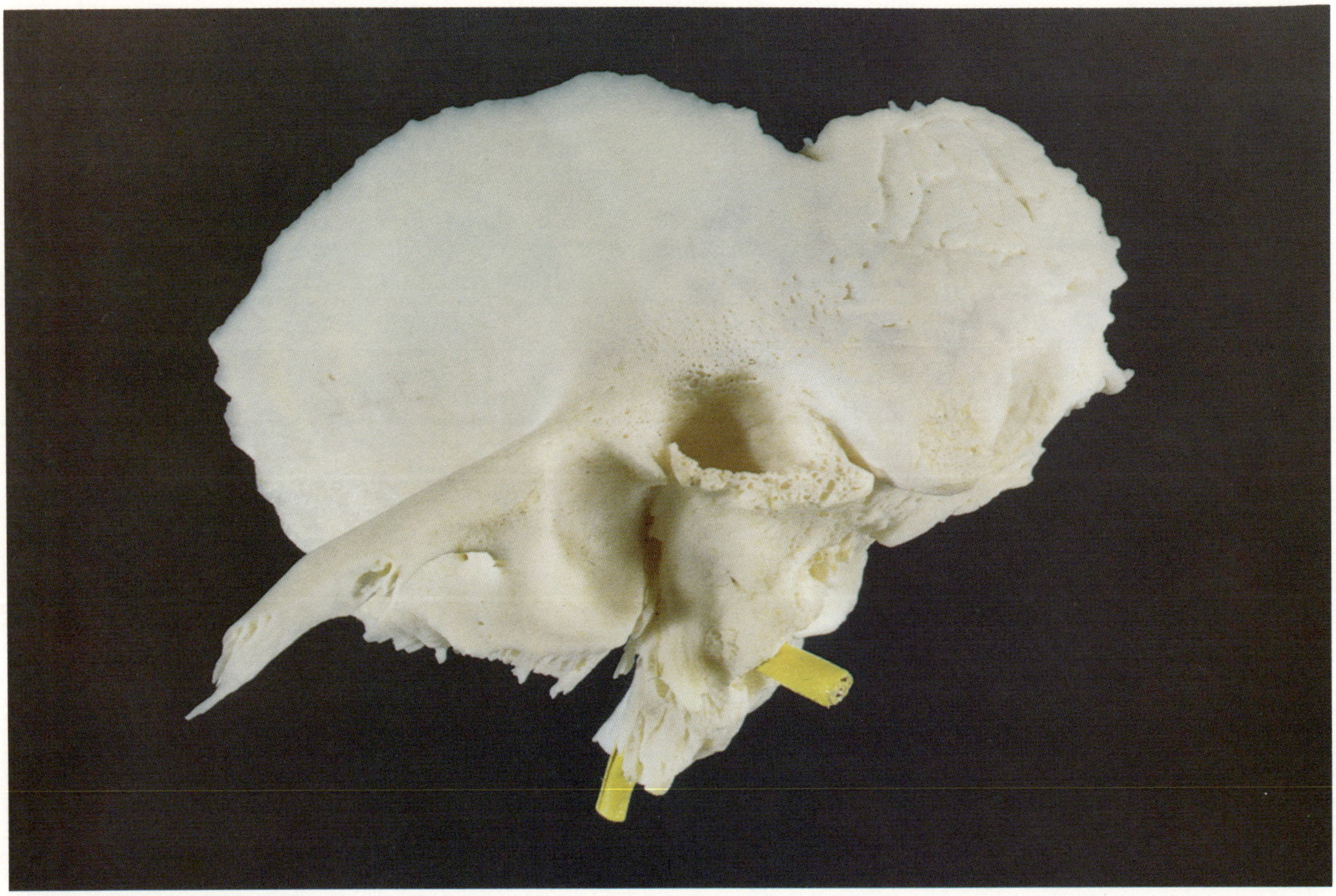

From left side and below.

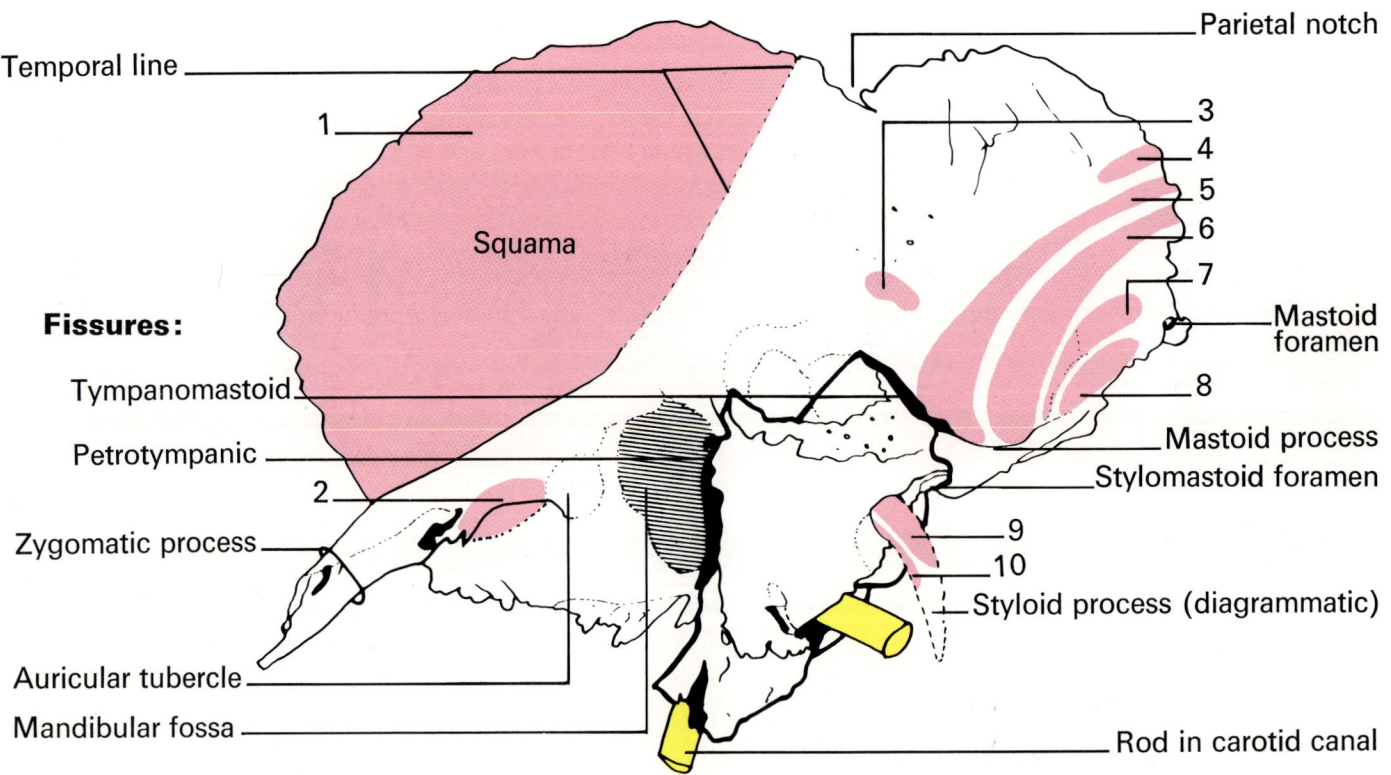

Tympanic portion outlined with heavy black line.

Temporal.

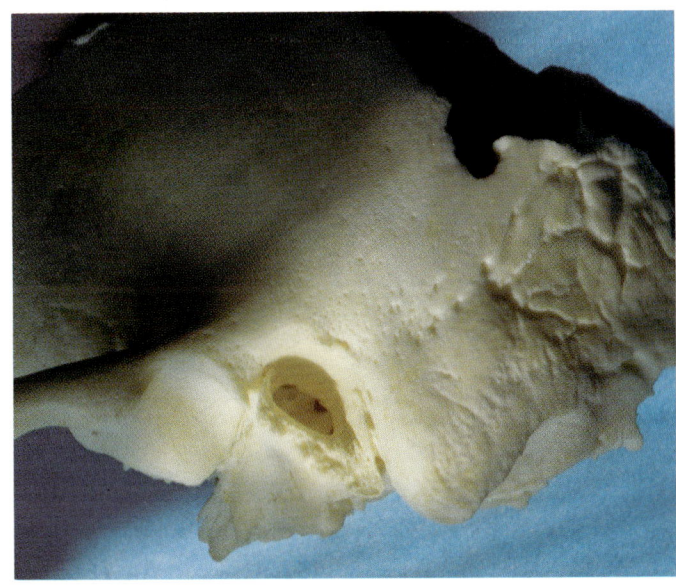

A. Closeup view left, external acoustic meatus. (Tympanic membrane absent.)

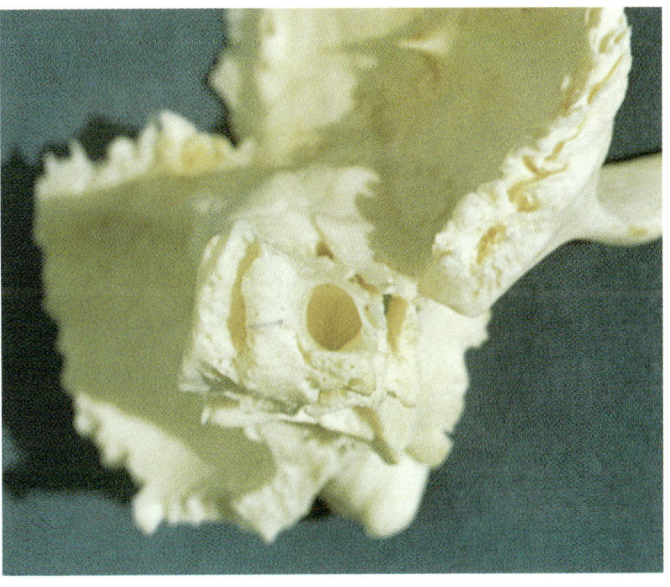

B. From front, apex and opening of carotid canal.

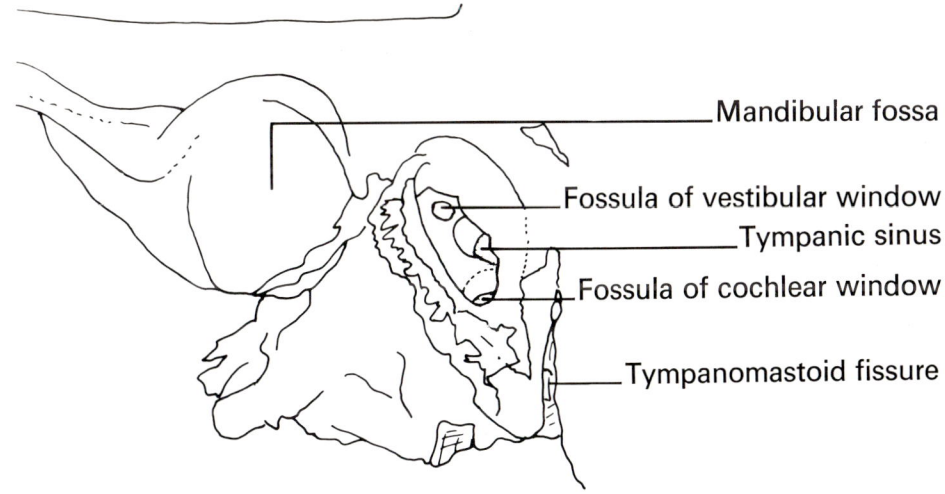

- Mandibular fossa
- Fossula of vestibular window
- Tympanic sinus
- Fossula of cochlear window
- Tympanomastoid fissure

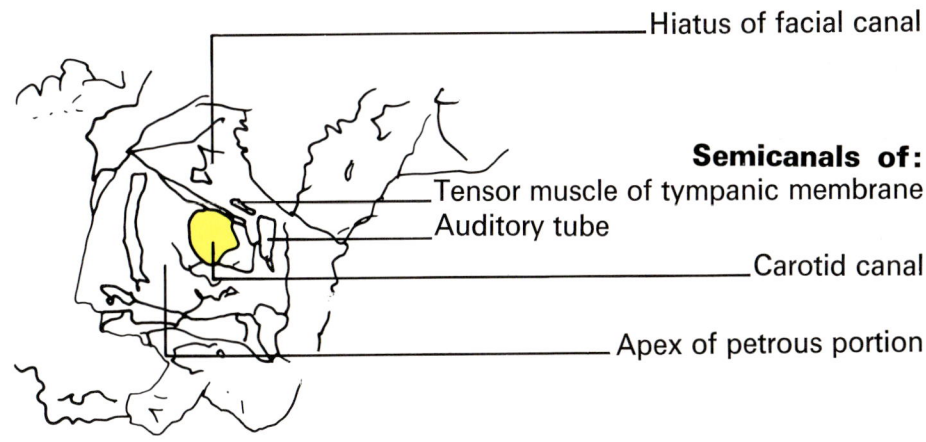

- Hiatus of facial canal

Semicanals of:
- Tensor muscle of tympanic membrane
- Auditory tube
- Carotid canal
- Apex of petrous portion

Key to numbers, opposite page
Muscle attachments:

1. Temporal
2. Masseter
3. Posterior auricular
4. Occipitofrontal
5. Sternocleidomastoid
6. Splenius of head
7. Long (of head)
8. Digastric
9. Stylohyoid
10. Styloglossus

Temporal.

Right temporal bone, from below.

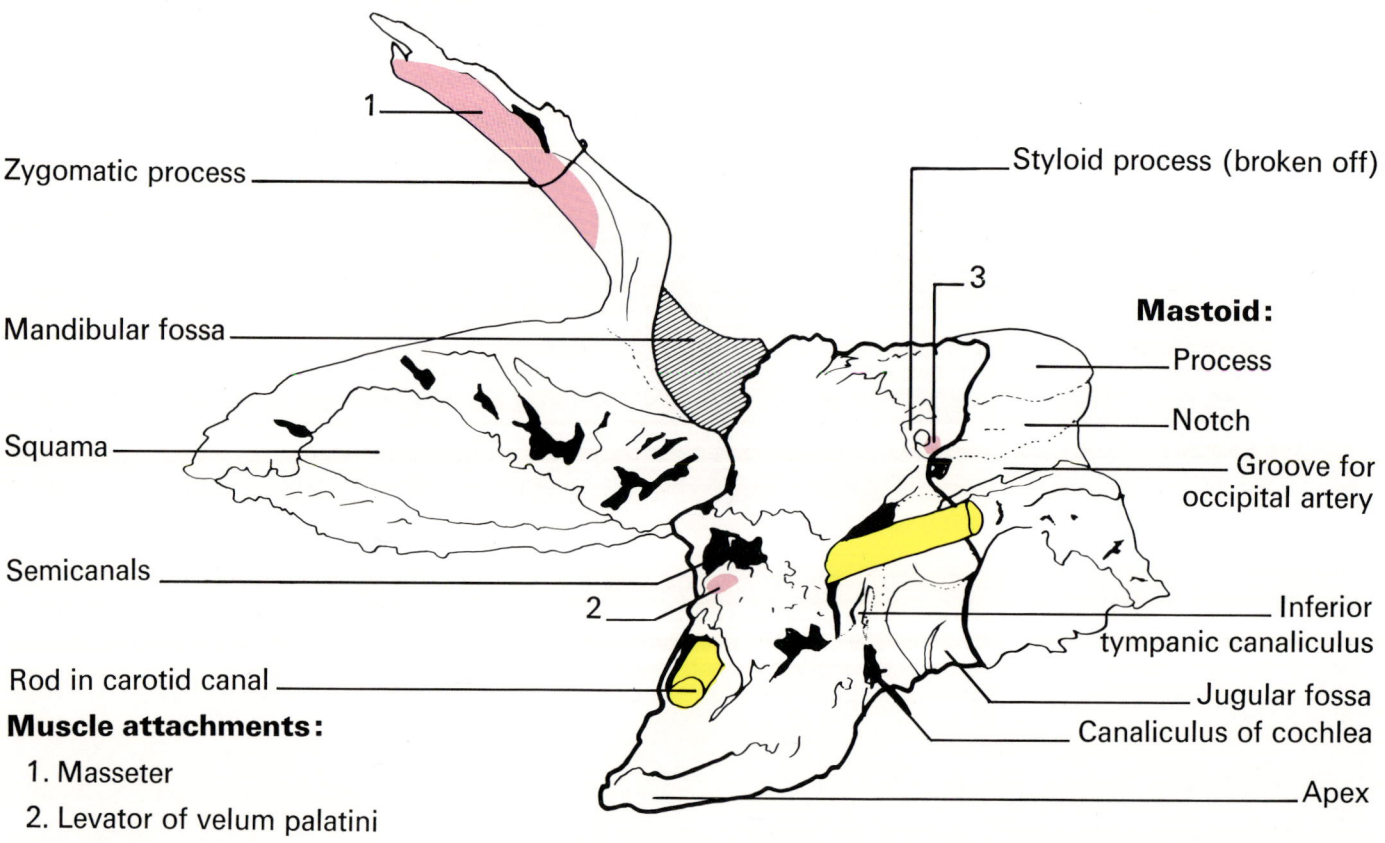

Zygomatic process

Mandibular fossa

Squama

Semicanals

Rod in carotid canal

Muscle attachments:
 1. Masseter
 2. Levator of velum palatini
 3. Stylopharyngeal

Styloid process (broken off)

Mastoid:
Process
Notch
Groove for occipital artery
Inferior tympanic canaliculus
Jugular fossa
Canaliculus of cochlea
Apex

Petrous portion outlined with heavy black line.

Temporal.

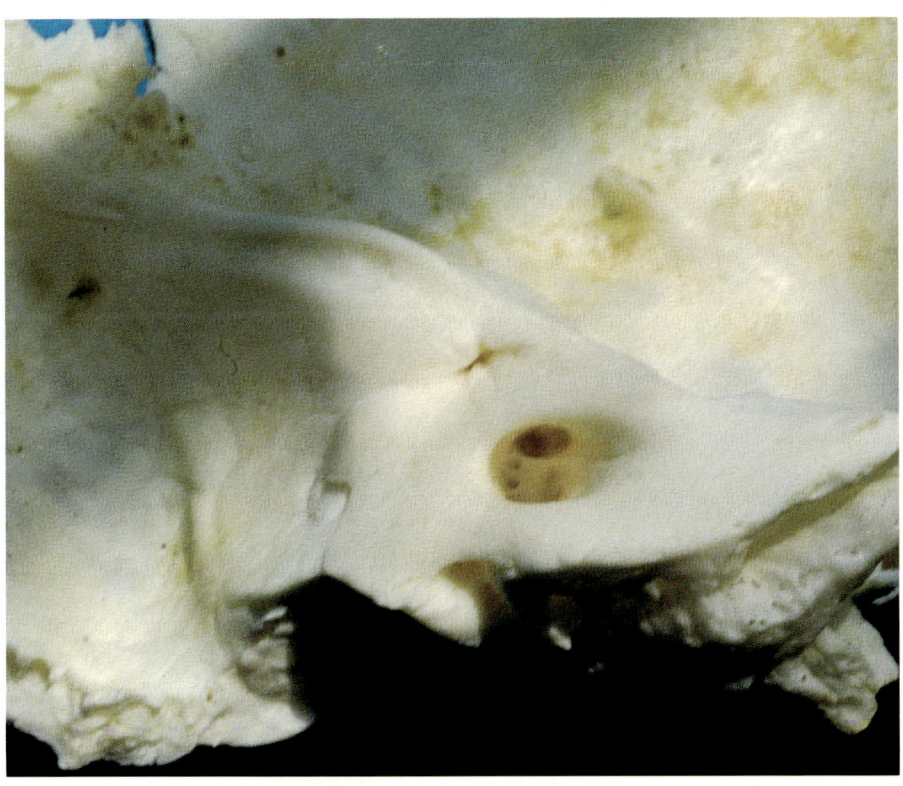

Left internal acoustic meatus.

- Subarcuate fossa
- **Vestibular area:**
  - Superior
  - Inferior
- Cochlear area
- Facial area
- Apex
- Transverse crest of internal acoustic meatus

- Groove for superior petrosal sinus
- **Vestibular area:**
  - Superior
  - Inferior
- Cochlear area
- Anterior surface
- Facial area
- Transverse crest
- Singular foramen
- Inferior margin of internal acoustic meatus

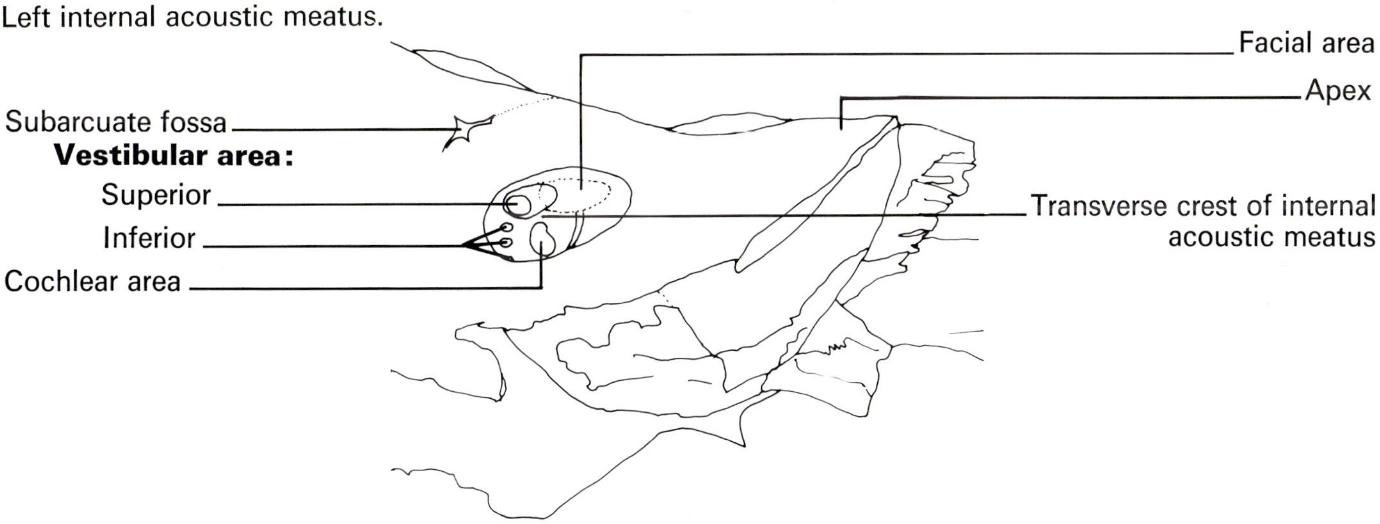

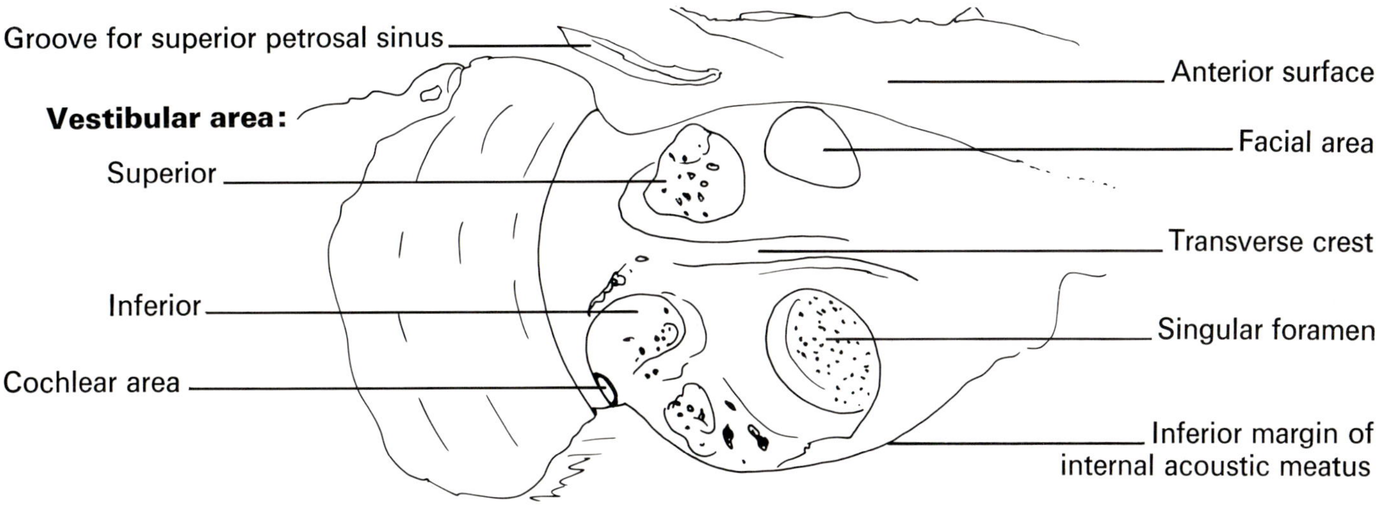

(Redrawn & relabeled from Surgical Anatomy of the Temporal Bone and Ear, 2nd ed., by Anson, B. J. and Donaldson, J. A. Courtesy of W. B. Saunders, publisher)

Temporal.

Left temporal bone, from behind, internal surface.
Yellow rod in carotid canal.

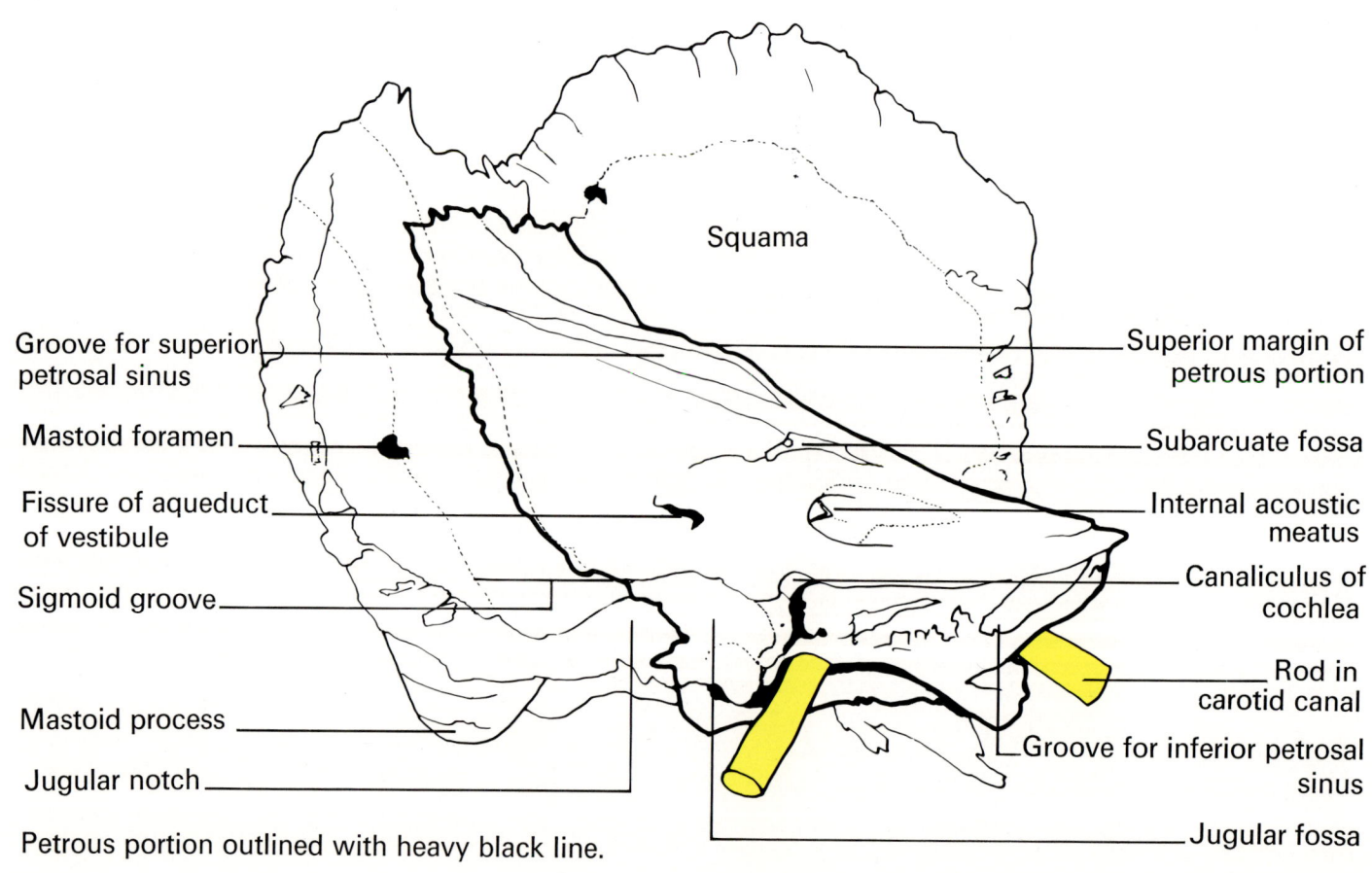

Temporal.

Left temporal bone, from above, internal surface.

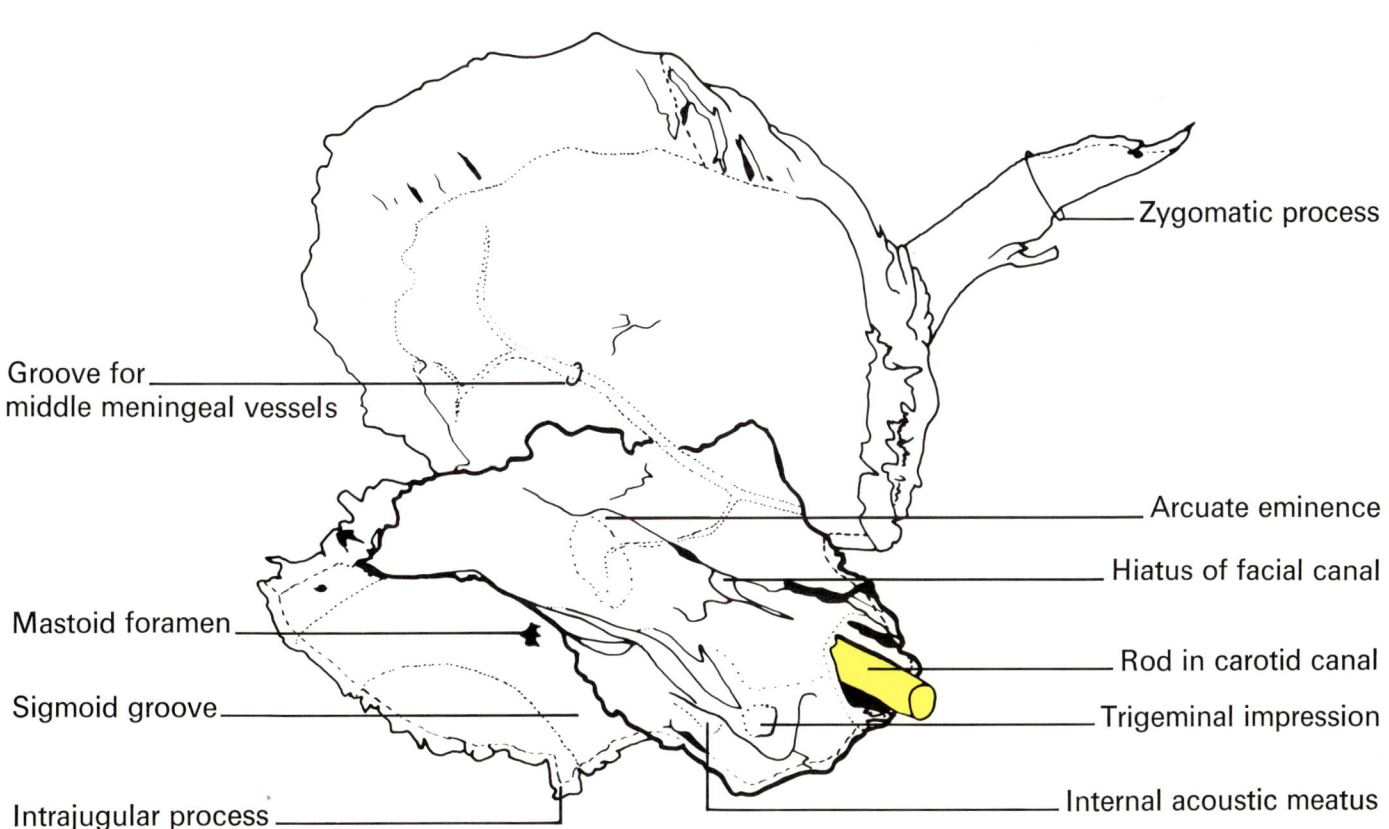

Temporal.

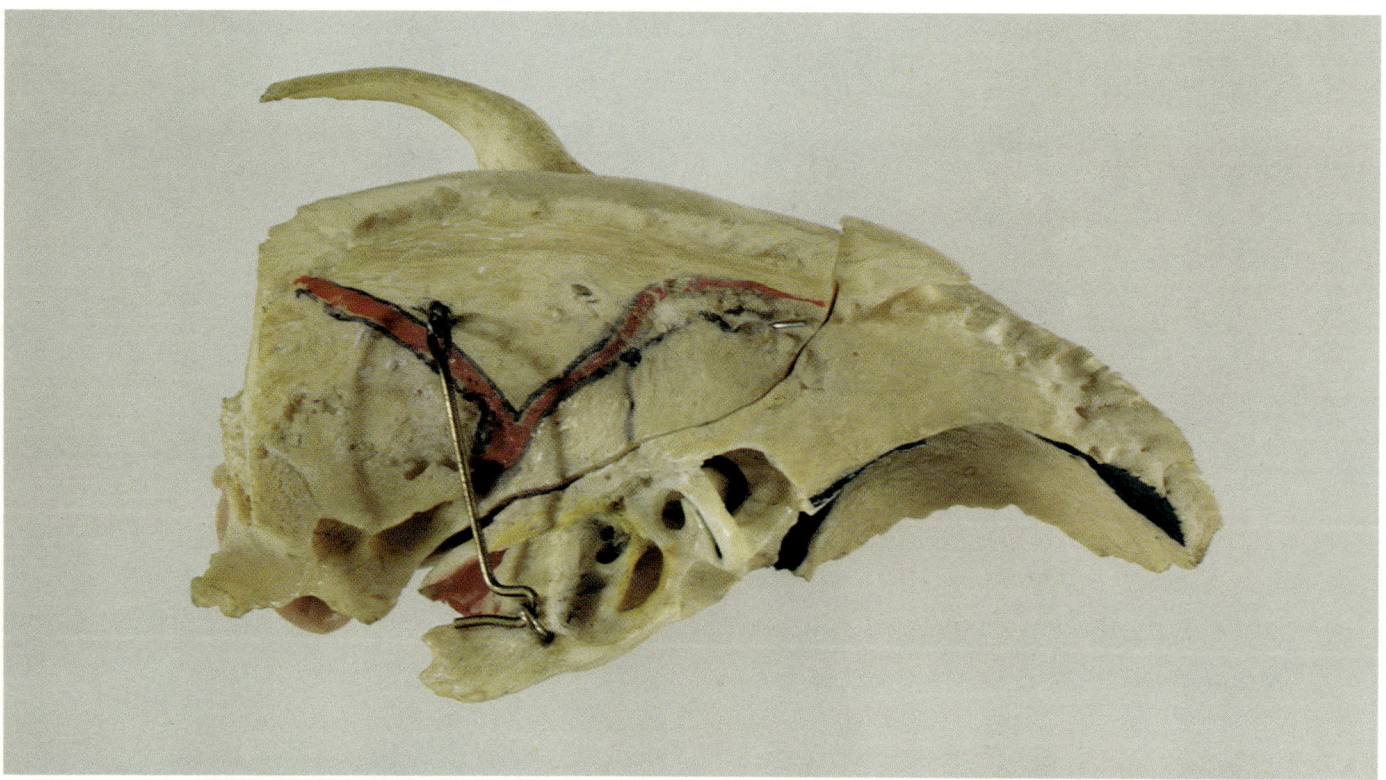

Right temporal bone, from above, semicircular canals exposed.

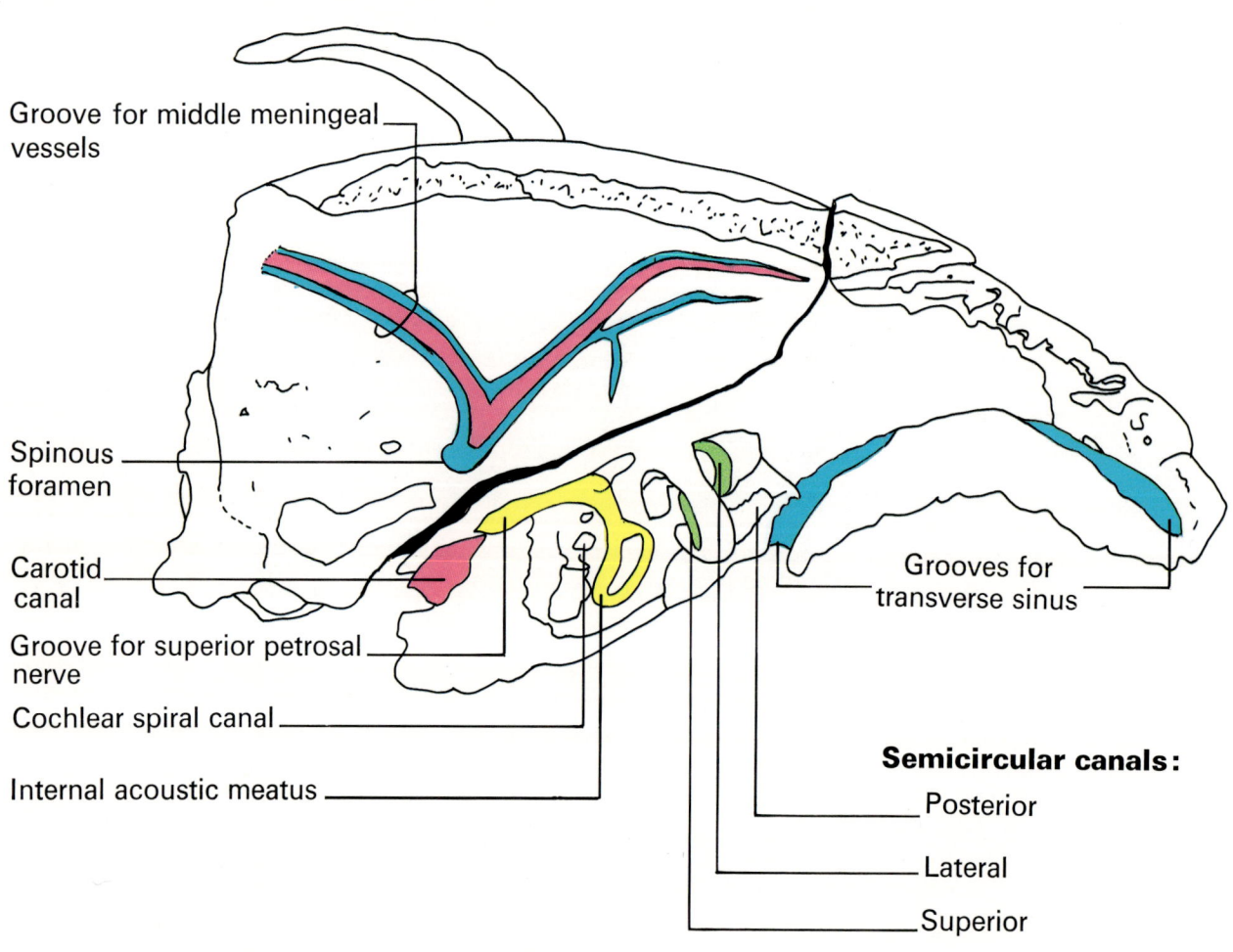

- Groove for middle meningeal vessels
- Spinous foramen
- Carotid canal
- Groove for superior petrosal nerve
- Cochlear spiral canal
- Internal acoustic meatus
- Grooves for transverse sinus

Semicircular canals:
- Posterior
- Lateral
- Superior

Temporal.

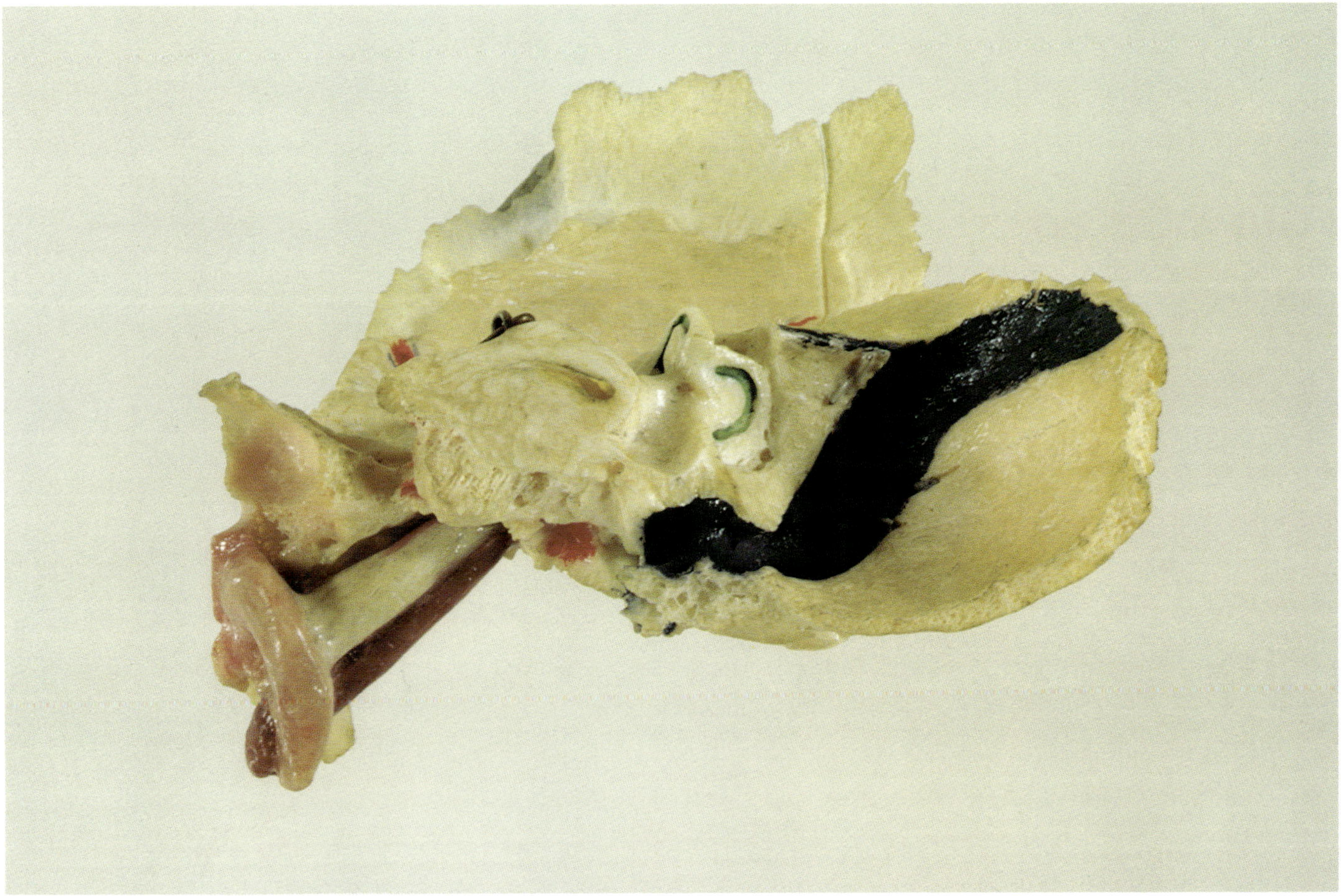

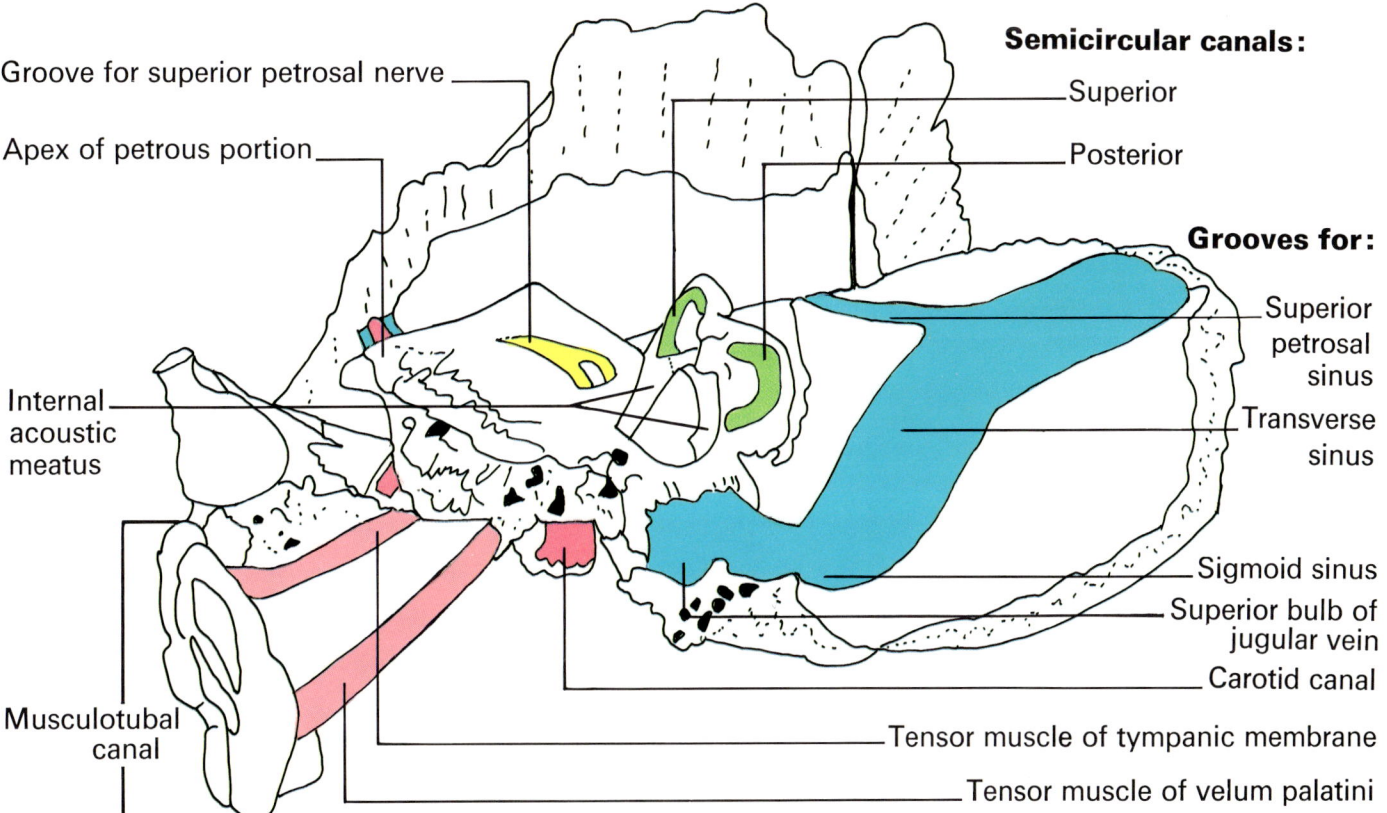

Right temporal, from behind.

Temporal.

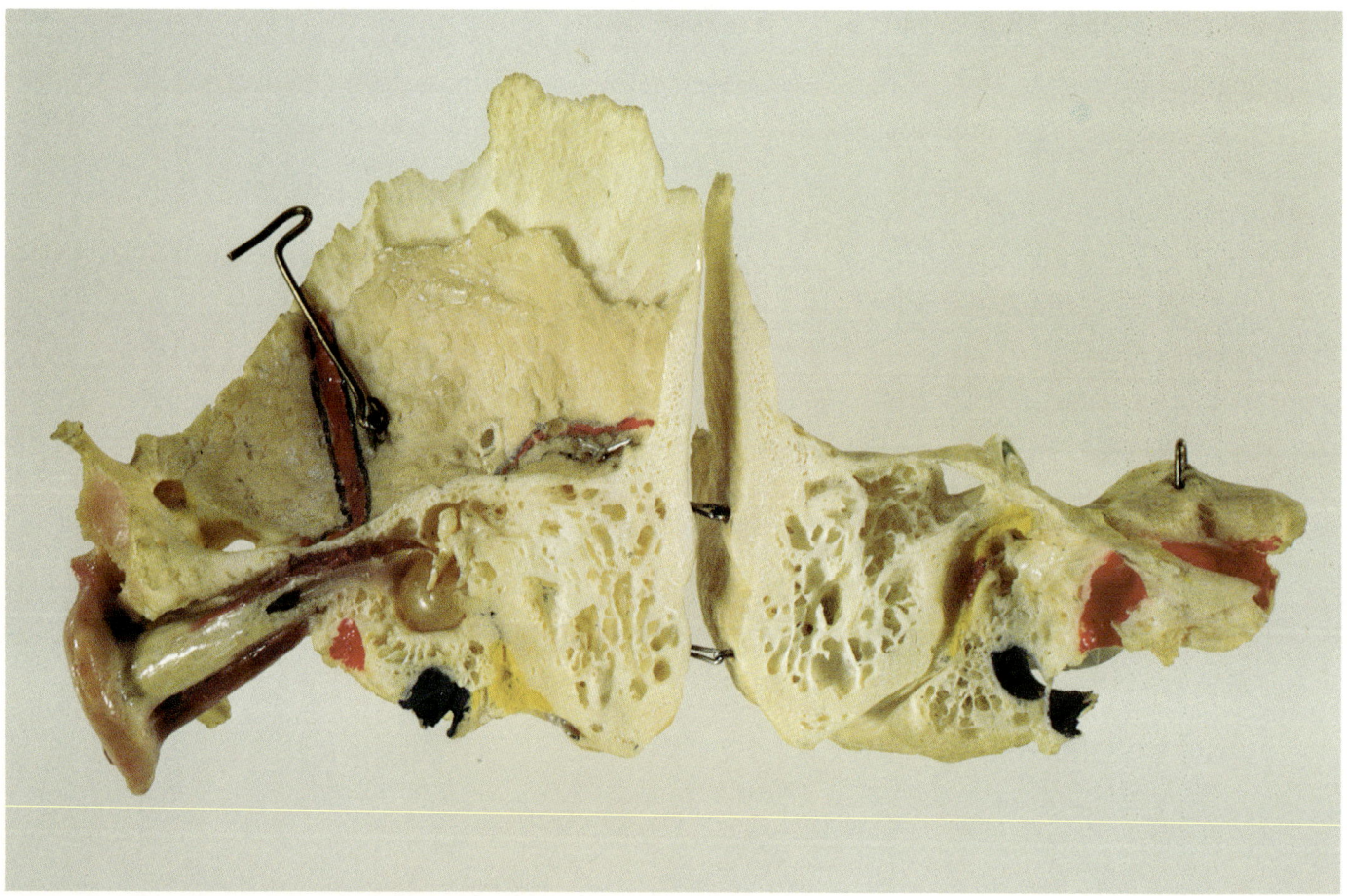

Sections separated, exposing facial canal and auditory ossicles.

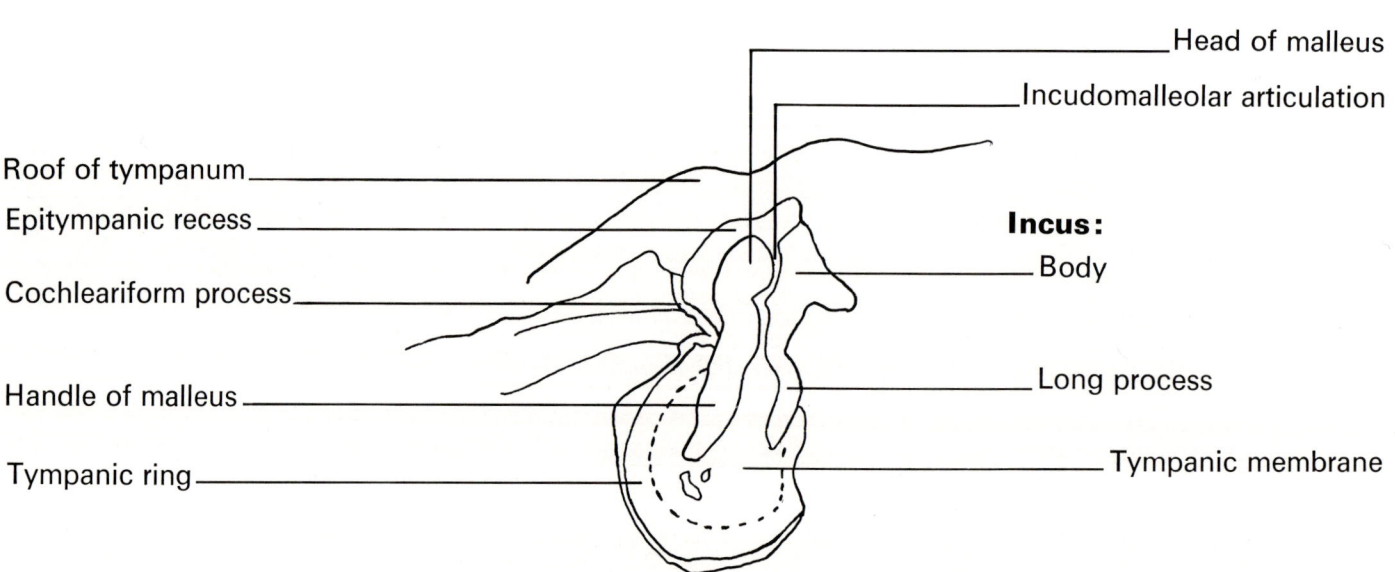

Line drawing of auditory ossicles magnified.

Temporal.

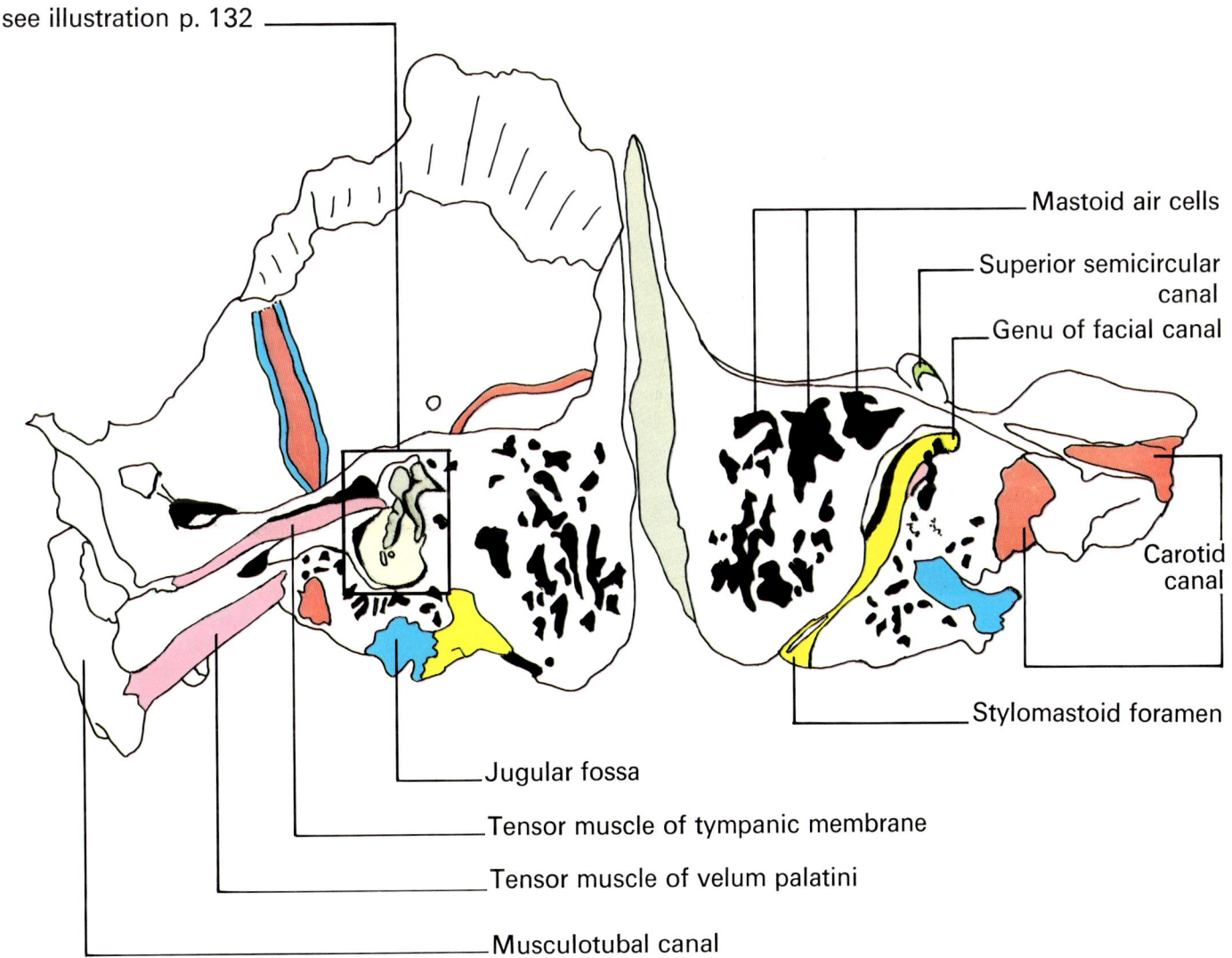

Temporal.

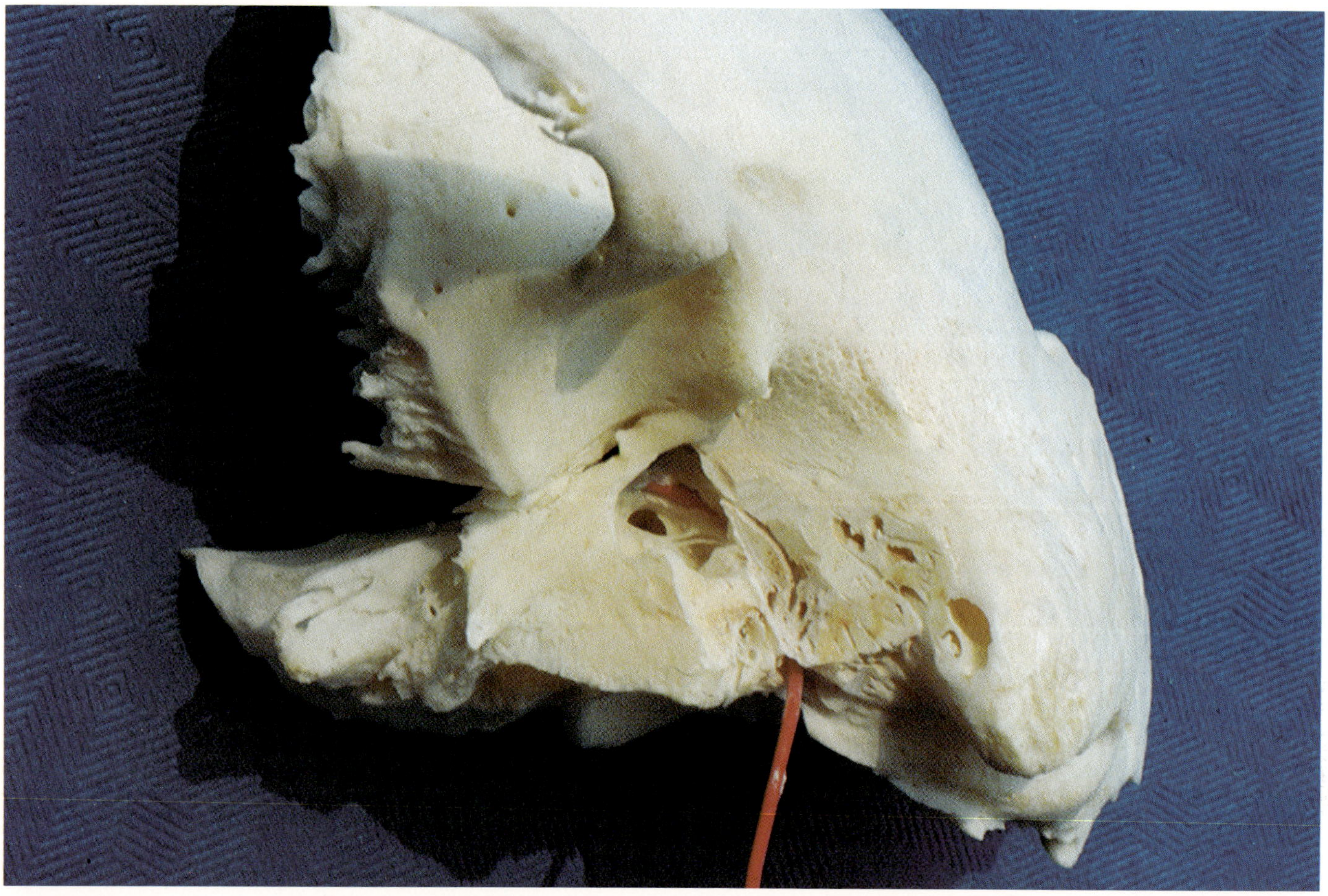

Left temporal bone, from side, and slightly below. Orange rod in facial canal which has been partially exposed.

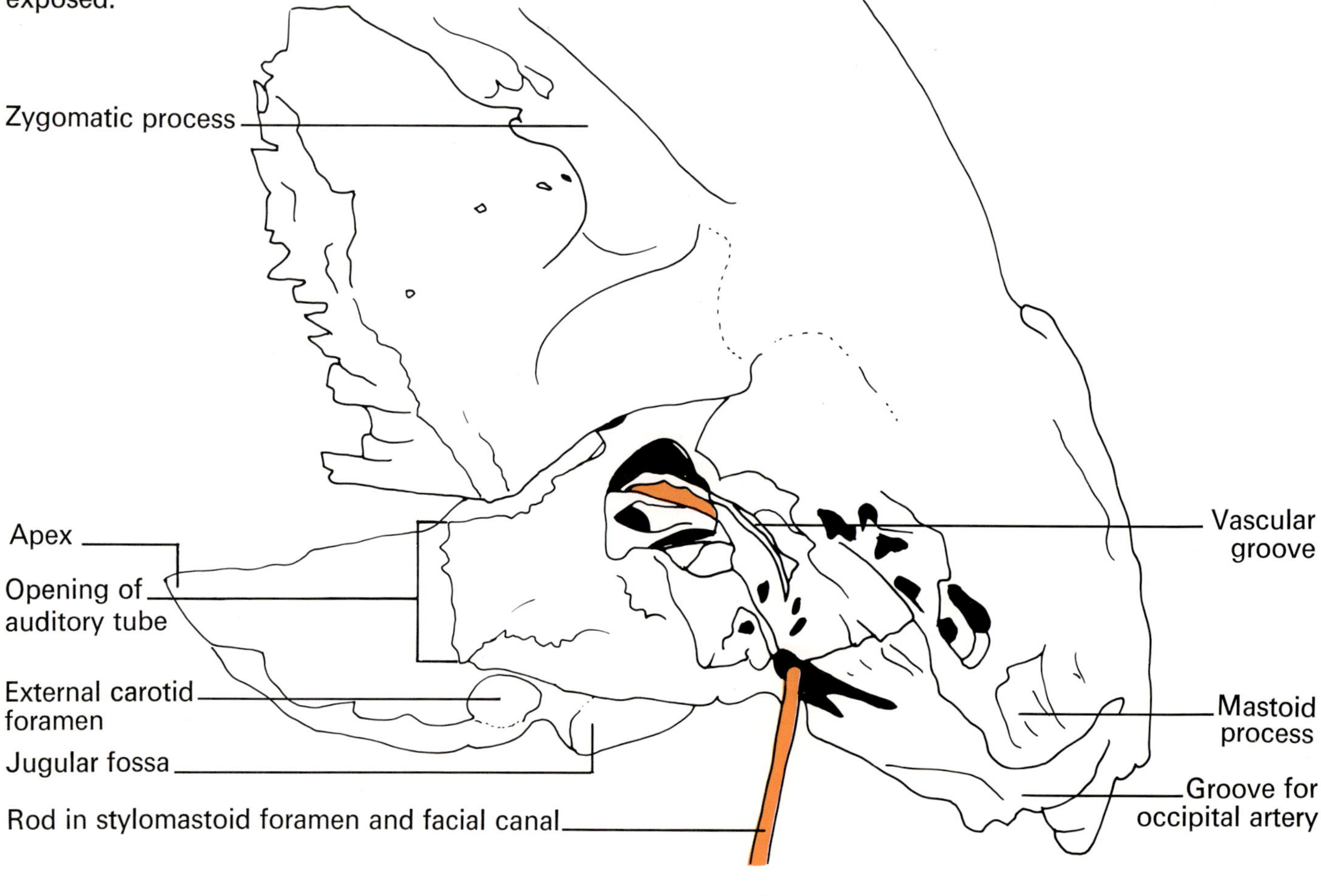

- Zygomatic process
- Apex
- Opening of auditory tube
- External carotid foramen
- Jugular fossa
- Rod in stylomastoid foramen and facial canal
- Vascular groove
- Mastoid process
- Groove for occipital artery

Temporal.

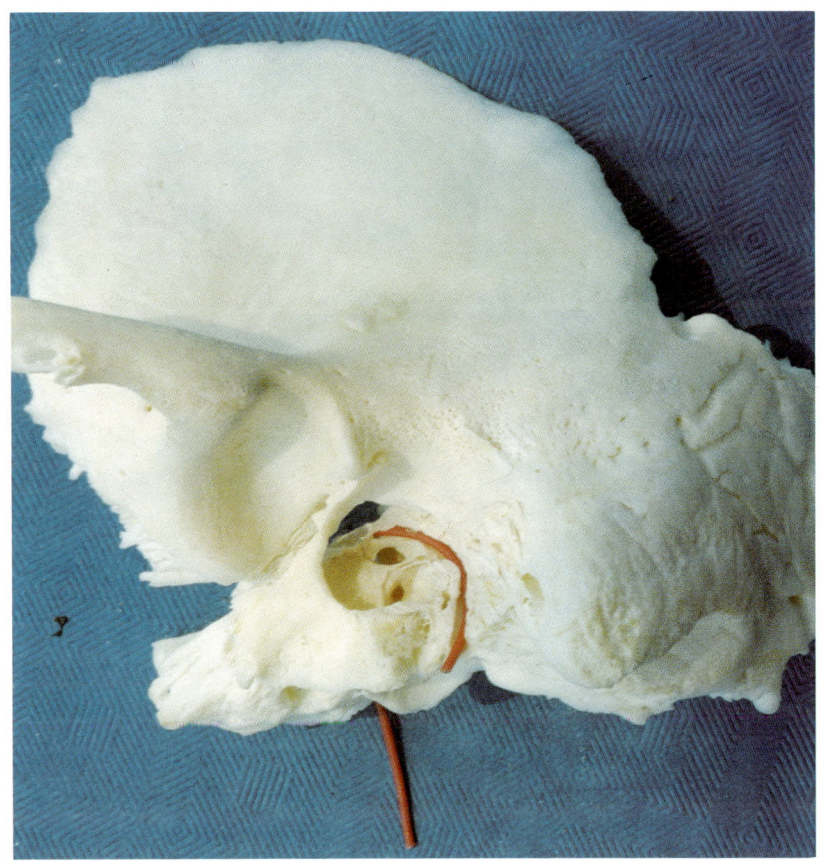

Facial canal exposed.

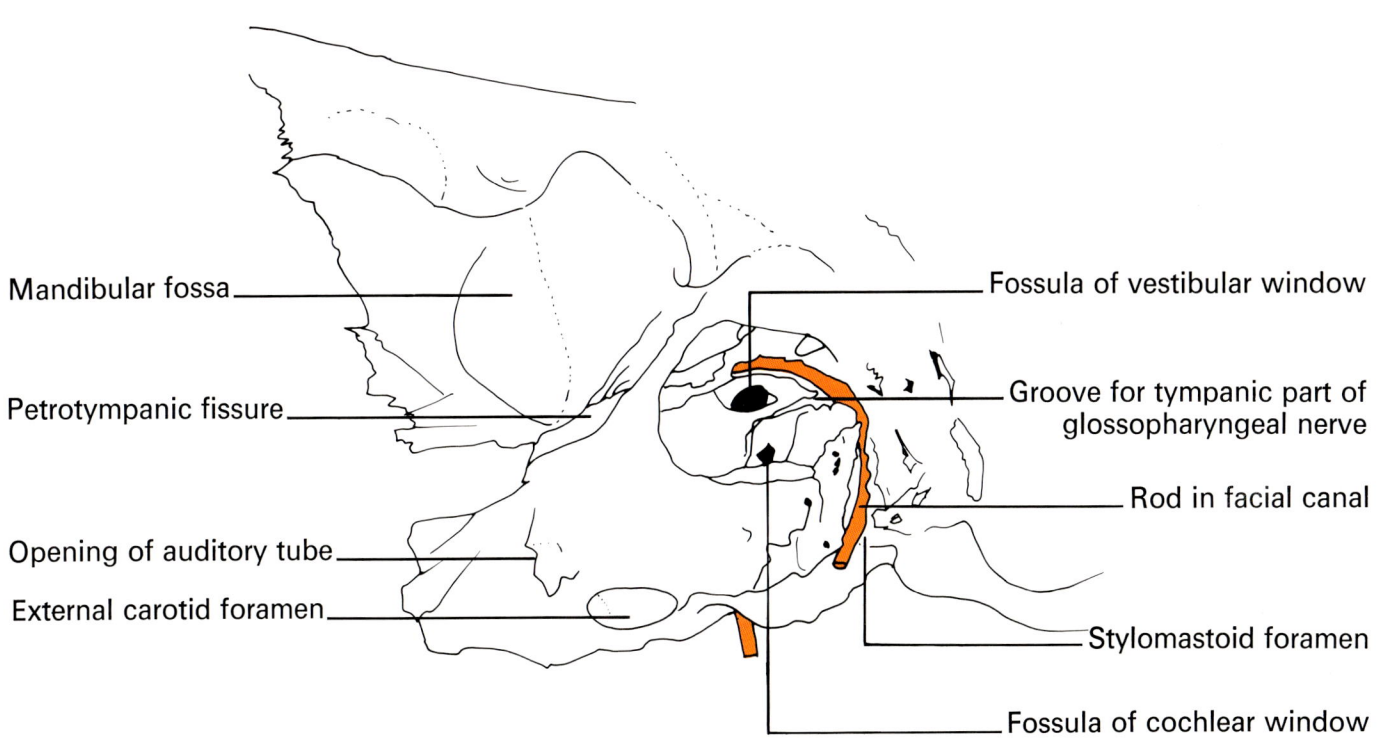

Temporal.

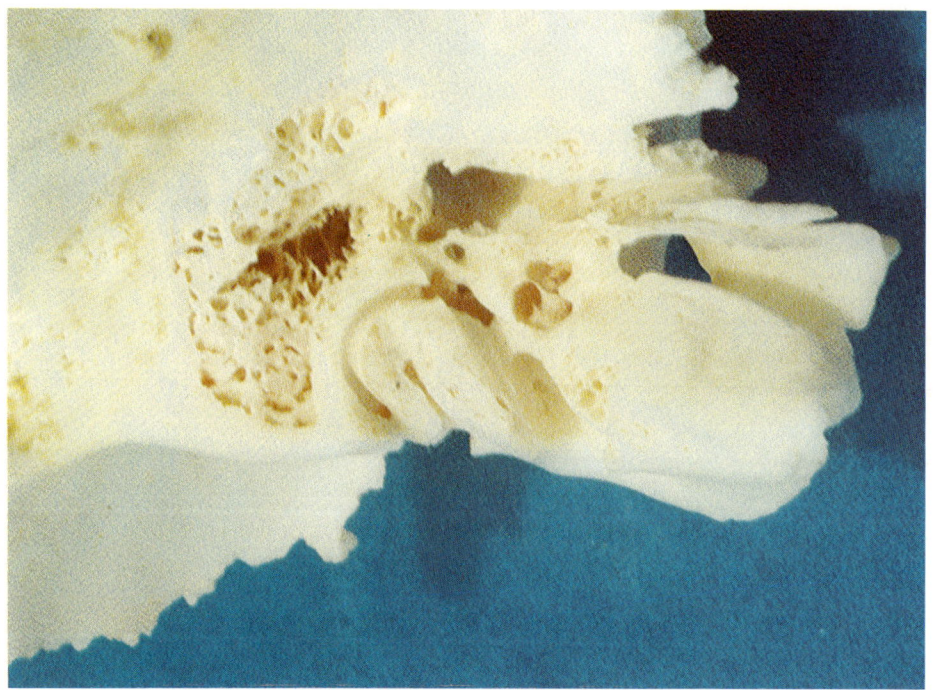

Left temporal bone, from above. Cochlear spiral canal, internal acoustic meatus and superior semicircular canal exposed.

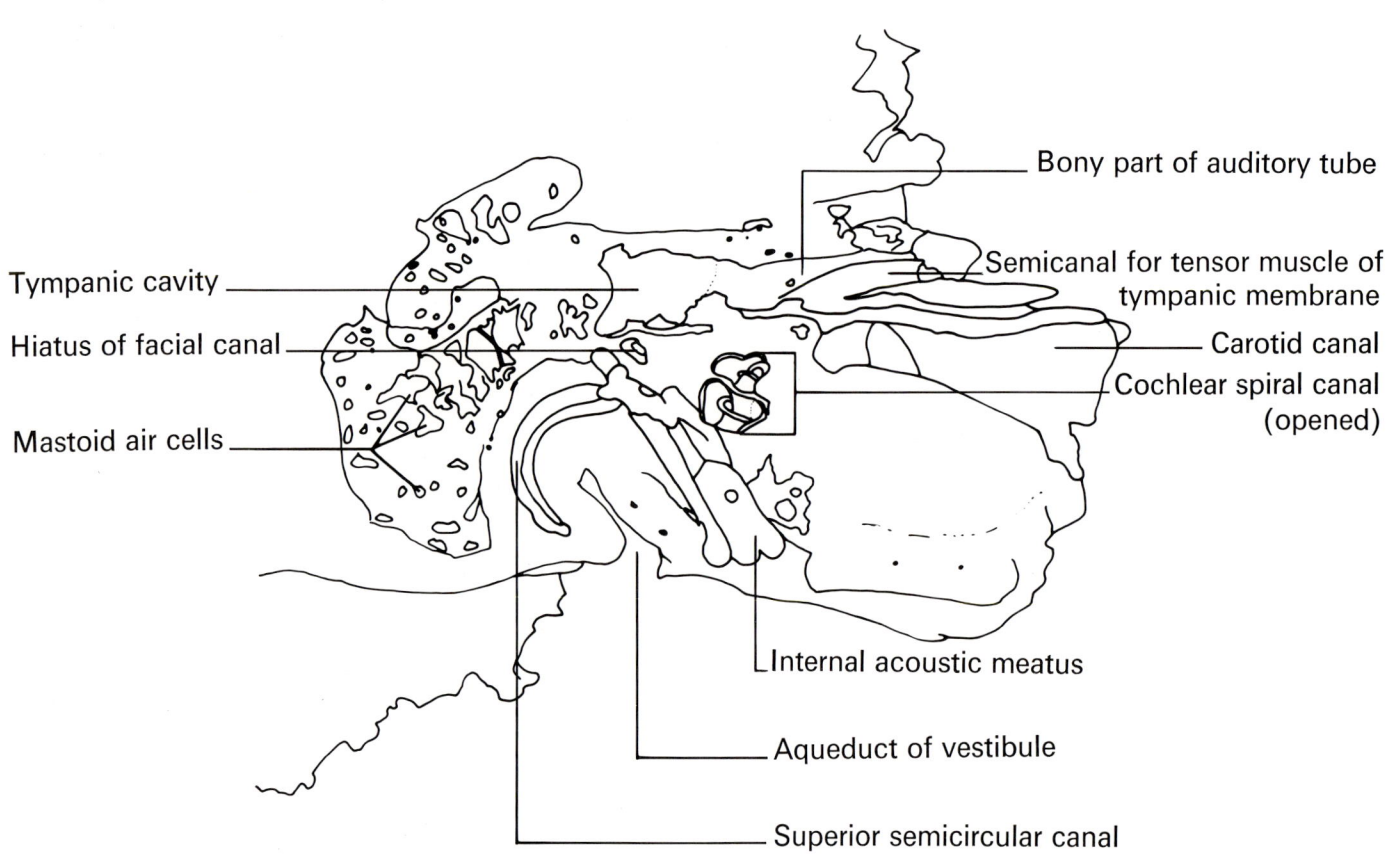

Temporal.

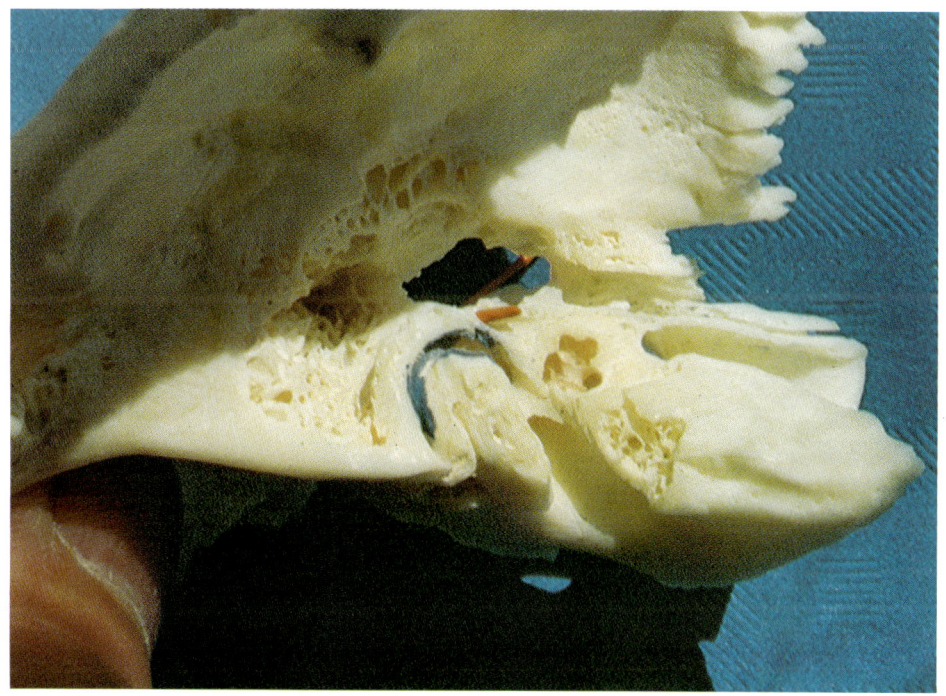

Left temporal bone from above. Orange rod in opening of facial canal.

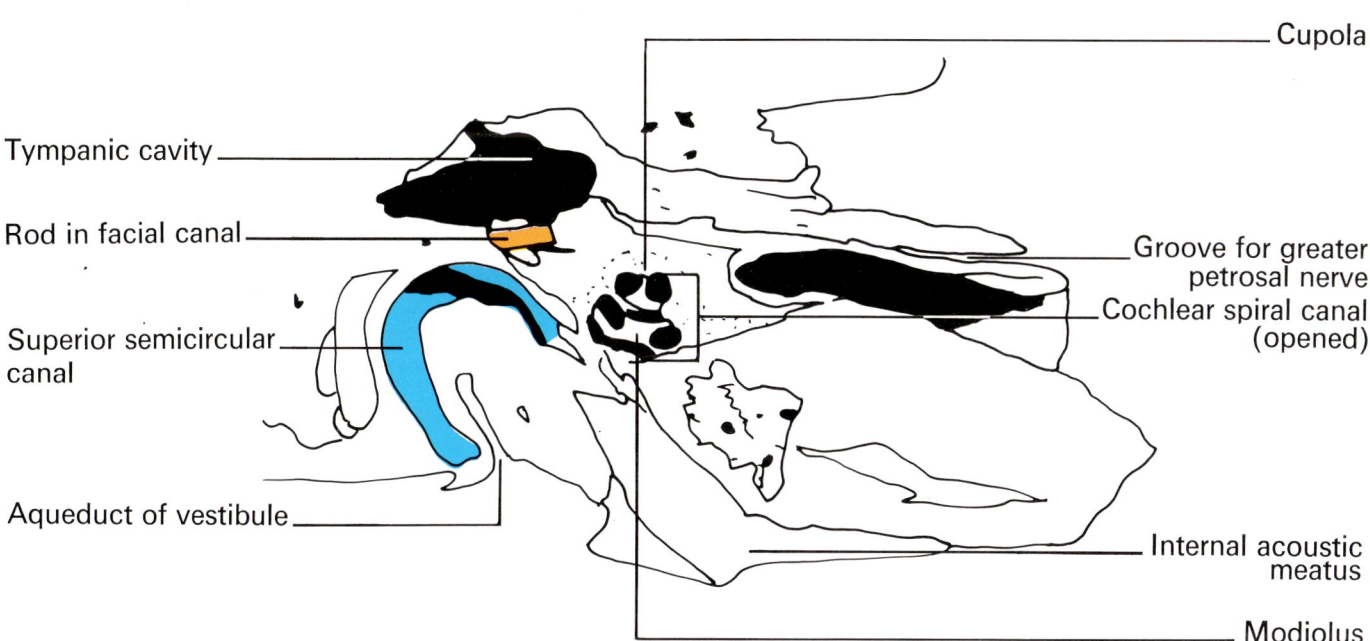

SECTION 7. Auditory ossicles.
Diagrammatic. A. Right side, from front.
B. and C. Composites of malleus, incus and stapes.

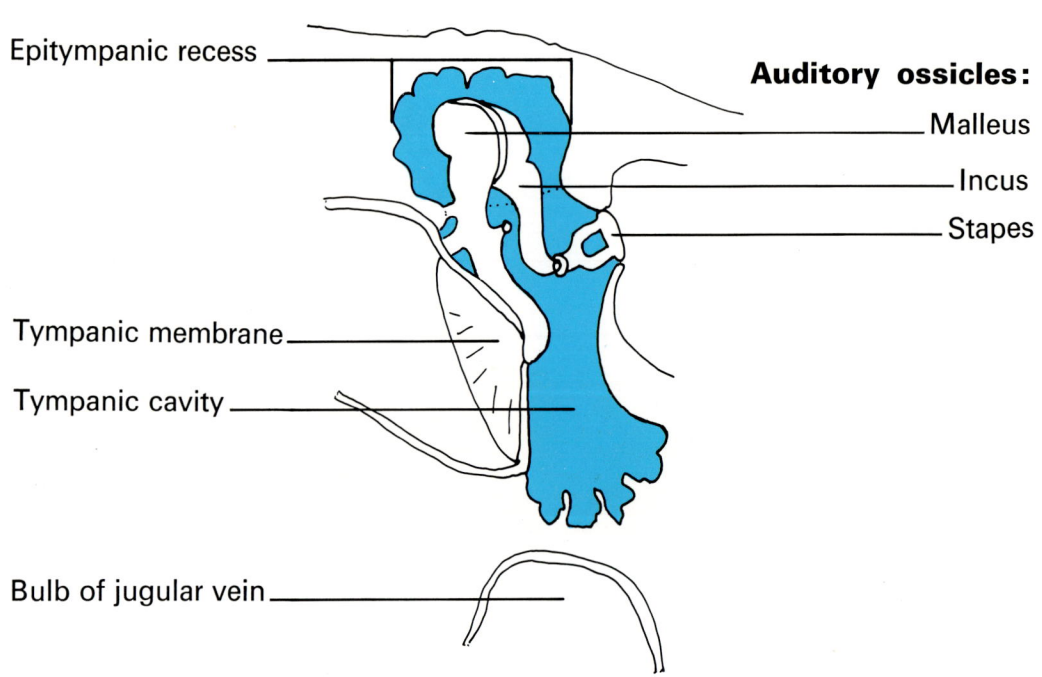

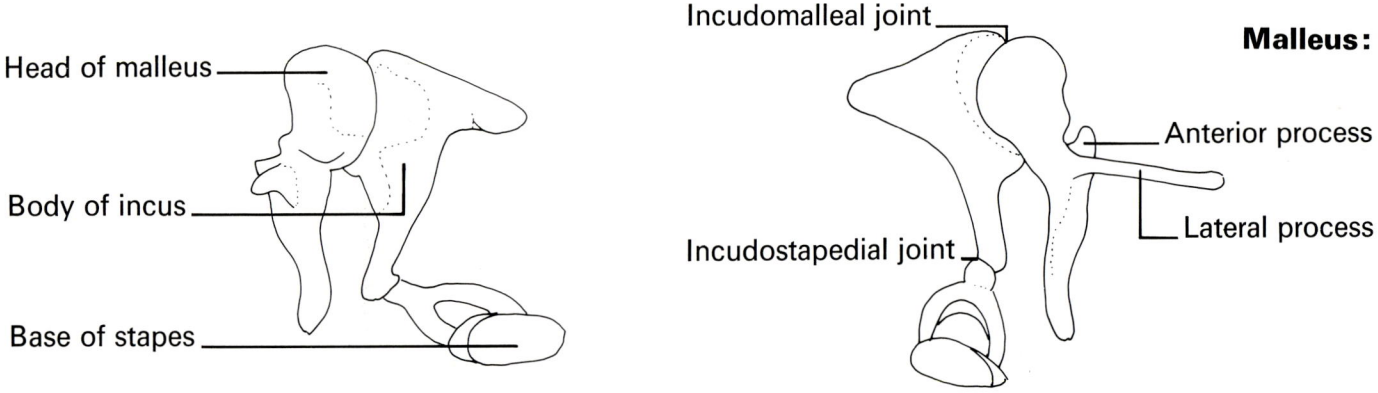

Auditory ossicles.

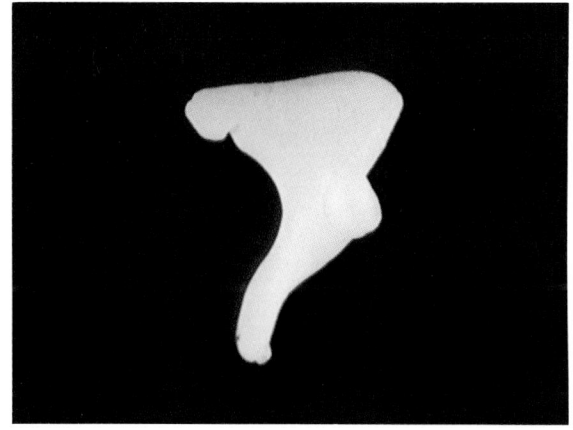
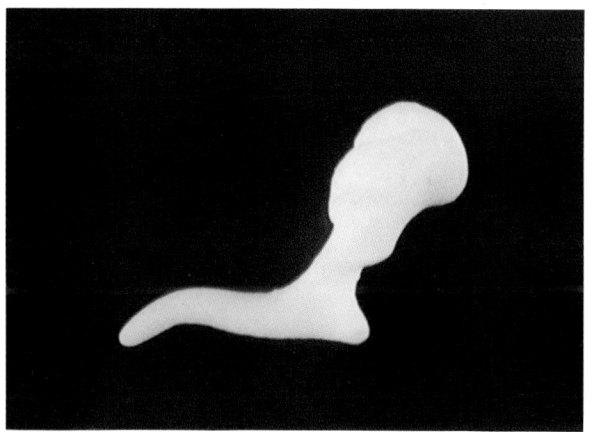
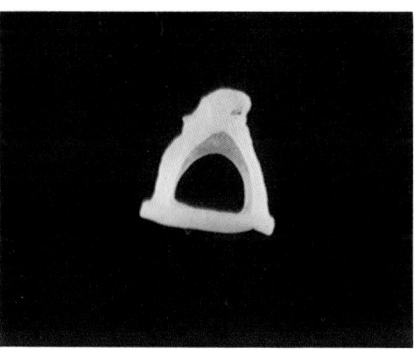

Incus, malleus and stapes, greatly enlarged.
Magnification 6.5X.

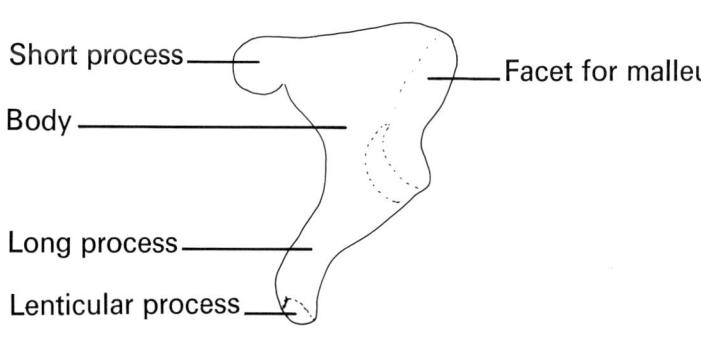

Incus

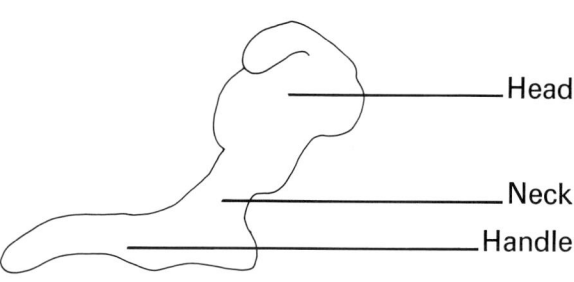

Malleus

Stapes

Auditory ossicles.
Ligaments and muscles.

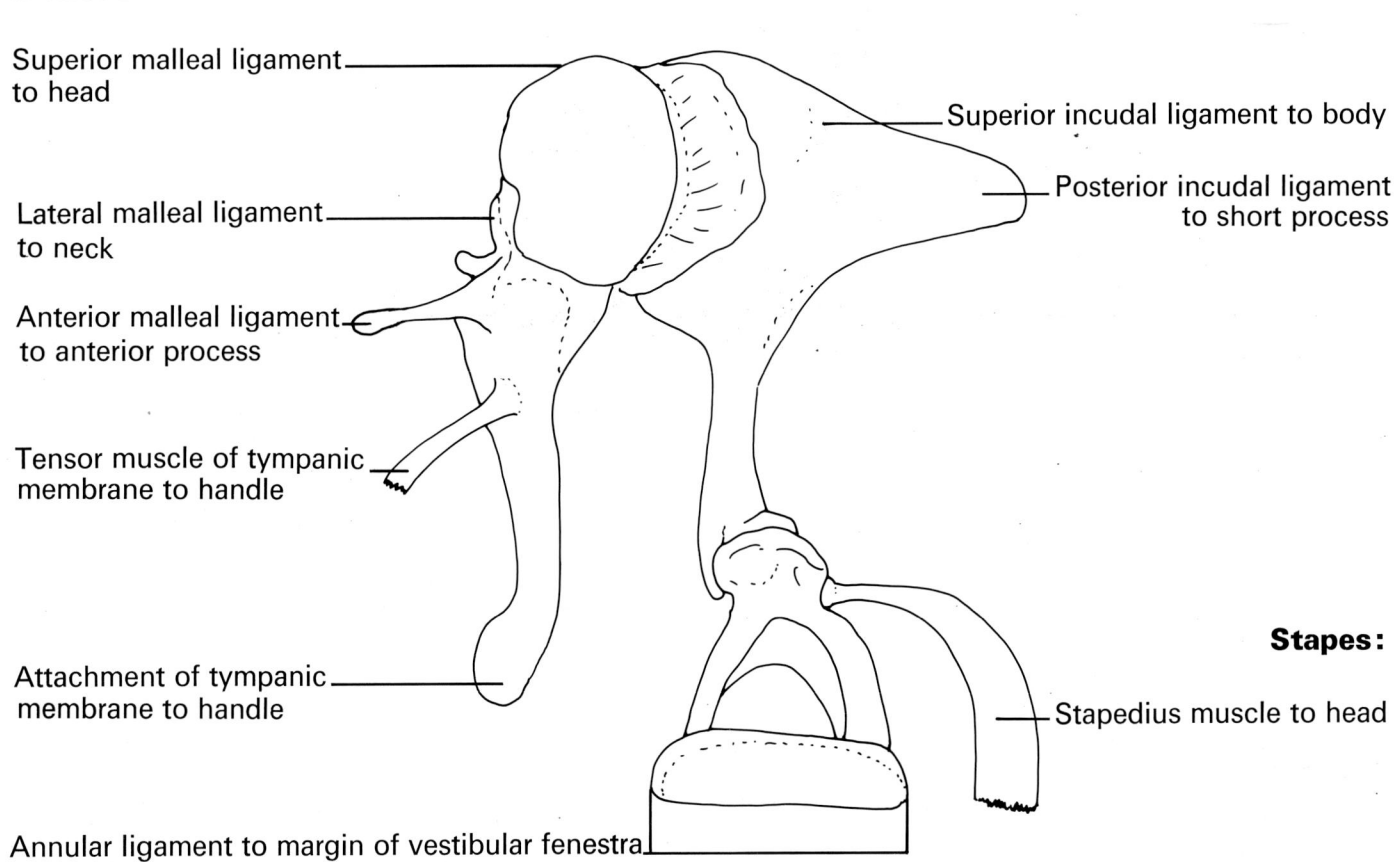

Malleus:

Superior malleal ligament to head

Lateral malleal ligament to neck

Anterior malleal ligament to anterior process

Tensor muscle of tympanic membrane to handle

Attachment of tympanic membrane to handle

Annular ligament to margin of vestibular fenestra

Incus:

Superior incudal ligament to body

Posterior incudal ligament to short process

Stapes:

Stapedius muscle to head

(Illustrations 140-145 redrawn & relabeled from Surgical Anatomy of the Temporal Bone and Ear, 2nd ed., by Anson, B. J. and Donaldson, J. A. Couritesy of W. B. Saunders, publisher)

SECTION 8. Tympanic cavity.

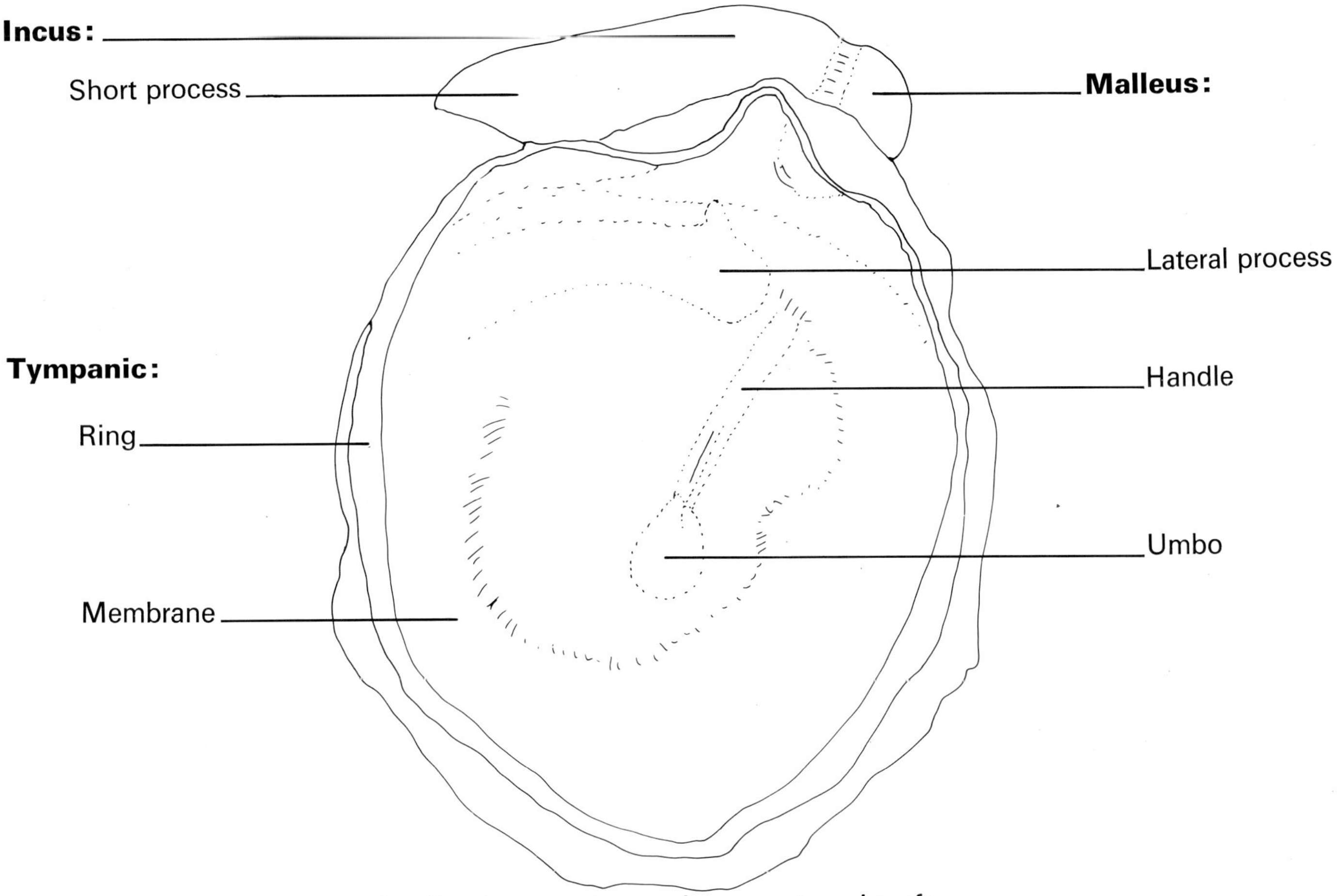

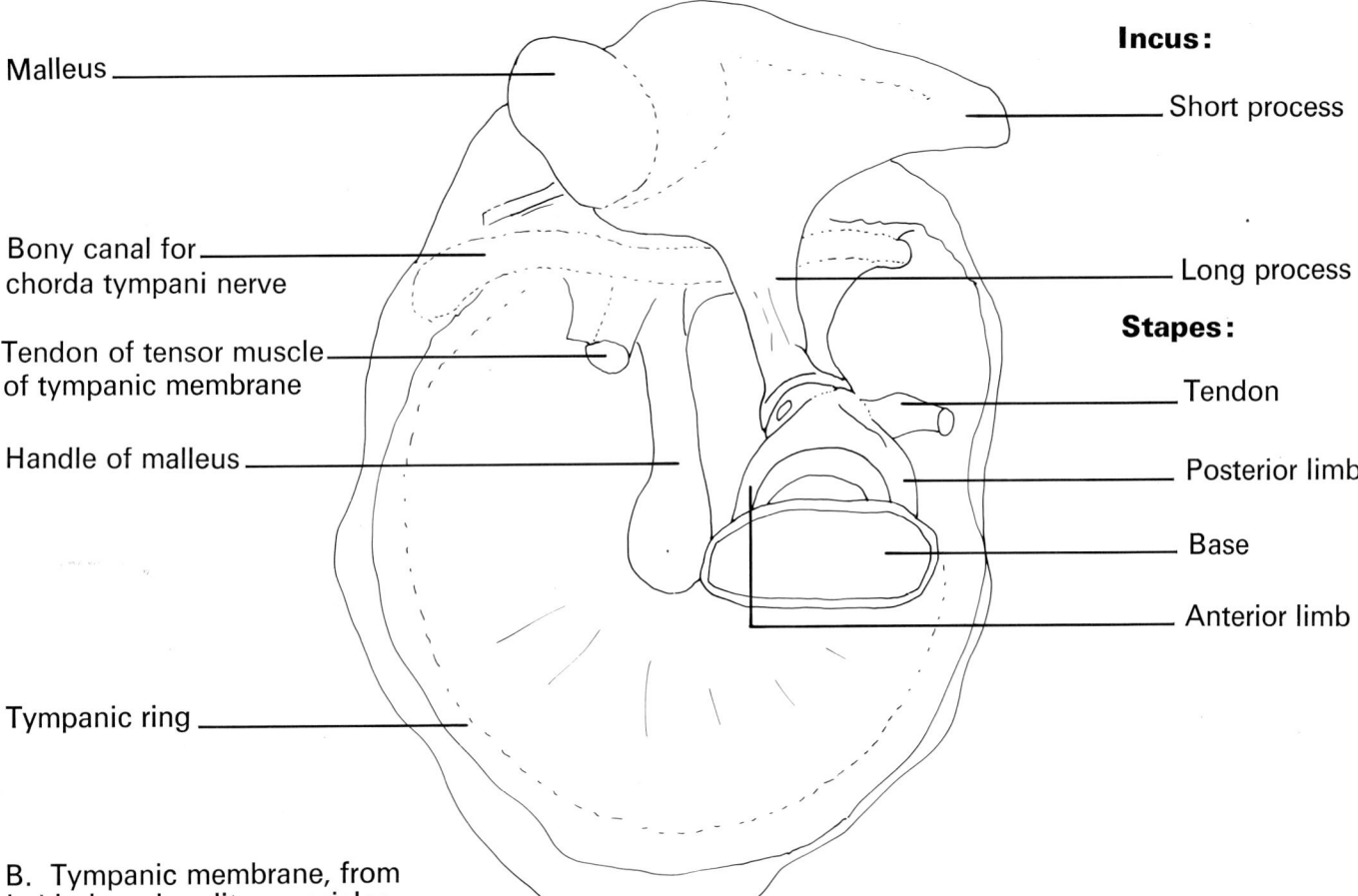

A. Right tympanic membrane, external surface.

B. Tympanic membrane, from behind, and auditory ossicles.

Tympanic cavity.

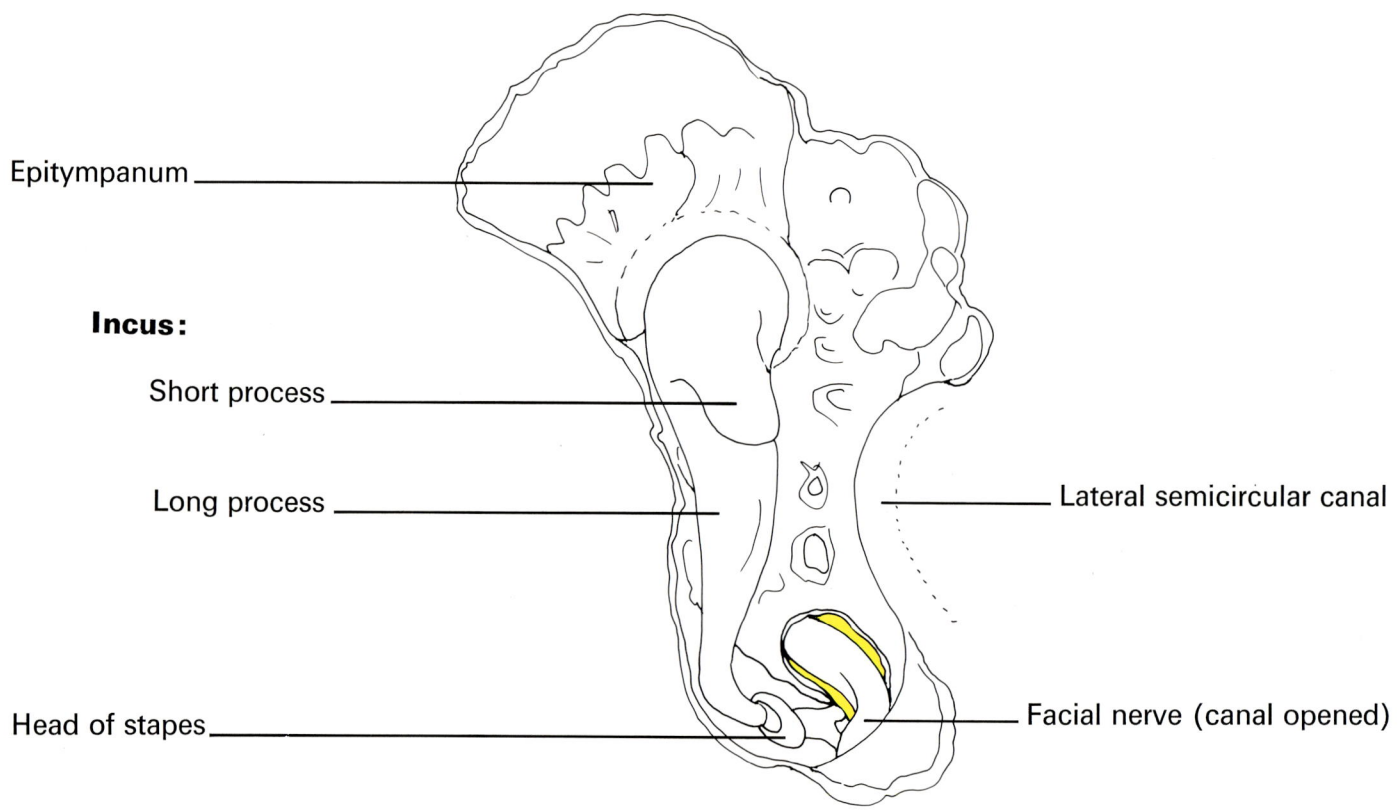

A. Left tympanic cavity, from side, facial canal exposed.

B. View, slightly superior, of incus, stapes and facial canal.

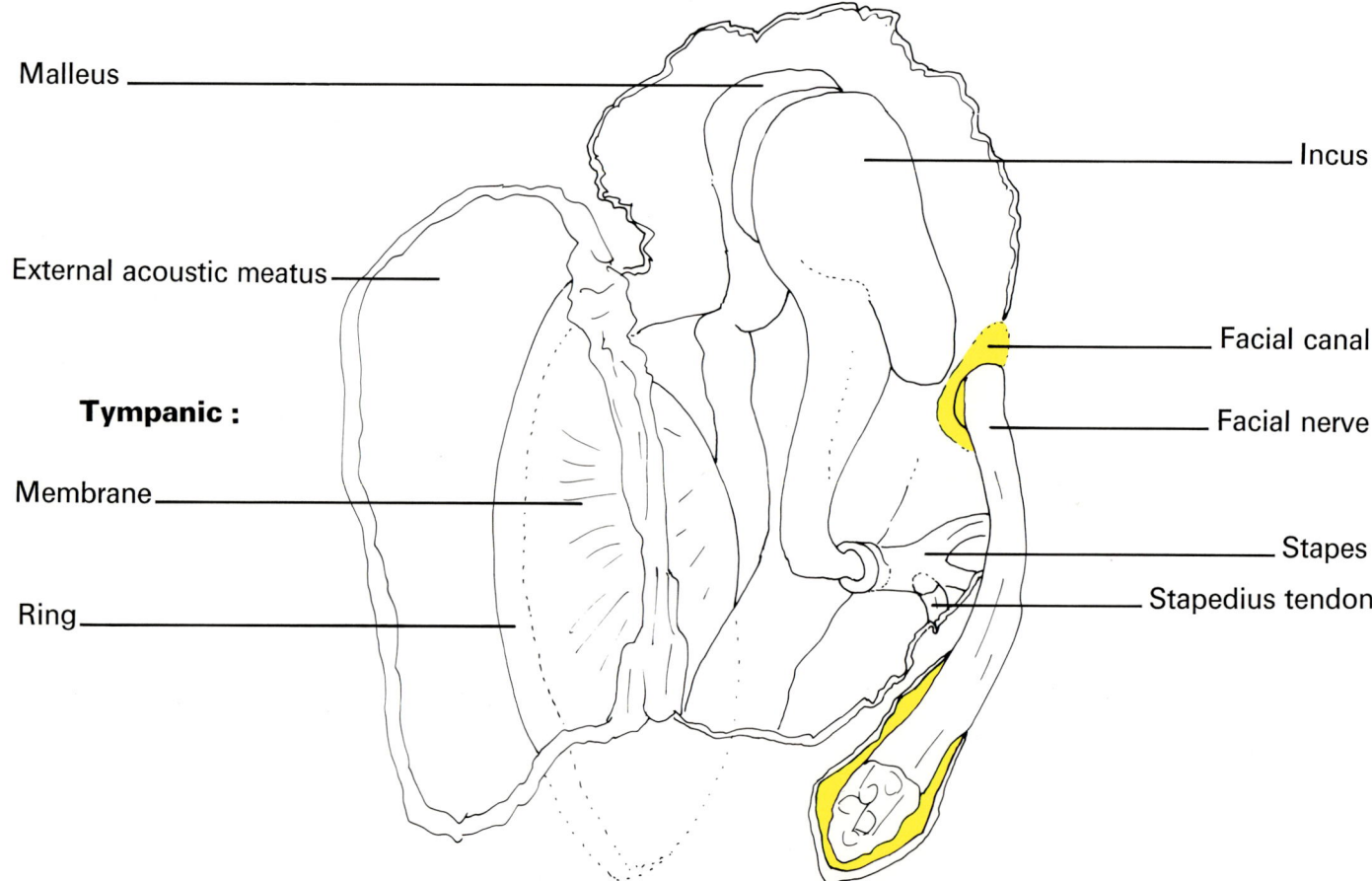

Tympanic cavity.

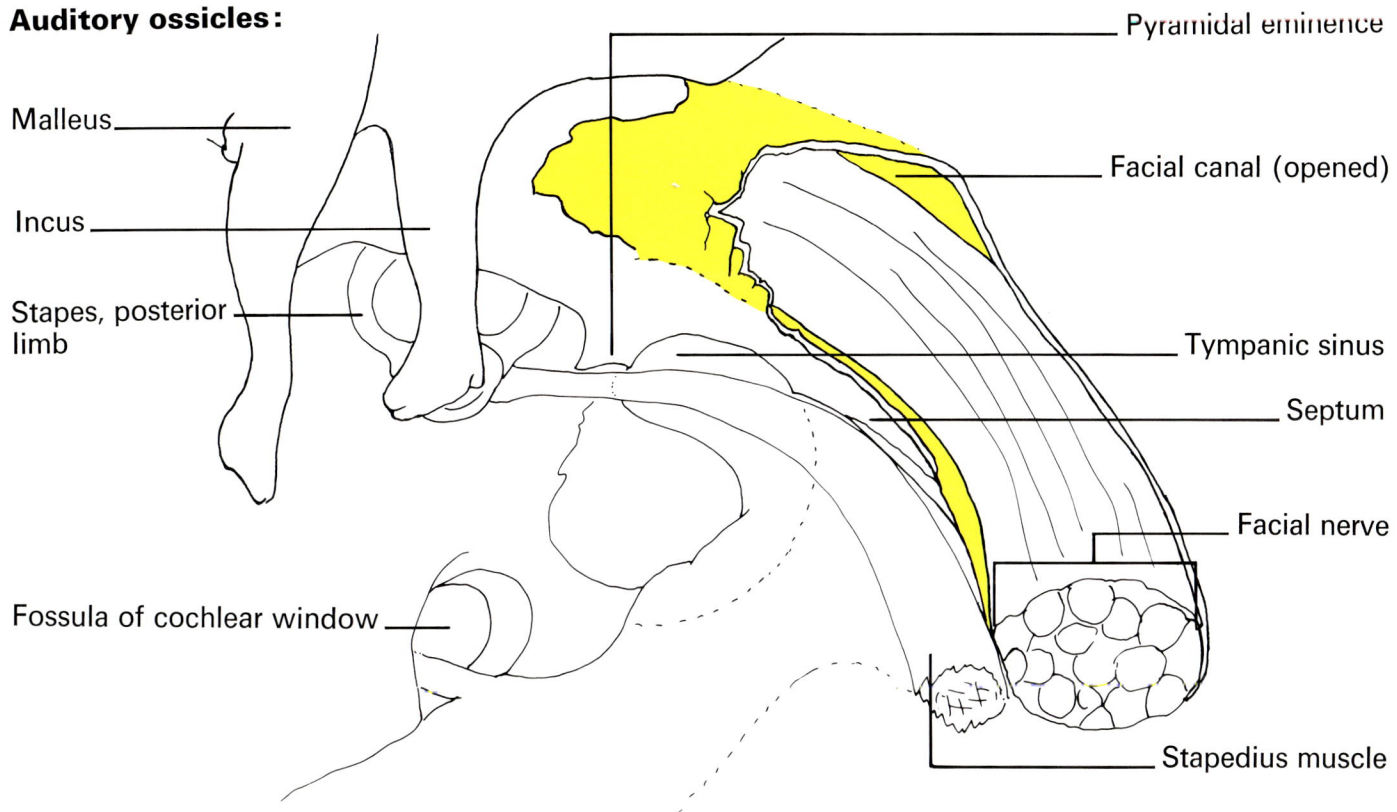

A. Facial nerve, canal and stapedius muscle in relation to auditory ossicles.

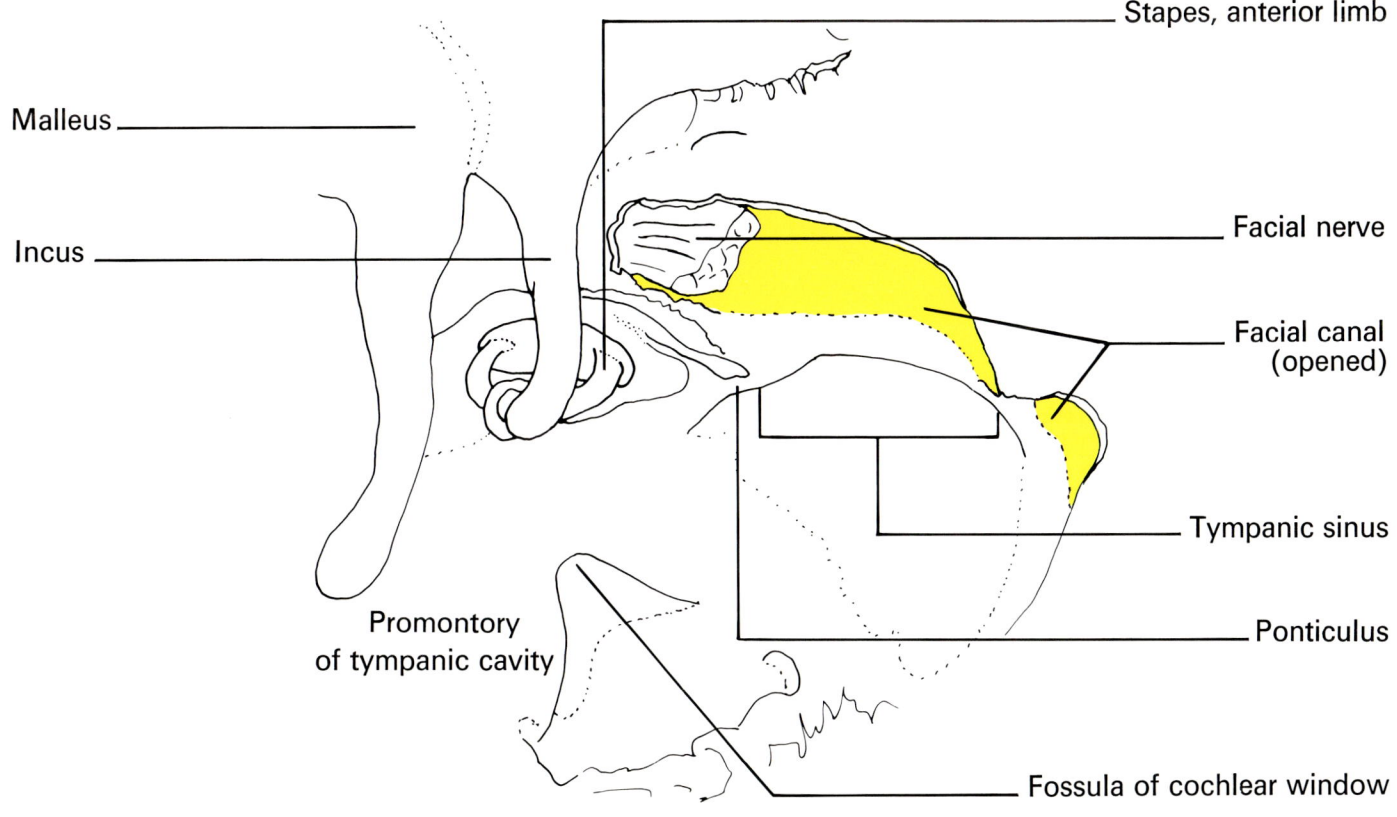

B. Facial nerve and stapedius muscle resected.

Tympanic cavity.

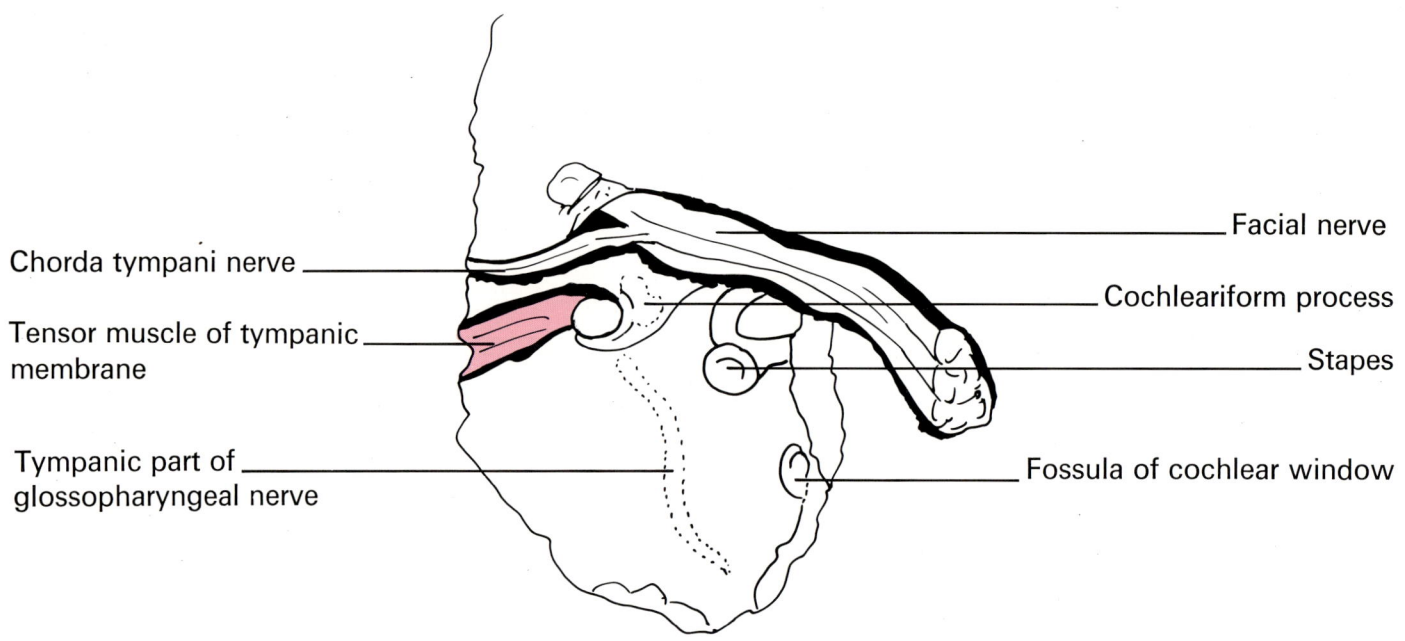

A. Stapes and adjacent nerves.
Closeup view of cochleariform process.

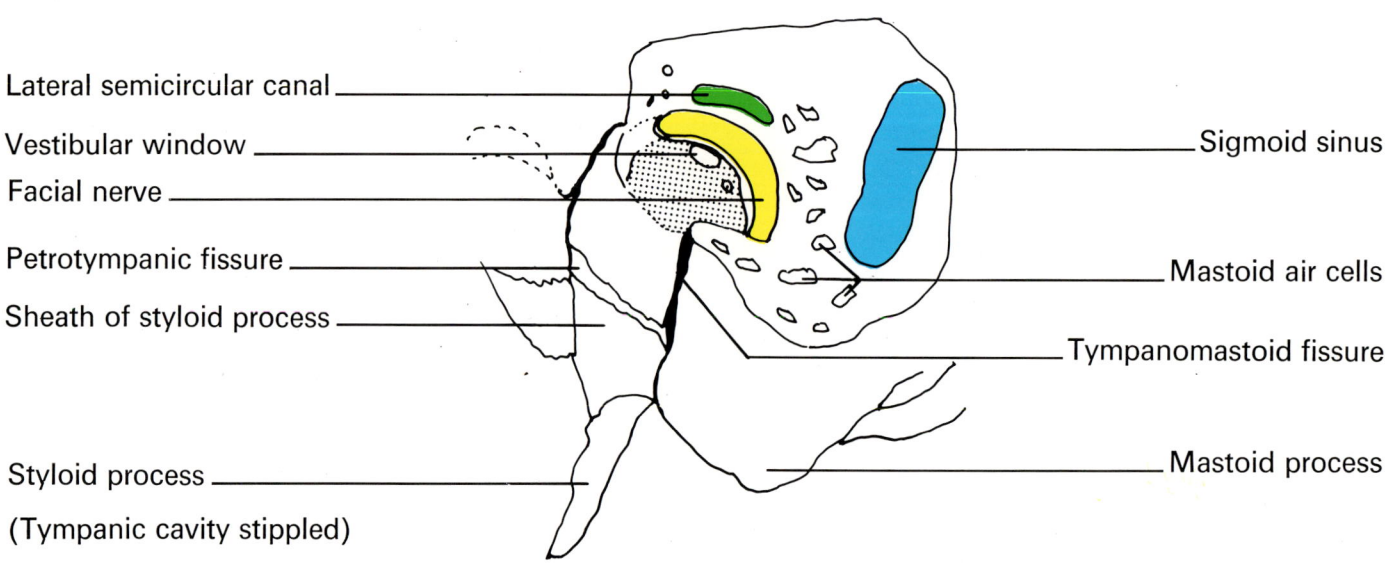

B. Relation of facial nerve to lateral semicircular canal and sigmoid sinus, diagrammatic.

Tympanic cavity.

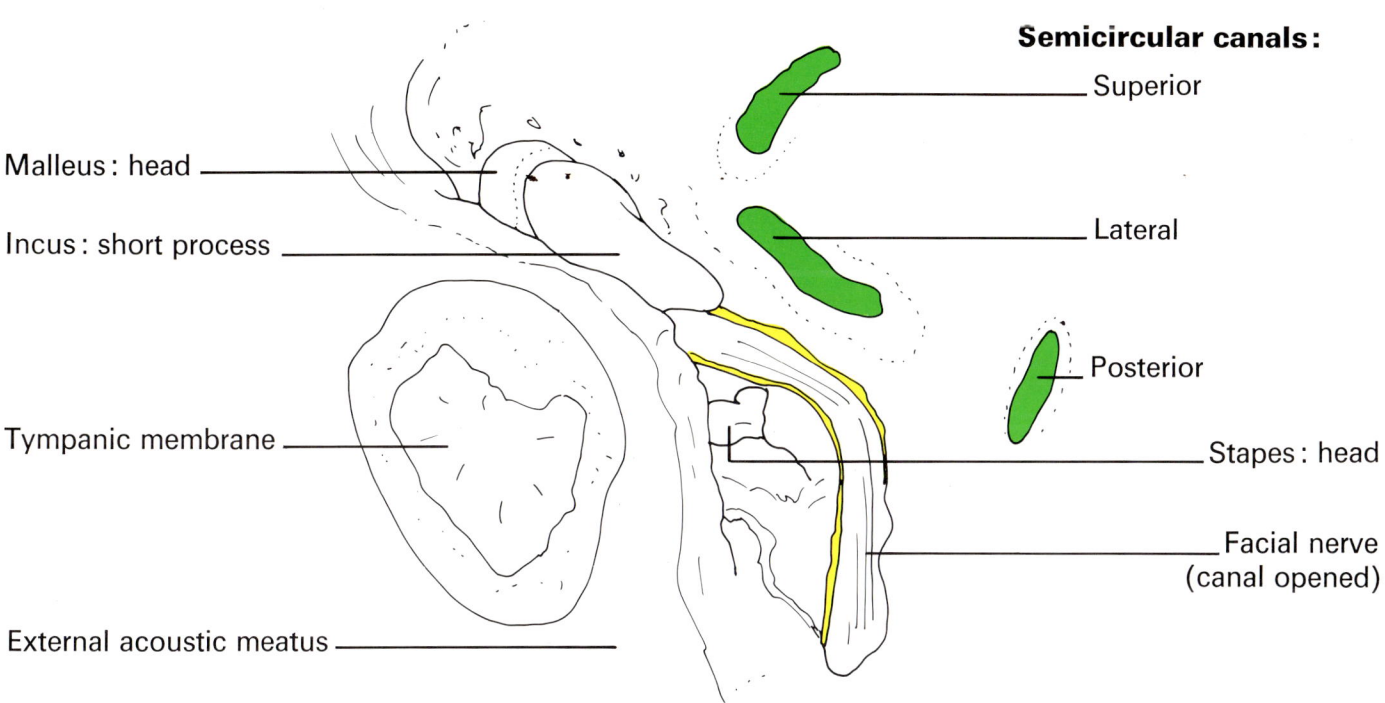

A. and B. Semicircular canals in relation to facial nerve and auditory ossicles.

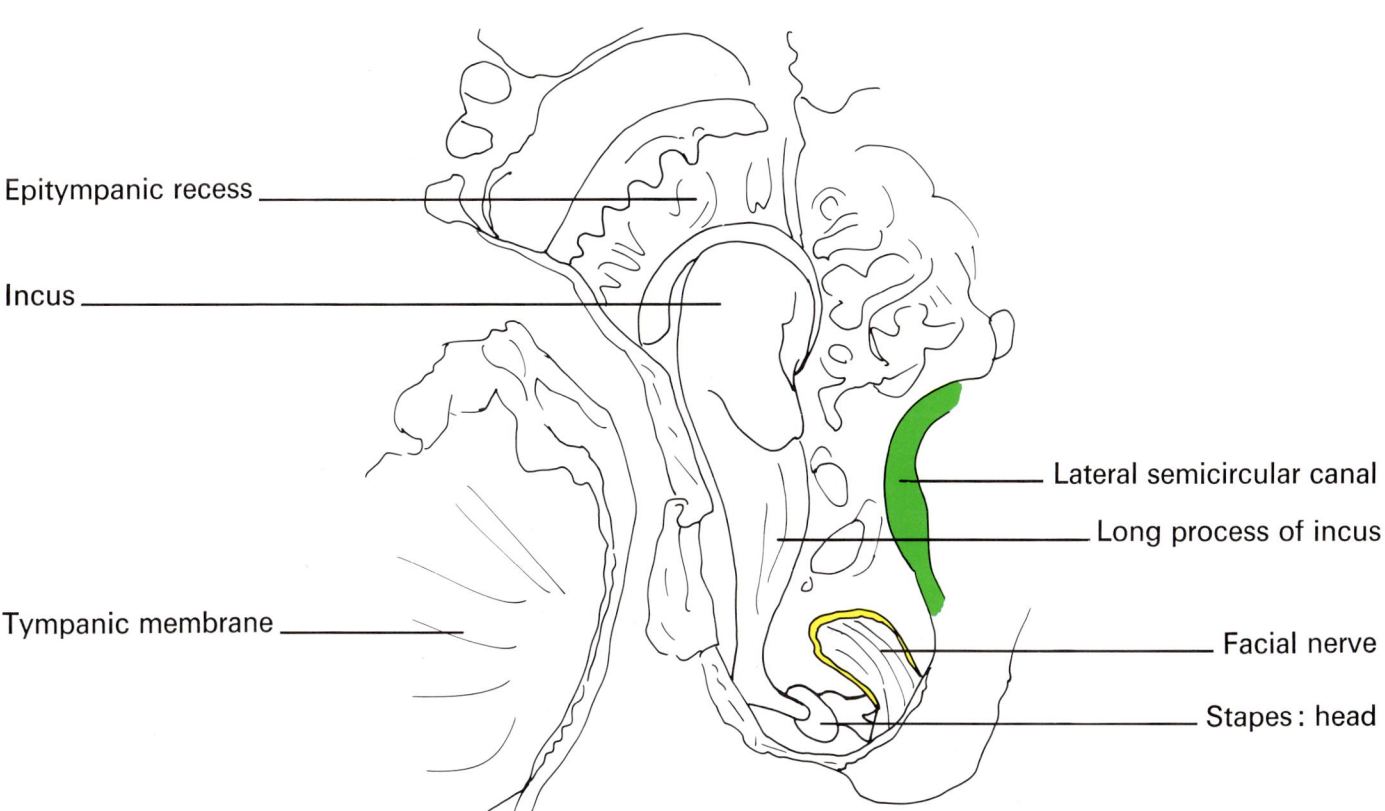

PART IV. INFANT SKULL AT BIRTH.

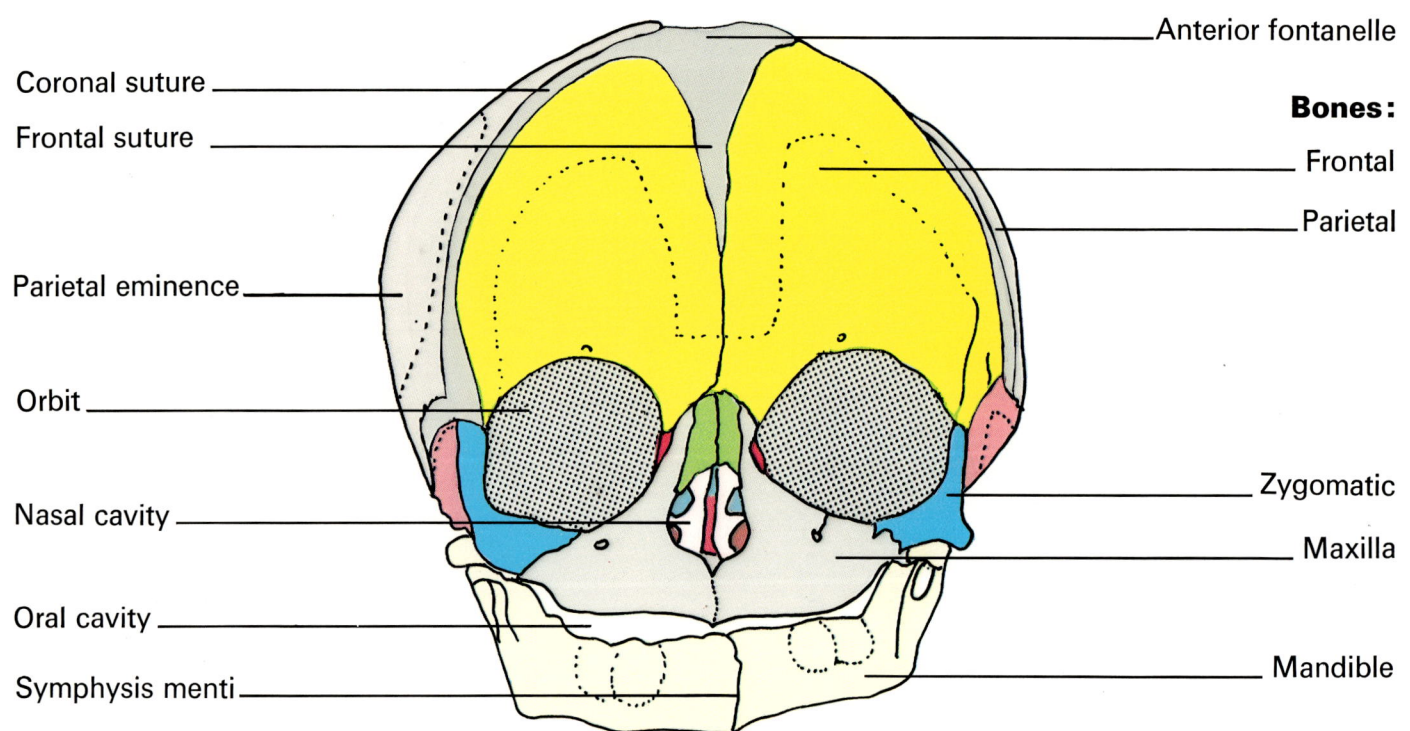

A. From front.

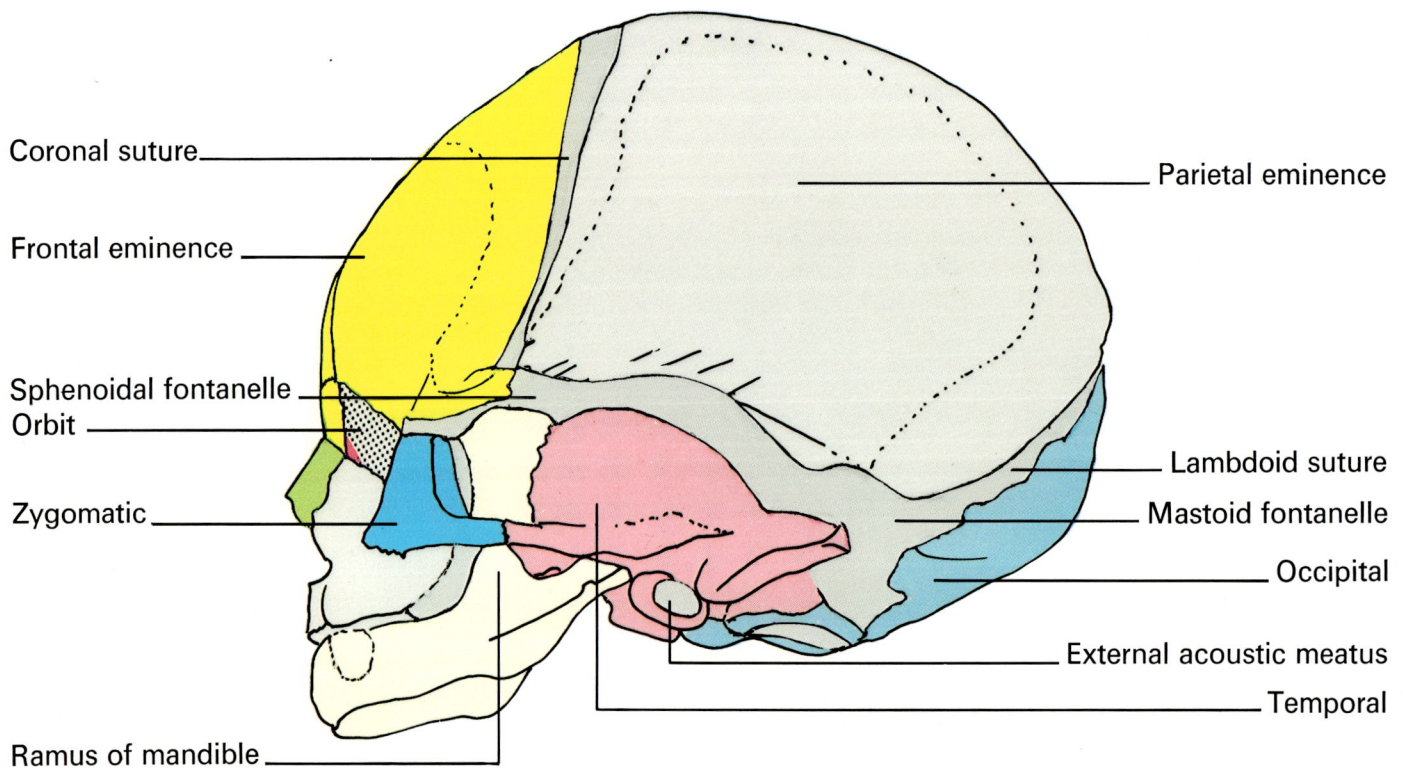

B. From left side.

(Redrawn from actual specimen.)

Infant skull at birth.

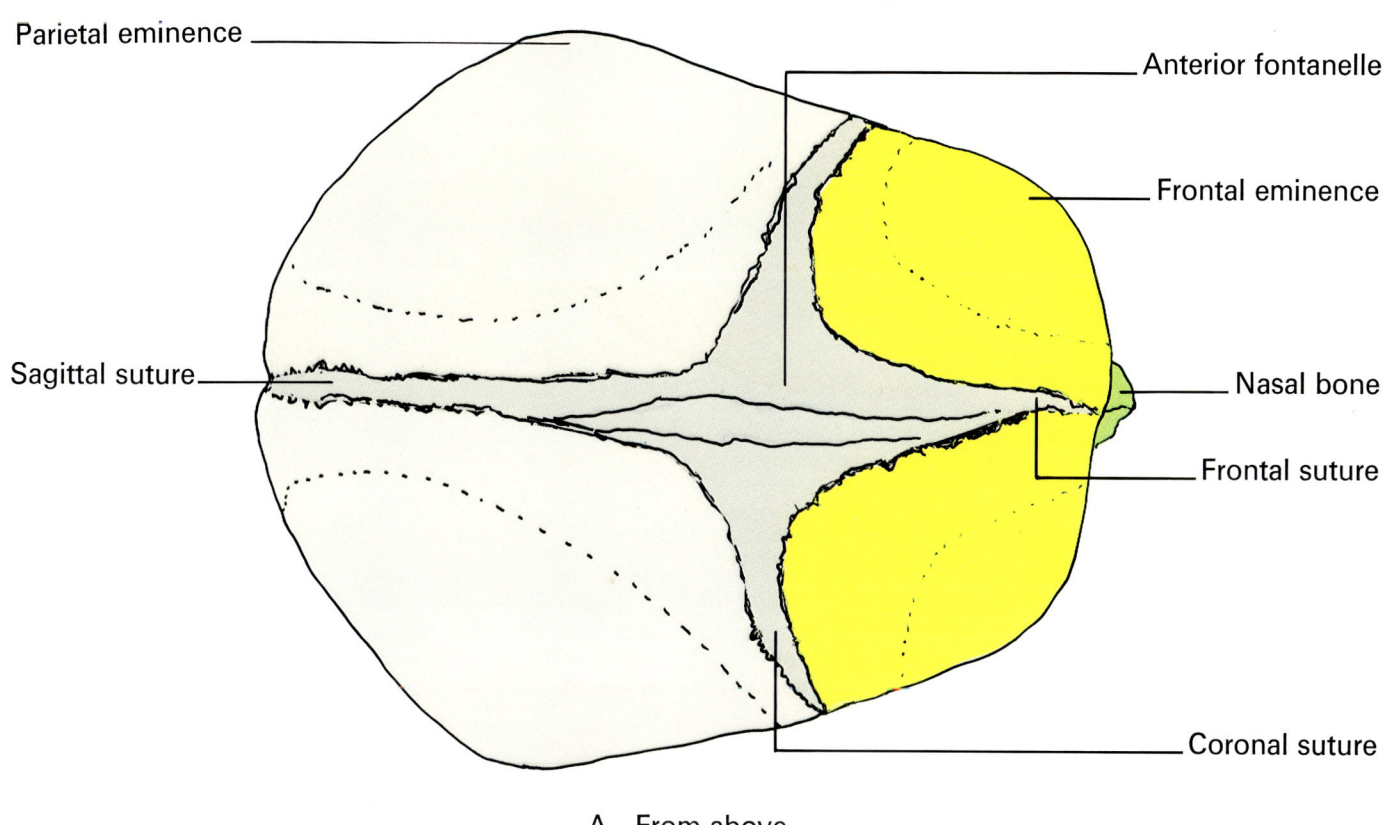

A. From above.

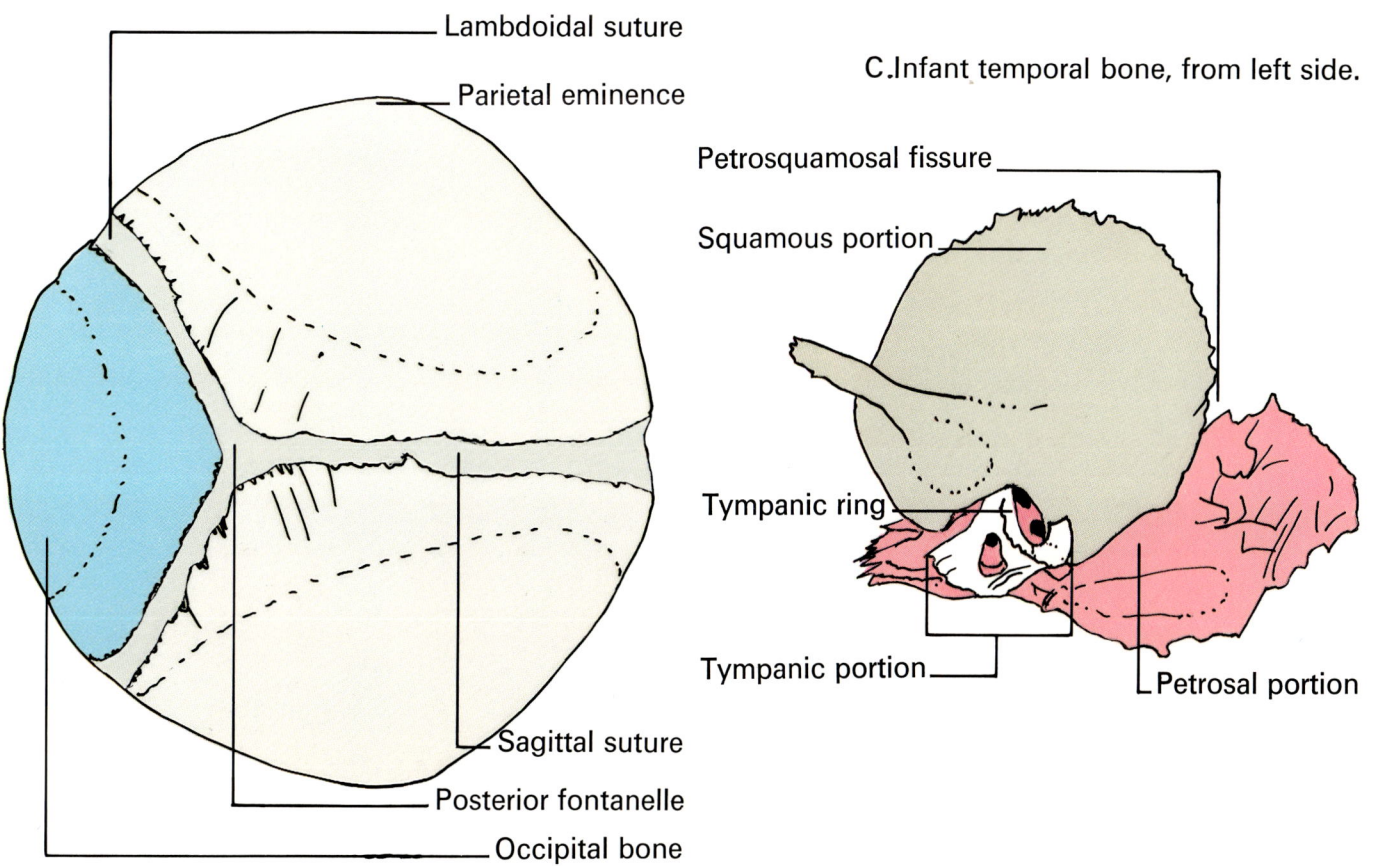

B. From above and from behind.

C. Infant temporal bone, from left side.

GLOSSARY

Ala
a general term for a winglike structure or process.

Ala cristae galli
a small winglike process on the anterior part of the crista galli of the ethmoid bone.

Ala of vomer   (*ala vomeris*)
one of the two lateral expansions on the superior border of the vomer, coming into contact with the sphenoidal process of the palatine bone and the vaginal process of the medial pterygoid plate.

Alar ligaments   (*ligamenta alaria*)
two strong bands that pass from the posterolateral part of the tip of the dens of the axis upward and laterally to the condyles of the occipital bone; they limit rotation of the head.

Alveolar arch of maxilla   (*arcus alveolaris maxillae*)
the inferior free border of the alveolar process of the maxilla

Alveolar foramina of maxilla   (*foramina alveolia maxillae*)
the openings of the alveolar canals at the deepest portion of the tooth sockets into maxilla.

Alveolar point
on the skull, the point of the alveolar process that projects most anteriorly in the midline of the maxilla.

Alveolar process
that portion of bone in either the maxilla or the mandible which surrounds and supports the teeth. In the maxilla, it is called processus alveolaris maxillae; in the mandible, pars alveolaris mandibulae.

Ampulla
a flask-like dilatation of a tubular structure.

Angle of mandible   (*angulus mandibulae*)
the angle created at the junction of the posterior edge of the ramus and the lower edge of the mandible.

Annular
shaped like a ring.

Anterior clinoid process   (*processus clinoideus anterior*)
the bony process found on the medial extremity of the posterior border of the small wing of the sphenoid bone.

Anterior cranial fossa   (*fossae cranii anterior*)
the anterior subdivision of the floor of the cranial cavity, supporting the frontal lobes of the brain, and composed of portions of three bones; the ethmoid, the frontal, the sphenoid.

Anterior ethmoidal foramen   (*foramen ethmoidale anterius*)
: an opening on the medial wall of the orbit, on the line of the frontoethmoidal suture, that transmits the anterior ethmoidal nerve and vessels.

Anterior fontanelle   (*fonticulus anterior*)
: the unossified area of the skull situated at the junction of the frontal, coronal, and sagittal sutures.

Anterior lacrimal crest   (*crista lacrimalis anterior*)
: the lateral margin of the groove on the posterior border of the frontal process of the maxilla.

Anterior process of malleus   (*processus anterior mallei*)
: a slender process that arises from the anterior aspect of the neck of the malleus, passes forward and downward to the petrotympanic fissure, and is attached to the petrous portion of the temporal bone by ligamentous fibers.

Anterior surface   (*facies anterior*)
: that surface which is toward the front of the body (on or nearest the ventral aspect) in man.

Anterior transverse line
: at the base of the skull, locates the oval foramen, lacerate foramen, nerve of pterygoid canal and synchondrosis between the sphenoid and basioccipital.

Apex of orbit
: is a small area posteriorly bounded by the roof of the orbit, the lateral wall of the orbit and it contains the optic foramen for transmission of optic nerve and ophthalmic artery.

Apex of petrous portion of temporal bone   (*apex partis petrosae ossis temporalis*)
: the truncated portion of the petrous part of the temporal bone that is directed anteriorly and medially and ends at the medial opening of the carotid canal.

Aqueduct of vestibule   (*aqueductus vestibuli*)
: a small canal extending from the vestibule of the inner ear to open onto the posterior part of the internal surface of the petrous part of the temporal bone. It lodges the endolymphatic duct and an arteriole and a venule.

Arch   (*arcus*)
: a general term to designate any structure having a curved or bowlike outline.

Arcuate eminence   (*eminentia arcuata*)
: an arched prominence on the internal surface of the petrous part of the temporal bone in the floor of the middle cranial fossa, marking the position of the superior semicircular canal. It is particularly prominent in young skulls.

Articular tubercle of temporal bone   (*tuberculum articulare ossis temporalis*)
: an enlargement of the inferior border of the zygomatic process of the temporal bone, forming the anterior boundary of the mandibular fossa and part of the anterior root of the zygoma; it gives attachment to the lateral ligament of the temporomandibular articulation.

Asterion
: the point on the surface of the skull where the lambdoid, parietomastoid, and occipitomastoid sutures meet.

Auditory ossicles   (*ear bones*)
1. Incus: the middle of the three ossicles of the ear, which, with the stapes, malleus, serves to conduct vibration from the tympanic membrane to the inner ear.
   called also: anvil
2. Malleus: the largest of the auditory ossicles, and the one attached to the tympanic mambrane; its club-shaped head articulates with the incus.
   called also: hammer
3. Stapes: the innermost of the auditory ossicles, shaped somewhat like a stirrup; it articulates by its head with the incus, and its base is inserted into the fenestra vestibuli.
   called also: stirrup

Auditory tube   (*tuba auditiva*)
: the channel that establishes communication between the tympanic cavity and the nasopharynx.

Basion
: it is the anterior midpoint of the foramen magnum.

Basisphenoid
: an embryonic bone that becomes the back part of the body of the sphenoid.

Bony labyrinth   (*labyrinthus osseus*)
: the bony part of the inner ear, consisting of vestibule, semicircular canals and cochlea.

Bony spiral lamina   (*lamina spiralis ossea*)
: a double plate of bone winding spirally around the modiolus and dividing the spiral canal of the cochlea incompletely into two parts, the scala tympani and the scala vestibuli.

Bregma
: the point on the surface of the skull where sagittal and coronal sutures intersect. It represents the position of the anterior fontanelle in the skull of the infant.

Calvarium   (*calvaria*)
: the domelike superior portion of the cranium, composed of the superior portions of the frontal, parietal, and occipital bones. It is transversed by three sutures, the coronal, sagittal and lambdoidal.

Canal   (*canalis*)
: a general term for a relatively narrow tubular passage or channel.

Canaliculus of chorda tympani   (*canaliculus chordae tympani*)
: a small canal that opens off the facial canal just before its termination, transmitting the chorda tympani nerve into the tympanic cavity.

Canaliculus of cochlea   (*canaliculus cochleae*)
: a small canal in the petrous part of the temporal bone that interconnects the scala tympani of the inner ear with the subarachnoid cavity; it lodges the perilymphatic duct and a small vein.

Canine fossa   (*fossa canina*)
: a wide depression on the external surface of the maxilla superolateral to the canine tooth socket; the levator anguli oris muscle arises from it.

Caroticoclinoid foramen   (*foramen caroticoclinoideus*)
: formed by the union of the anterior with the middle clinoid process, inconstant.

Carotid canal   (*canalis caroticus*)
: a passage in the petrous portion of the temporal bone, beginning on the inferior surface just anterior to the jugular foramen, and running anteromedially for about 2 cm.; it is seen interiorly in the floor of the middle cranial fossa, where it meets the carotid groove on the body of the sphenoid bone. It lodges the internal carotid artery.

Carotid groove   (*sulcus caroticus ossis spehnoidalis*)
: the sulcus on the side of the body of the sphenoid bone that lodges the internal carotid artery and the cavernous sinus.

Choana
: the paired openings between the nasal cavity and the nasopharynx. It is located behind and above the hard palate. It is bounded by the horizontal part of the palatine, the body and medial pterygoid plate of the sphenoid. The choanae are separated by the vomer and are funnel shaped openings.

Clivus
: a bony surface in the posterior cranial fossa, sloping upward from the foramen magnum to the dorsum sellae, the lower part being formed by a portion of the basilar part of the occipital bone and the upper part by a surface of the body of the sphenoid bone.

Cochlea
: the essential organ of hearing: a spirally wound tube, resembling a snail shell, which forms part of the inner ear. Its base lies against the lateral end of the internal acoustic meatus and its apex is directed anterolaterally.

Common tendinous ring   (*anulus tendineus communis*)
: the annular ligament of origin common to the recti muscles of the eye, attached to the edge of the optic canal and the inner part of the superior orbital fissure.

Conchal crest of maxilla   (*crista conchalis maxillae*)
: an oblique ridge on the nasal surface of the body of the maxilla, just anterior to the lacrimal sulcus, which articulates with the inferior nasal concha.

Conchal crest of palatine bone   (*crista conchalis ossis palatini*)
: a sharp transverse ridge, near the posterior edge of the palatine bone, which articulates with the inferior concha.

Condylar canal   (*canalis condylaris*)
: an opening sometimes present in the floor of the condylar fossa for the transmission of a vein from the transverse sinus.

Condylar Fossa   (*fossa condylaris*)
: either of two pits situated on the lateral portions of the occipital bone, one on either side of the foramen magnum, posterior to the occipital condyle.

Condylar process of mandible   (*processus condylaris mandibulae*)
: the posterior process on the ramus of the mandible that articulates with the mandibular fossa of the temporal bone.

Condyle   (*condylus*)
: a rounded projection on a bone, usually for articulation with another.

Condyloid tubercle
: an eminence on the condylar process of the mandible for attachment of the lateral ligament of the temporomandibular articulation.

Coronal suture   (*sutura coronalis*)
: the line of junction of the frontal bone with the two parietal bones.

Coronoid process of mandible   (*processus coronoideus mandibulae*)
: the anterior part of the upper end of the ramus of the mandible, to which the temporal muscle is attached.

Corpus
: main portion of an anatomic part.

Cranial bones (*ossa cranii*)
>the bones of the cranium, or skull, including the occipital, sphenoidal, temporal, parietal, frontal, ethmoidal.

Cranial fossa (*fossa cranii*)
>any one of the depressions on the floor of the cranial cavity.

Cranium
>the skeleton of the head, variously construed as including all of the bones of the head, all of them except the mandible, or the eight bones which form the vault that lodges the brain.

Cribriform plate of ethmoid bone (*lamina cribosa ossis ethmoidalis*)
>the horizontal plate of the ethmoid bone that forms the roof of the nasal cavity; it is perforated by many foramina for the passage of the olfactory nerves.

Crista galli
>a thick triangular process projecting upward from the cribriform plate of the ethmoid bone; the falx cerebri attaches to it.

Crown (*corona dentis*)
>the portion of a tooth that is covered by enamel and is separated from the root or roots at the cementoenamel junction.

Crus
>a limb, leg or process.

Cupola
>a small inverted cup or dome-shaped cap over some structure.

Dacryon
>the point on the skull formed by the junction of the anterior border of lacrimal and frontal bone.

Digastric fossa (*fossa digastrica*)
>a depression on the internal surface of the body of the mandible on each side of the symphysis to which the anterior belly of the digastric is attached.

Diploë
>the loose osseous tissue between the two tables of the cranial bones.

Dorsum sellae
>the quadrilateral plate on the sphenoid bone that forms the posterior boundary of the sella turcica; the posterior clinoid processes project from its superior extremity, and it is continuous with the clivus.

Emissary foramen
>any foramen in a cranial bone that gives passage to an emissary vein.

Epitympanic recess   (*recessus epitympanicus*)
>the upper portion of the tympanic cavity, extending above the level of the tympanic membrane and containing the greater part of the incus and the upper half of the malleus.

Ethmoid bone   (*os ethmoidale*)
>the cubical bone located between the orbits and consisting of the lamina cribosa, the lamina perpendicularis and the paired lateral masses.

Ethmoidal bulla   (*bulla ethmoidalis ossis ethmoidalis*)
>a rounded projection of the ethmoid bone into the lateral wall of the middle nasal meatus just below the middle nasal concha, enclosing a large ethmoid air cell.

Ethmoidal foramina   (*foramina ethmoidalia*)
>these foramina are formed in the frontoethmoidal suture. The nasociliary nerve and ethmoidal vessels are transmitted via the anterior ethmoidal foramina. The posterior ethmoidal foramina transmit ethmoidal nerves and vessels.

Ethmoidal groove   (*sulcus ethmoidalis ossis nasalis*)
>a groove that extends the entire length of the posteromedial surface of the nasal bone and lodges the external nasal branch of the anterior ethmoid nerve.

Ethmoidal labyrinth   (*labyrinthus ethmoidalis*)
>either of the paired lateral masses of the ethmoid bone, consisting of numerous thin-walled cellular cavities, the ethmoid cells.

Ethmoidal notch of frontal bone   (*incisura ethmoidalis ossis frontalis*)
>a space between the orbital parts of the frontal bone, in which the ethmoid bone is lodged.

External acoustic meatus   (*meatus acusticus externus*)
>the narrow passage of the external ear leading to the tympanic membrane. It is bounded by the tympanic part of the temporal bone, mastoid process posteriorly, and the posterior root of the zygomatic arch superiorly.

External occipital crest   (*crista occipitalis externa*)
>a variable crest of the bone that sometimes extends from the external occipital protuberance toward the foramen magnum.

External occipital protuberance   (*protuberantia occipitalis externa*)
>a prominence at the center of the outer surface of the squama of the occipital bone which gives attachment to the ligamentum nuchae.

Facial bones   (*ossa faciei*)
: the bones that consitute the facial part of the skull, including the palatine and zygomatic bones, the mandible, and the maxilla, the lacrimal and nasal bones, the inferior nasal concha, and the vomer. Some authorities also include the ethmoid bone.

Facial canal   (*canalis facialis*)
: a canal in the temporal bone for the facial nerve, beginning in the internal acoustic meatus and passing anterolaterally dorsal to the vestibule of the inner ear for about 2 mm.. Turning sharply backward at the genu of the facial canal, it runs along the medial wall of the tympanic cavity, then turns inferiorly and reaches the exterior of the petrous part of the bone at the stylomastoid foramen.

Fissure   (*fissura*)
: any cleft or groove or sulcus.

Fissure of aqueduct of vestibule   (*aperatura externa aqueductus vestibuli*)
: the external opening for the aqueduct of the vestibule, located on the internal surface of the petrous part of the temporal bone, lateral to the opening for the internal acoustic meatus.

Floor of orbit
: made up of the orbital surface of the maxilla, orbital process of zygomatic and orbital process of palatine. Along the medial margin the lacrimal sulcus is found. It harbors the nasolacrimal canal. Lateral to the sulcus a small depression serves for attachment of the inferior oblique eye muscle. Close to the middle of the floor the infraorbital groove is formed which leads anteriorly to the infraorbital canal. The latter transmits infraorbital nerves and vessels.

Fontanelle   (*fonticuli cranii*)
: a soft spot, such as one of the membrane-covered spaces remaining in the incompletely ossified skull of a fetus or infant.

Foramen
: a natural opening or passage especially one into or through bone.

Foramen cecum of frontal bone   (*foramen cecum ossis frontalis*)
: a blind opening formed between the frontal crest and the crista galli; it sometimes transmits a vein from the nasal cavity to the superior sagittal sinus.

Foramen lacerum
: an irregular gap formed at the junction of the base of the great wing of the sphenoid bone, the tip of the petrous part of the temporal bone, and the basilar part of the occipital bone; in life its inferior aspect is filled with fibrocartilage, superior to which the internal carotid artery lies.

Foramen magnum
 the large opening in the anterior and inferior part of the occipital bone, interconnecting the vertebral canal and the cranial cavity. It is bounded laterally by the occipital condyles. It transmits: medulla oblongata; accessory nerves; vertebral, anterior, and posterior spinal arteries; and ligaments connecting axis and occipital bone.

Foramen rotundum (*foramen rotundum ossis sphenoidalis*)
 a round opening in the medial part of the great wing of the sphenoid bone that transmits the maxillary branch of the trigeminal nerve.

Foramen of Vesalius
 an opening occasionally found medial to the oval foramen of the sphenoid, for the passage of a vein from the cavernous sinus.

Fossa
 a trench or channel; a general term for a hollow or depressed area.

Fossula
 a small fossa, a general term for a slight depression in the surface of a structure.

Fossula of cochlear window (*fossula fenestrae cochleae*)
 a depression on the medial wall of the tympanic cavity, at the bottom of which is the cochlear window.

Fossula of vestibular window (*fossula fenestrae vestibuli*)
 a depression on the medial wall of the tympanic cavity, at the bottom of which is the vestibular window.

Fovea
 a small pit or depression.

Frontal angle of parietal bone (*angulus frontalis ossis parietalis*)
 the anterosuperior angle of the parietal bone, which is membranous at birth and forms part of the anterior fontanelle.

Frontal bone (*os frontale*)
 a single bone that closes the front part of the cranial cavity and forms the skeleton of the forehead; it is developed from the two halves, the line of separation sometimes persisting in adult life.

Frontal crest (*crista frontalis*)
 a median ridge on the internal surface of the frontal bone, extending upward from the foramen cecum to unite with the sulcus for the superior sagittal sinus.

Frontal eminence (*tuber frontale*)
 one of the slightly rounded prominences on the frontal bone on either side above the eyes, forming the most prominent portions of the forehead.

Frontal notch   (*incisura frontalis*)
: a notch located in the supraorbital margin of the frontal bone medial to the supraorbital notch or foramen, for transmission of branches of the supraorbital nerve and vessels; frequently converted into a foramen by a bridge of osseous tissue.

Frontal sinus   (*sinus frontalis*)
: one of the paired paranasal sinuses located in the frontal bone, and communicating by way of the nasofrontal duct with the middle meatus of the nasal cavity on the same side.

Frontal squama   (*squama frontalis*)
: the broad, curved portion of the frontal bone, situated above the supraorbital margin and forming the forehead.

Frontal suture   (*sutura frontalis*)
: the usually transient line of junction between the right and left halves of the frontal bone. The inferior part often persists in the adult. If the entire suture persists, it is called the metopic suture.

Frontomaxillary suture   (*sutura frontomaxillaris*)
: the line of junction between the frontal bone and the frontal process of the maxilla.

Frontonasal suture   (*sutura frontonasalis*)
: the line of junction between the frontal and the two nasal bones.

Frontozygomatic suture   (*sutura frontozygomatica*)
: the line of junction between the zygomatic bone and the zygomatic process of the frontal bone.

Glabella
: the most prominent point in the midsagittal plane between the eyebrows.

Granular foveolae   (*foveolae granulares*)
: small pits on the internal surface of the cranial bones on either side of the groove for the superior sagittal sinus; they are occupied by the arachnoid granulations.

Great palatine canal   (*canalis palatinus major*)
: a passage in the sphenoid and palatine bones for the greater palatine vessels and nerve.

Greater palatine foramen   (*foramen palatinum majus*)
: the inferior opening in the great palatine canal, found laterally on the horizontal plate of each palatine bone opposite the root of each third molar tooth; it transmits a palatine nerve and artery.

Greater palatine groove of maxilla (*sulcus palatinus major maxillae*)
the sulcus on the nasal surface of the maxilla which, along with the corresponding one on the perpendicular plate of the palatine bone, forms the canal for the greater palatine nerve.

Greater palatine groove of palatine bone (*sulcus palatinus major ossis palatini*)
a vertical groove on the maxillary surface of the perpendicular plate of the palatine bone; it articulates with the maxilla to form the canal for the greater palatine nerve.

Greater wing of sphenoid bone (*ala major ossis sphenoidalis*)
a large wing-shaped process arising from either side of the body of the sphenoid bone; its cerebral surface forms the anterior part of the floor of the middle cranial fossa, and its orbital surface forms the chief part of the lateral wall of the orbit.

Groove
a shallow linear depression, especially one appearing during embryonic development or persisting in definitive bone.

Groove for auditory tube (*sulcus tubae auditivae*)
a groove on the medial part of the base of the spine of the sphenoid bone; it lodges a portion of the cartilaginous part of the auditory tube. At the bottom of the sulcus the petrosphenoidal fissure is found.

Groove for occipital artery (*sulcus arteriae occipitalis*)
the groove just medial to the mastoid notch on the temporal bone, lodging the occipital artery.

Groove for superior petrosal sinus (*sulcus sinus petrosi superioris*)
a small posterolaterally directed sulcus that runs along the internal surface of the petrous part of the temporal bone on the angle separating the posterior and middle cranial fossae; it lodges the superior petrosal sinus.

Groove of lacrimal bone (*sulcus lacrimalis ossis lacrimalis*)
a deep vertical groove on the anterior part of the lateral surface of the lacrimal bone, which with the maxilla forms the fossa for the lacrimal sac.

Hamular groove of medial pterygoid plate (*sulcus hamuli pterygoidei*)
a smooth groove on the lateral surface of the medial pterygoid plate of the sphenoid bone, in the angle at the base of the pterygoid hamulus; it lodges the tendon of the tensor veli palatini muscle.

Hamulus
a general term denoting a hook-shaped process.

Hiatus
a general term for a gap, cleft or opening.

Hiatus of facial canal (*hiatus canalis nervi petrosi majoris*)
a canal in the petrous part of the temporal bone for the greater petrosal nerve, and a branch of the middle meningeal artery.

Highest nuchal line (*linea nuchae suprema*)
a sometimes indistinct line arching upward from the external occipital protuberance and running toward the lateral angle of the occipital bone; the epicranial aponeurosis attaches to it.

Hypoglossal canal (*canalis hypoglossi*)
an opening in the lateral part of the occipital bone at the base of the condyle, which transmits the hypoglossal nerve and a branch of the posterior meningeal artery.

Hypophyseal fossa (*fossa hypophysialis*)
a deep depression in the middle of the sella turcica of the sphenoid bone, lodging the hypophysis cerebri.

Incisive canal (*canalis incisivus*)
one of the small canals opening into the incisive fossa of the hard palate, and transmitting small vessels and nerves from the floor of the nose into the front part of the roof of the mouth.

Incisura
a notch, indentation or depression, chiefly on the edge of a bone or other structure.

Incudal fossa (*fossa incudis*)
a groove in the posterior wall of the tympanic cavity, lodging the short limb of the incus.

Incus
the middle of the three ossicles of the ear, which, with the stapes and malleus serve to conduct vibrations from the tympanic membrane to the inner ear.

Inferior meatus of nose (*meatus nasi inferior*)
the space beneath the inferior nasal concha, into which the nasolacrimal duct opens.

Inferior nasal concha (*concha nasalis inferior*)
a thin bony plate with curved margins, articulating with the ethmoid, maxilla, lacrimal and palatine bones, forming the lower part of the lateral wall of the nasal cavity.

Inferior nuchal line (*linea nuchae inferior*)
the lowest of the three nuchal lines found on the outer surface of the occipital bone, extending laterally from the middle of the external occipital crest to the jugular process.

Inferior orbital fissure   (*fissura orbitalis Inferior*)
: a cleft in the inferolateral wall of the orbit bounded by the great wing of the sphenoid and the orbital process of the maxilla; it transmits the infraorbital and zygomatic nerves and the infraorbital vessels.

Inferior surface   (*facies inferior*)
: that surface which is lower, directed away from the head, in man.

Infraorbital canal   (*canalis infraorbitalis*)
: a passage beneath the orbital surface of the maxilla, continuous posteriorly with the infraorbital sulcus, and opening anteriorly on the anterior surface of the body of the maxilla, in the infraorbital foramen. It contains the infraorbital vessels and nerve.

Infraorbital fissure
: it is an opening in the posterior lateral part of the orbit. Its boundaries are formed by the maxilla, orbital process of palatine, great wing of sphenoid bone and a small portion of the zygomatic bone. The fissure transmits the zygomatic and sphenopalatine branch of the maxillary nerve and the clinically important infraorbital vessels and veins. The latter connects with the inferior ophthalmic vein and the pterygoid venus plexus.

Infraorbital foramen   (*foramen infraorbitale*)
: the opening of the orbital canal on the anterior surface of the maxilla, giving passage to the infraorbital nerve and vessels.

Infratemporal crest   (*crista infratemporalis*)
: a crest separating the temporal surface of the great wing of the sphenoid bone into a temporal portion above and an infratemporal portion below. The crest joins the anterior root of the zygomatic process.

Infratemporal fossa   (*fossa infratemporalis*)
: The fossa contains two fissures which intersect at right angles. Superiorly, the horizontal, infraorbital fissure; medially, the vertical, pterygopalatine fissure.

  Boundaries: the infratemporal surface of the maxilla, ridge of the zygomatic process, articular tubercle of temporal, spine and great wing of sphenoid, inferior surface of the temporal squama, the lateral pterygoid plate and alveolar border of maxilla.

  Contents: the oval and spinous foramen open into it. It also contains the pterygoid and temporalis muscle, the mandibular and maxillary nerves and vessels.

Inion
: the most prominent point of the external occipital protuberance.

Innominate foramen
>   an occasional opening in the temporal bone for passage of the small superficial petrosal nerve.

Internal acoustic meatus   (*meatus acusticus internus*)
>   the opening on the posterior surface of the petrous part of the temporal bone through which the facial, intermediate, and vestibulocochlear nerves, and the labyrinthine artery pass.

Interparietal bone   (*os interparietale*)
>   the part of the squama of the occipital bone that lies superior to the highest nuchal line when this portion remains separate throughout life.

Jugular foramen   (*foramen jugulare*)
>   the opening formed by the jugular notches on the temporal and occipital bones.
>   1. the anterior portion of the foramen transmits the inferior petrosal sinus;
>   2. the intermediate portion, the glossopharyngeal, vagus and accessory nerves;
>   3. the posterior portion, the transverse sinus and several meningeal branches from the occipital and ascending pharyngeal arteries.

Jugular fossa of temporal bone   (*fossa jugularis ossis temporalis*)
>   a prominent depression on the inferior surface of the petrous part of the temporal bone, forming the major part of the jugular notch; it forms the anterior and lateral wall of the jugular foramen and lodges the superior bulb of the internal jugular vein.

Jugum
>   a general term for a depression or ridge connecting two structures.

Jugum sphenoidale
>   the portion of the body of the sphenoid bone that connects the lesser wings.

Lacrimal bone   (*os lacrimale*)
>   a thin scalelike bone at the anterior part of the medial wall of the orbit, articulating with the frontal and ethmoid bones and the maxilla and inferior nasal concha.

Lacrimal sulcus of lacrimal bone   (*sulcus lacrimalis ossis lacrimalis*)
>   a deep vertical groove on the anterior part of the lateral surface of the lacrimal bone, which with the maxilla forms the fossa for the lacrimal sac.

Lambda
>   the point on the skull where sagittal and lambdoidal suture intersect. It represents the position of the posterior fontanelle in the skull of the infant.

Lambdoidal suture   (*sutura lambdoidea*)
> the line of junction between the occipital and parietal bones, shaped like the Greek letter lambda. The lambdoidal sutures may contain one or several sutural bones.

Lateral pterygoid plate   (*lamina lateralis processus pterygoidei*)
> either of a pair of bony plates projecting downward from the roots of the greater wings of the sphenoid bone and forming the medial wall of the ipsilateral infratemporal fossa.

Lateral surface   (*facies lateralis*)
> a surface nearer to, or directed toward the side of the body.

Lateral wall of orbit
> it is composed of the orbital process of the zygomatic and the orbital surface of the greater wing of sphenoid. Inferiorly, it is bounded by the inferior orbital fissure.

Lesser palatine canal   (*canales palatini minores*)
> openings in the palatine bone that branch off the great palatine canal to carry the lesser and middle palatine nerves and vessels to the roof of the mouth.

Lesser palatine foramina   (*foramina palatina minora*)
> the openings of the palatine canals behind the palatine crest and the greater palatine foramina.

Lesser wing of sphenoid bone   (*ala minor ossis sphenoidalis*)
> the thin triangular plate of bone that extends horizontally and laterally from either side of the anterior part of the body of the sphenoid bone; it articulates with the frontal bone and helps form the roof of the orbit and the floor of the anterior cranial fossa.

Lingula of mandible   (*lingula mandibulae*)
> the sharp medial boundary of the mandibular foramen, to which is attached the sphenomandibular ligament.

Mandible   (*mandibula*)
> the horseshoe-shaped bone forming the lower jaw; the largest and strongest bone of the face, presenting a body and a pair of rami, which articulate with the skull at the temporomandibular joints. The median ridge on the mandible indicates the position of the fetal symphysis menti. The incisive fossa of the mandible is located below the incisive teeth. Inferior to the second premolar tooth is located the mental foramen which transmits mental nerve and vessels. The oblique line of the mandible is continuous with the anterior border of the ramus and runs anteriorly to the mental tubercle. Posterior border of ramus runs from the angle of the mandible to the condyle.

Mandibular canal   (*canalis mandibulae*)
: a canal that traverses the ramus and body of the mandible between the mandibular and mental foramina, transmitting the inferior alveolar vessels and nerve.

Mandibular foramen   (*foramen mandibulae*)
: the opening on the medial surface of the ramus of the mandible, leading into the mandibular canal.

Mandibular fossa   (*fossa mandibularis*)
: a prominent depression on the inferior surface of the squamous part of the temporal bone at the base of the zygomatic process. It is divided into two parts by the petrotympanic fissure. It is located between the articular tubercle and the tympanic part of the temporal. The anterior, larger, part of the fossa articulates with the condyles of the mandible. The posterior part may contain portions of the parotid gland. It is a non-articular surface.

Manubrium
: a handle-like structure or part.

Mastoid canaliculus   (*canaliculus mastoideus*)
: a minute passage beginning in the lateral wall of the jugular fossa of the temporal bone and passing into the temporal bone. The tympanic branch of the vagus nerve passes through it to exit via the tympanomastoid fissure.

Mastoid cells   (*cellulae mastoideae*)
: the air spaces of the mastoid process of the temporal bone.

Mastoid fontanelle   (*fonticulus mastoideus*)
: the unossified area of the skull at the junction of the lambdoidal, parietomastoid, and occipitomastoid sutures.

Mastoid foramen   (*foramen mastoideum*)
: a prominent opening in the temporal bone posterior to the mastoid process and near its occipital articulation; an artery and vein usually pass through it.

Mastoid groove
: located medial to the mastoid notch and lodges the occipital artery.

Mastoid notch   (*incisura mastoidea ossis temporalis*)
: a deep groove on the medial surface of the mastoid process of the temporal bone, which gives origin to the posterior belly of the digastric muscle.

Mastoid process of temporal bone   (*processus mastoideu ossis temporalis*)
: a conical process projecting forward and downward from the external surface of the petrous part of the temporal bone just posterior to the external acoustic meatus. It serves for attachment of three muscles, namely: sternocleidomastoid, splenius capitis, longissimus capitis.

Maxilla
    the irregularly shaped bone that with its fellow forms the upper jaw; it assists in the formation of the orbit, the nasal cavity, and the palate, and lodges the upper teeth.

Maxillary sinus (*sinus maxillaris*)
    one of the paired paranasal sinuses, located in the body of the maxilla on either side and communicating with the middle meatus of the nasal cavity on the same side.

Meatus
    an opening or passage.

Medial pterygoid plate (*lamina medialis processis pterygoidei*)
    either of a pair of bony plates projecting downward from the roots of the greater wings of the sphenoid bone and forming the lateral boundary of the ipsilateral posterior aperture of the nasal cavity and the most posterior part of the lateral wall of the nasal cavity.

Medial surface (*facies medialis*)
    a surface nearer to or directed toward the midline of the body.

Medial wall of orbit
    it is nearly vertical and formed by the frontal process of maxilla, lacrimal, orbital plate of ethmoid and a small part of the body of the sphenoid. The lacrimal groove lodges the lacrimal sac and posteriorly to the groove the posterior lacrimal crest serves for attachment of portions of the orbicularis oculi muscle (lacrimal part).

Mental foramen (*foramen mentale*)
    an opening on the lateral part of the body of the mandible, opposite the second bicuspid tooth, for passage of the mental nerve and vessels.

Mental spine (*spina mentalis*)
    any of the small bony projections (usually four in number) located on the internal surface of the mandible, near the lower end of the midline, serving for attachment of the genioglossal and geniohyoid muscles.

Middle clinoid process (*processus clinoideus medius*)
    either of two small inconstant eminences on the internal surface of the sphenoid bone, one on either side of the anterior part of the hypophyseal fossa.

Middle cranial fossa (*fossa cranii media*)
    the middle subdivision of the cranial cavity, supporting the temporal lobes of the brain and the pituitary gland; it is composed of the body and greater wings of the sphenoid bone and the squamous and petrous portions of the temporal bone.

Middle meatus of nose   (*meatus nasi medius*)
: the space beneath the middle nasal concha, with which the anterior ethmoidal cells and frontal and maxillary sinuses communicate.

Middle nasal concha   (*concha nasalis media*)
: the lower of two bony plates projecting from the inner wall of the ethmoid labyrinth and separating the superior from the middle meatus of the nose.

Modiolus
: is the central axis of the cochlea.

Musculotubal canal   (*canalis musculotubarius*)
: the combined canals of the auditory tube and the tensor tympani muscle in the temporal bone.

Mylohyoid fossa of mandible   (*fovea sublingualis*)
: a depression on the inner surface of the body of the mandible, lodging a portion of the sublingual gland.

Mylohyoid line of mandible   (*linea mylohyoidea mandibulae*)
: a ridge on the inner surface of the mandible from the base of the symphysis to the ascending ramus behind the last molar tooth; it affords attachment to the mylohyoid muscle and superior constrictor of the pharynx.

Mylohyoid groove of mandible   (*sulcus mylohyoideus mandibulae*)
: a groove on the medial surface of the ramus of the mandible, passing downward and forward from the mandibular foramen and lodging the mylohyoid artery and nerve.

Nasal bone   (*os nasale*)
: either of the two small, oblong bones that together form the bridge of the nose.

Nasal cavity   (*cavum nasi osseum*)
: the space between the floor of the cranium and the roof of the mouth, extending between the pharynx posteriorly and the external nose anteriorly, and divided by a median septum.

Nasal foramina   (*foramina nasalis*)
: openings on the outer surface of each nasal bone for the transmission of blood vessels.

Nasal groove   (*sulcus ethmoidalis ossis nasalis*)
: a groove that extends the entire length of the posteromedial surface of the nasal bone and lodges the external nasal branch of the anterior ethmoid nerve.

Nasal notch of maxilla   (*incisura nasalis maxillae*)
: the large notch in the anterior border of the maxilla that forms the lateral and inferior margins of the anterior nasal aperature.

Nasal septum   (*septum nasi osseum*)
: the bone of the skull interposed between the openings of the nose, consisting of the vomer and the perpendicular plate of the ethmoid bone.

Nasion
: a point on the skull representing the midpoint of the frontonasal suture.

Nasolacrimal canal   (*canalis nasolacrimalis*)
: a canal formed by the lacrimal sulcus of the maxilla, lacrimal bone, and inferior nasal concha; it contains the nasolacrimal duct.

Nasopalatine groove
: a furrow on the lateral surface of the vomer for the nasopalatine nerve and vessels.

Occipital bone   (*os occipitale*)
: a single trapezoid-shaped bone situated at the posterior and inferior part of the cranium, articulating with the two parietal and two temporal bones, the sphenoid bone, and the atlas; it contains a large opening, the foramen magnum.

Occipital condyle   (*condylus occipitalis*)
: one of two oval processes on the lateral portions of the occipital bone, on either side of the foramen magnum, for articulation with the atlas.

Occipital squama   (*squama occipitalis*)
: the largest of the four parts of the occipital bone, extending from the posterior edge of the foramen magnum to the lambdoid suture, its external surface bearing the external occipital protuberance and nuchal lines.

Occipitomastoid suture   (*sutura occiptomastoidea*)
: it is a continuation of the lambdoidal suture joining the occipital bone and the mastoid portion of the temporal bone.

Olfactory foramen
: any one of the many openings of the cribriform plate of the ethmoid bone.

Olfactory groove
: a groove situated in the cribriform plate of the ethmoid bone; it lodges the olfactory bulb of the brain.

Opisthion
: it is the posterior midpoint of the foramen magnum.

Optic canal   (*canalis opticus*)
: one of the paired openings in the sphenoid bone where the small wings are attached to the body of the bone at the apex of the orbit; each canal transmits one of the optic nerves and the opthalmic artery of that side.

Optic groove   (*sulcus chiasmatis*)
: a furrow on the superior surface of the sphenoid bone located just anterior to the tuberculum sellae, and lodging the optic chiasm.

Oral cavity   (*cavum oris*)
: the cavity of the mouth and the associated structures.

Orbit   (*orbita*)
: a bony cavity that contains the eyeball and its associated muscles, vessels, and nerves. The orbits are conventionally divided into roof, floor, medial wall, lateral wall and apex. The medial walls are virtually parallel with each other. The lateral walls diverge forming nearly an angle of 45°. The ethmoid, frontal, lacrimal, nasal, palatine, sphenoid, and zygomatic bones, and the maxilla contribute to its formation.

Oval foramen of sphenoid bone   (*foramen ovale basis cranii*)
: an opening in the posterior of the medial portion of the great wing of the sphenoid bone; it transmits the mandibular branch of the trigeminal nerve and some vessels.

Palatine bone   (*os palatinum*)
: one of the two irregularly shaped bones forming the posterior part of the hard palate, the lateral wall of the nasal fossa between the medial pterygoid plate and the maxilla, and the posterior part of the floor of the orbit.

Palatine grooves of maxilla   (*sulci palatini maxillae*)
: the laterally placed furrows, between the palatine spines on the inferior surface of the hard palate, that lodge the palatine vessels and nerves.

Parietal bone   (*os parietale*)
: either of the two quadrilateral bones forming part of the superior and lateral surfaces of the skull, and joining each other in the midline at the sagittal suture.

Parietal eminence   (*tuber parietale*)
: the somewhat laterally bulging prominence just superior to the superior temporal line on the external surface of the parietal bone.

Parietal foramen   (*foramen patietale*)
: an opening on the posterior part of the superior portion of the parietal bone near the sagittal suture, for passage of a vein and arteriole.

Parietomastoid suture   (*sutura parietomastoidea*)
: it is a nearly horizontal suture, continuation of the squamosal suture. It unites the mastoid process of the temporal bone with the adjacent mastoid angle of the parietal bone.

Petrosal
: pertaining to the petrous portion of the temporal bone.

Petrosal fossa   (*fossula petrosa*)
: a small depression on the under surface of the petrous portion of the temporal bone. A small ridge separates the jugular fossa from the external carotid foramen.

Petrotympanic fissure   (*fissura petrotympanica*)
: a narrow transversely running slit just posterior to the articular surface of the mandibular fossa of the temporal bone; an arteriole and the chorda tympani nerve pass through it, and it lodges a portion of the malleus.

Petrous
: resembling a rock; hard, stony.

Petrous portion of temporal bone   (*pars petrosa ossis temporalis*)
: a pyramid of dense bone located at the base of the cranium; it houses the organ of hearing.

Pharyngeal canal   (*canalis palatovaginalis*)
: a narrow canal located in the roof of the nasal cavity between the inferior surface of the body of the sphenoid bone and the sphenoidal process of the palatine bone; it opens posteriorly into the nasal cavity and anteriorly into the pterygopalatine fossa.

Pharyngeal raphe   (*raphe pharyngis*)
: a band of connective tissue extending downward from the base of the skull along the posterior wall of the pharynx in the median plane, and giving attachment to the constrictor muscles of the pharynx.

Pharyngeal tubercle   (*tuberculum pharyngeum*)
: a midline eminence on the inferior surface of the basilar part of the occipital bone, for attachment of the pharynx, the superior constrictor muscle and the pharyngeal raphe.

Planum nuchale
: the occipital bone beneath the highest nuchal line. It extends from the inion to the foramen magnum and is divided by the median nuchal line. The median nuchal line serves for attachment of the ligamentum nuchae.

Planum occipitale
: it is the portion of the occipital bone above the highest nuchal line.

Ponticulus   (*ponticulus promontorii*)
: a ridge on the median wall of the tympanic cavity.

Posterior clinoid process   (*processus clinoideus posterior*)
> either of two tubercles found on the superior angle of either side of the dorsum sellae of the sphenoid bone, and giving attachment to the tentorium of the cerebellum.

Posterior cranial fossa   (*fossa cranii posterior*)
> the posterior subdivision of the floor of the cranial cavity, lodging the cerebellum, pons, medulla oblongata; it is formed by portions of the sphenoid, temporal, parietal, and occipital bones.

Posterior ethmoidal foramen   (*foramen ethmoidale posterius*)
> a small opening on the medial wall of the orbit, on the line of the frontoethmoidal suture, that transmits the posterior ethmoidal nerve and vessels.

Posterior fontanelle   (*fonticulus posterior*)
> the unossified area of the skull at the junction of the sagittal and lambdoidal sutures.

Posterior surface   (*facies posterior*)
> that surface which is toward the back of the body (on or nearest the dorsal aspect) in man.

Posterior transverse line
> at the base of the skull. Passes between the styloid and mastoid processes and locates the stylomastoid foramen, jugular foramen and hypoglossal canal.

Process   (*processus*)
> a prominence or projection; a general term for such a mass projecting from a larger structure.

Promontory of tympanic cavity   (*promontorium tympani*)
> the prominence on the medial wall of the tympanic cavity formed by the first turn of the cochlea.

Pterion
> it is the point at the end of the sphenoparietal suture. It is located slightly superior to the level of the zygomatic process of the frontal bone, approximately 3 centimeters posteriorly.

Pterygoid
> shaped like a wing.

Pterygoid canal   (*canalis pterygoideus*)
> a horizontally running canal that passes forward through the base of the medial pterygoid plate of the sphenoid bone to open into the posterior wall of the pterygopalatine fossa just medial and inferior to the foramen rotundum; it transmits the pterygoid vessels and nerves.

Pterygoid fissure   (*fissura pterygoidea*)
: a fissure on the inferior portion of each pterygoid process where the pyramidal process of the palatine bone is inserted between the diverging medial and lateral pterygoid plates.

Pterygoid fossa of sphenoid bone   (*fossa pterygoidea ossis sphenoidalis*)
: the posteriorly facing fossa which is formed by the divergence of the medial and lateral pterygoid plates of the sphenoid bone, and lodges the origins of the internal pterygoid muscle and tensor veli palatini muscle.

Pterygoid fovea   (*fovea pterygoidea mandibulae*)
: a depression on the inner side of the neck of the condyloid process of the mandible, for attachment of the external pterygoid muscle.

Pterygoid hamulus   (*hamulus pterygoideus*)
: a hooklike process on the distal end of the medial pterygoid plate of the sphenoid bone. It serves as a pully for the tensor veli palatini muscle.

Pterygoid process of sphenoid   (*processus pterygoideus ossis sphenoidalis*)
: either of two processes on the sphenoid bone descending from the points of junction of the great wings and body of the bone, and each consisting of a lateral and a medial plate.

Pterygoid tuberosity of mandible   (*tuberositas pterygoidea mandibulae*)
: a roughened area on the inner side of the angle of the mandible for the insertion of the internal pterygoid muscle.

Pterygopalatine fissure   (*fissura pterygomaxillaris*)
: it is a cleft formed by the separation of the pterygoid process of the sphenoid from the maxilla. It connects the pterygopalatine and infratemporal fossa. It transmits maxillary vessels.

Pterygopalatine fossa   (*fossa pterygopalatina*)
: it is a space formed by the junction of the pterygomaxillary fissure in the inferior part of the apex of the orbit and the junction at the inferior orbital fissure. The boundaries are formed by the body, great wing and pterygoid process of sphenoid, the infratemporal surface of the maxilla and parts of the palatine (orbital, sphenoidal process and portions of the vertical part of the bone). Five foramina open into this fossa. The foramen rotundum, pterygoid canal and pharyngeal canal on the posterior wall. The sphenopalatine foramen and the pterygopalatine canal on the medial wall. Contents of fossa: Maxillary nerve and artery, pterygopalatine ganglion.

Pterygospinous process   (*processus pterygospinosus*)
    a small spine on the posterior edge of the lateral pterygoid plate of the sphenoid bone, giving attachment to the pterygospinous ligament.

Pyramidal eminence   (*eminentia pyramidalis*)
    the hollow elevation in the inner wall of the middle ear, which contains the stapedius muscle.

Pyramidal process of palatine bone   (*processus pyramidalis ossis palatini*)
    a strong process projecting downward, backward, and laterally from the lateral part of the posterior margin of the palatine bone and helping to form the pterygoid fossa.

Recess   (*recessus*)
    a small empty space, hollow or cavity.

Roof of orbit
    is made up of the orbital plate of the frontal and lesser wing of the sphenoid. Medially, the trochlear fovea serves for attachment of the cartilaginous pulley of the superior oblique muscle of the eye. The lacrimal fossa is located more laterally. It harbors the lacrimal gland. Lesser wing of sphenoid and frontal bone are joined by a suture posteriorly.

Roof of tympanum   (*tegmen tympani*)
    the thin layer of translucent bone, on the petrous part of the temporal bone on the floor of the middle cranial fossa, separating the tympanic antrum from the cranial cavity.

Rostrum
    a beaklike appendage or part.

Sagittal groove
    a groove on the inner surface of the skull for the superior longitudinal sinus.

Scaphoid fossa of sphenoid bone   (*fossa scaphoidea ossis sphenoidalis*)
    it is a depression located at the base and lateral side of the medial pterygoid plate of the sphenoid bone. It serves for attachment of the tensor veli palatini.

Sella
    a saddle shaped depression.

Semicanal   (*semicanalis*)
    a channel which is open on one side.

Semicanal of auditory tube   (*semicanalis tubae auditivae*)
    a small canal in the temporal bone opening on the inferior surface of the skull just posterior and superior to the spinous foramen. It constitutes the inferior part of the musculotubal canal and lodges the auditory tube.

Semicanal of tensor tympani muscle   (*semicanalis musculi tensoris tympani*)
: a small canal hidden in the temporal bone, constituting the superior part of the musculotubal canal, and lodging the tensor tympani muscle.

Sheath of styloid process   (*vagina processus styloidei*)
: a ridge on the lower surface of the temporal bone, partly enclosing the base of the styloid process.

Singular foramen   (*foramen singulare*)
: the opening in the inferior vestibular area of the fundus of the internal acoustic meatus that gives passage to the nerves of the ampulla of the posterior semicircular duct.

Sphenoidal fontanelle   (*fonticulus sphenoidalis*)
: the unossified area at the junction of the parietal and frontal bones, the greater wing of the sphenoid, and the squamous part of the temporal bones.

Sphenoid bone   (*os sphenoidale*)
: a single irregular, wedge-shaped bone at the base of the skull, forming a part of the floor of the anterior, middle, and posterior cranial fossae.

Sphenoid sinus   (*sinus sphenoidalis*)
: an air cavity of variable size and shape situated in the anterior part of the body of the sphenoid bone; separated from its fellow of the opposite side by a septum, and opening into the nasal cavity above the superior nasal concha on the same side.

Sphenopalatine foramen   (*foramen sphenopalatinum*)
: an opening on the medial wall of the pterygopalatine fossa, interconnecting this fossa with the nasal cavity, and transmitting the sphenopalatine artery and nasal nerves.

Sphenopalatine notch of palatine bone   (*incisura sphenopalatina ossis palatini*)
: a notch between the orbital and sphenoid processes of the palatine bone; it is converted into a foramen by the under surface of the sphenoid bone.

Spine   (*spina*)
: a thornlike process or projection.

Spine of sphenoid bone   (*spina ossis sphenoidalis*)
: a small bony process projecting downward from the inferior aspect of the great wing of the sphenoid bone; it serves for attachment of the sphenomandibular and pterygospinous ligaments.

Spinous foramen   (*foramen spinosum*)
>an opening in the great wing of the sphenoid bone, near its posterior angle, for the middle meningeal vessels.

Spiral canal of modiolus   (*canalis spiralis modioli*)
>a canal following the course of the bony spiral lamina of the cochlea and containing the spiral ganglion of the cochlear division of the vestibulocochlear nerve.

Squama
>a scale or plate-like structure.

Stephanion
>the point on the side of the cranium at which the coronal suture meets the superior temporal line.

Stylomandibular ligament   (*ligamentum stylomandibulare*)
>an aponeurotic band attached superiorly to the tip of the styloid process of the temporal bone and inferiorly to the angle and posterior margin of the ramus of the mandible.

Stylomastoid foramen   (*foramen stylomastoideum*)
>a foramen on the inferior part of the temporal bone between the styloid and mastoid processes, for the facial nerve and the stylomastoid artery.

Subarcuate fossa of temporal bone   (*fossa subarcuate ossis temporalis*)
>a small fossa on the internal surface of the petrous part of the temporal bone just below the arcuate eminence. It lodges a piece of dura and transmits a small vein.

Submandibular fossa   (*fovea submandibularis*)
>a depression on the medial aspect of the body of the mandible, lodging a small portion of the submandibular gland.

Sulcus
>a groove, trench, or furrow.

Superior nasal concha   (*concha nasalis superior*)
>the upper of two bony plates projecting from the inner wall of the ethmoid labyrinth and forming the upper boundary of the superior meatus of the nose.

Superior nasal meatus   (*meatus nasi superior*)
>the narrow cavity below the superior nasal concha, with which the posterior ethmoidal cells communicate.

Superior nuchal line   (*linea nuchae superior*)
>a curved line on the outer surface of the occipital bone, extending from the external occipital protuberance toward the lateral angle, and giving attachment medially to the trapezius muscle and laterally to the sternocleidomastoid muscle.

Superior orbital fissure   (*fissura orbitalis superior*)
    an elongated cleft between the small and great wings of the sphenoid bone, which transmits various nerves and vessels. Through this fissure the oculomotor, the trochlear, the ophthalmic division of the trigeminal and abducens nerves enter the orbital cavity; also sympathetic fibers from the cavernous plexus and the orbital branches of the middle cerebral artery. Leaving through this fissure are the recurrent branch from the lacrimal artery to the dura mater and the superior ophthalmic vein.

Superior surface   (*facies superior*)
    that surface which is upper or higher (toward the head, in man).

Superior temporal line of parietal bone   (*linea temporalis superior ossis parietalis*)
    a curved line on the external surface of the parietal bone, above and parallel to the inferior temporal line, giving attachment to the temporal fascia.

Suprainterparietal bone
    a sutural bone sometimes occurring at the posterior part of the sagittal suture.

Supraorbital foramen   (*foramen supraorbitalis*)
    an opening in the frontal bone in the supraorbital margin, giving passage to the supraorbital artery and nerve.

Supraorbital notch   (*incisura supraorbitalis*)
    a palpable notch in the frontal bone at the junction of the medial one-third and lateral two-thirds of the supraorbital margin, for transmission of the supraorbital nerve and vessels to the forehead. In life it is bridged by fibrous tissue, which is sometimes ossified, forming a bony aperture (foramen supraorbitalis).

Sutural bones   (*ossa suturarum*)
    small irregular bones in the sutures between the bones of the skull.

Suture   (*sutura*)
    a type of fibrous joint in which the opposed surfaces are closely united, as in the skull.

Symphysis
    a site or line of union; used in official anatomical nomenclature to designate a type of cartilaginous joint in which the apposed bony surfaces are firmly united by a plate of fibrocartilage.

Styloid process of temporal bone   (*processus styloideus ossis temporalis*)
> it originates between the lamina of the vaginal process of the tympanic part of the temporal. It serves for attachment of three muscles and two ligaments. The muscles are: stylopharyngeal, styloglossus and stylohyoid. The ligaments are: stylomandibular and stylohyoid.

Table   (*tabula*)
> a flat layer or surface.

Temporal bone   (*os temporale*)
> one of the two irregular bones forming part of the lateral surfaces and base of the skull, and containing the organs of hearing.

Temporal fossa   (*fossa temporalis*)
> the area on the outside of the cranium outlined posteriorly and superiorly by the temporal lines, anteriorly by the frontal and zygomatic bones, laterally by the zygomatic arch, and inferiorly by the infratemporal crest. The fossa communicates with the orbital cavity through the inferior orbital and sphenomaxillary fissures. Contents of the fossa: temporalis muscle, vessels and nerves, zygomaticotemporal nerve.

Temporal lines
> these lines demarcate the superior limit of the temporal fossa. They extend from the zygomatic process of the frontal bone across the frontal and parietal bones and then curve anteriorly to be continuous with the posterior root of the zygomatic arch. They are continuous with the supramastoid crest.

Trigeminal impression of temporal bone   (*impressio trigemini ossis temporalis*)
> the shallow impression in the floor of the middle cranial fossa on the petrous part of the temporal bone, lodging the semilunar ganglion of the trigeminal nerve.

Tuber
> a swelling, protuberance.

Tubercle
> a nodule, or small eminence, such as a rough, rounded eminence on a bone.

Tubercle of sella turcica   (*tuberculum sellae turcicae*)
> a transverse ridge on the upper surface of the body of the sphenoid bone, it is in front of the sella turcica, back of the sulcus chiasmatis, and between the anterior clinoid processes.

Tuberosity   (*tuberositas*)
> an elevation or protuberance.

Tympanic antrum   (*antrum mastoideum*)
>an air space in the mastoid portion of the temporal bone, communicating with the tympanic cavity and the mastoid cells.

Tympanic canaliculus   (*canaliculus tympanicus*)
>a small opening on the inferior surface of the petrous part of the temporal bone in the floor of the petrosal fossa; it transmits the tympanic branch of the glossopharyngeal nerve and a small artery.

Tympanic cavity   (*auris media*)
>the space immediately medial to the tympanic membrane; it contains the auditory ossicles and connects with the mastoid cells and auditory tube.

Tympanic membrane   (*membrana tympani*)
>the membrane separating the middle from the external ear.

Tympanic opening of auditory tube   (*ostium tympanicum tubae auditivae*)
>the opening of the auditory tube on the carotid wall of the typanic cavity.

Tympanic plate
>a bony plate which forms the floor and sides of the meatus acusticus internus.

Tympanic portion of temporal bone   (*pars tympanica ossis temporalis*)
>the part of the temporal bone that forms the anterior and inferior walls and part of the posterior wall of the external auditory meatus.

Tympanic ring   (*anulus tympanicus*)
>the bony ring forming part of the temporal bone at the time of birth and developing into the tympanic part of the bone.

Tympanic sinus   (*sinus tympani*)
>a deep fossa on the medial wall of the tympanic cavity; it is bounded behind by the pyramidal eminence, below by the subiculum promontorii, and it goes over in front into the fossula fenestrae cochleae.

Tympanomastoid fissure   (*fissura tympanomastoidea*)
>an external fissure on the inferior and internal aspect of the skull between the tympanic portion and the mastoid process of the temporal bone; the auricular branch of the vagus nerve often passes through it.

Umbo of tympanic membrane   (*umbo membranae tympani*)
>the slight projection at the center of the outer surface of the tympanic membrane, corresponding to the point of attachment of the tip of the handle of the malleus.

Vertex of bony cranium   (*vertex cranii ossei*)
: the highest point of the skull; although its position varies somewhat in different skulls, it is generally located on the sagittal suture, usually near the midpoint of the suture.

Vestibular fenestra   (*fenestra vestibuli*)
: an oval opening in the inner wall of the middle ear, which is closed by the base of the stapes.

Vomer
: a plowshare.

Vomer bone   (*vomer*)
: the unpaired flat bone that forms the inferior and posterior part of the nasal septum.

Zygoma
: a bolt or bar (Gr.)

Zygomatic arch   (*arcus zygomaticus*)
: the arch formed by the articulation of the broad temporal process of the zygomatic bone and the slender zygomatic process of the temporal bone, giving attachment to the masseter muscle and serving as a line of demarcation between the temporal and infratemporal fossae.

Zygomatic bone   (*os zygomaticum*)
: the quadrangular bone of the cheek, articulating with the frontal bone, the maxilla, the zygomatic process of the temporal bone, and the great wing of the sphenoid bone.

Zygomaticoorbital foramen   (*foramen zygomaticoorbitale*)
: either of the two openings on the orbital surface of each zygomatic bone, which transmit branches of the zygomatic branch of the trigeminal nerve and branches of the lacrimal artery.

MUSCLES OF THE SKULL

Buccinator muscle (*musculus buccinator*)
    origin: buccinator ridge of mandible, alveolar process of maxilla, pterygomandibular ligament;
    insertion: orbicular muscle of mouth at angle of mouth;
    innervation: buccal branch of facial;
    action: compresses cheek and retracts angle of the mouth.

Depressor muscle of angle of mouth (*musculus depressor anguli oris*)
    origin: lower border of mandible;
    insertion: angle of mouth;
    innervation: facial;
    action: pulls down angle of mouth.

Depressor muscle of lower lip (*musculus depressor labii inferioris*)
    origin: anterior portion of lower border of mandible;
    insertion: orbicular muscle of mouth and skin of lower lip;
    innervation: facial;
    action: depresses lower lip.

Depressor muscle of nasal septum (*musculus depressor septi nasi*)
    origin: incisor fossa of maxilla;
    insertion: ala and septum of nose;
    innervation: facial;
    action: contracts nostril and depresses ala.

Digastric muscle (*musculus digastricus*)
    origin: anterior belly, digastric fossa on inner surface of lower border of mandible near symphysis; posterior belly, mastoid notch of temporal bone;
    insertion: intermediate tendon on hyoid bone;
    innervation: anterior belly, mylohyoid; posterior belly, digastric branch of facial;
    action: elevates hyoid bone, lowers jaw.

Genioglossus muscle (*musculus genioglossus*)
    origin: mental spine of mandible;
    insertion: hyoid bone and under surface of tongue;
    innervation: hypoglossal;
    action: protrudes and depresses tongue.

Geniohyoid muscle (*musculus geniohyoideus*)
    origin: mental spine of mandible;
    insertion: body of hyoid bone;
    innervation: a branch of first cervical nerve through hypoglossal;
    action: elevates, draws hyoid forward.

Greater zygomatic muscle (*musculus zygomaticus major*)
    origin: zygomatic bone in front of temporal process;
    insertion: angle of mouth;
    innervation: facial;
    action: draws angle of mouth backward and upward.

Inferior oblique muscle of eyeball (*musculus obliquus inferior bulbi*)
    origin: orbital plate of maxilla;
    insertion: sclera;
    innervation: oculomotor;
    action: rotates eyeball upward and outward.

Lateral pterygoid muscle (*musculus pterygoideus lateralis*)
    origin: upper head, lateral surface of greater wing of sphenoid and infratemporal crest;
          lower head, lateral surface of lateral pterygoid plate;
    insertion: neck of condyle of mandible, temporomandibular joint capsule;
    innervation: mandibular division of trigeminal;
    action: protrudes mandible, open jaws, moves mandible from side to side.

Lesser zygomatic muscle (*musculus zygomaticus minor*)
    origin: zygomatic bone near maxillary suture;
    insertion: orbicular muscle of mouth and levator muscle of upper lip;
    innervation: facial;
    action: draws upper lip upward and laterally.

Levator muscle of angle of mouth (*musculus levator anguli oris*)
    origin: canine fossa of maxilla;
    insertion: orbicular muscle of mouth and skin at angle of mouth;
    innervation: facial;
    action: raises angle of mouth.

Levator muscle of upper eyelid (*musculus levator palpebrae superioris*)
    origin: upper border of optic foramen;
    insertion: tarsal plate of upper eyelid;
    innervation: oculomotor;
    action: raises upper lid.

Levator muscle of upper lip (*musculus levator labii superioris*)
    origin: lower orbital margin;
    insertion: muscle of upper lip;
    innervation: facial nerve;
    action: raises upper lip.

Levator muscle of upper lip and ala of nose (*musculus levator labii superioris alaeque nasi*)
    origin: nasal process of maxilla;
    insertion: cartilage of ala nasi and upper lip;
    innervation: infraorbital branch of facial;
    action: raises upper lip and dilates nostril.

Levator muscle of velum palatini (*musculus levator veli palatini*)
    origin: apex of petrous portion of temporal bone and cartilaginous part of auditory tube;
    insertion: aponeurosis of soft palate;
    innervation: pharyngeal plexus of vagus;
    action: raises soft palate.

Longissimus muscle of head (*musculus longissimus capitis*)
    origin: transverse processes of four or five upper thoracic vertebrae, articular processes of three or four lower cervical vertebrae;
    insertion: mastoid process of temporal bone;
    innervation: branches of cervical;
    action: draws head backward, rotates head.

Long muscle of head (*musculus longus capitis*)
    origin: transverse processes of third to sixth cervical vertebrae;
    insertion: basal portion of occipital bone;
    innervation: branches from first, second, and third cervical;
    action: flexes head.

Masseter muscle (*musculus masseter*)
    origin: superficial part, zygomatic process of maxilla and lower border of zygomatic arch;
    deep part, lower border and medial surface of zygomatic arch;
    insertion: superficial part, angle and ramus of mandible;
    deep part, upper half of ramus and lateral surface of coronoid process of mandible;
    innervation: mandibular division of trigeminal;
    action: raises mandible, closes jaws.

Medial pterygoid muscle (*musculus pterygoideus medialis*)
    origin: lateral pterygoid plate, tuberosity of maxilla;
    insertion: medial surface of ramus and angle of mandible;
    innervation: mandibular division of trigeminal;
    action: closes jaws.

Mental muscle (*musculus mentalis*)
- origin: incisive fossa of mandible;
- insertion: skin of chin;
- innervation: facial;
- action: wrinkles skin of chin.

Mylohyoid muscle (*musculus mylohyoideus*)
- origin: mylohyoid line of mandible:
- insertion: body of hyoid bone and median raphe;
- innervation: mylohyoid branch of trigeminal;
- action: elevates hyoid bone, supports floor of mouth.

Nasal muscle (*musculus nasalis*)
- origin: maxilla;
- insertion: alar part, ala of nose; transverse part, by aponeurotic expansion with fellow of opposite side;
- innervation: facial;
- action: alar part, aids in widening nostril; transverse part, depresses cartilage of nose.

Occipitofrontal muscle (*musculus occipitofrontalis*)
- origin: frontal belly, galea aponeurotica; occipital belly, highest nuchal line of occipital bone;
- insertion: frontal belly, skin of eyebrows and root of nose; occipital belly, galea aponeurotica;
- innervation: frontal belly, temporal branch of facial; occipital belly, posterior auricular branch of facial;
- action: frontal belly, raises eyebrows; occipital belly, draws scalp backward.

Orbicular muscle of eye (*musculus orbicularis oculi*)
- origin: orbital part, medial margin of orbit, including frontal process of maxilla; palpebral part, medial canthus, medial palpebral ligament; lacrimal part, posterior lacrimal crest;
- insertion: orbital part, near origin after encircling orbit; palpebral part, lateral canthus; lacrimal part, joins palpebral portion;
- innervation: facial;
- action: closes eyelids, wrinkles forehead, compresses lacrimal sac.

Orbicular muscle of mouth (*musculus orbicularis oris*)
- origin: labial part, fibers restricted to the lips; marginal part, fibers blending with those of adjacent muscle;
- insertion: forms part of lip;
- innervation: facial
- action: protrudes lips.

Orbital muscle (*musculus orbitalis*)
- origin: orbital periosteum
- insertion: fascia of inferior orbital fissure;
- innervation: sympathetic fibers;
- action: protrudes eye.

Platysma muscle (*platysma*)
- origin: fascia of the cervical region;
- insertion: mandible and skin around the mouth;
- innervation: cervical branch of the facial nerve;
- action: wrinkles skin of neck and depresses jaw.

Posterior auricular muscle (*musculus auricularis posterior*)
- origin: mastoid process;
- insertion: cartilage of ear;
- innervation: facial;
- action: draws auricle backward.

Procerus muscle (*musculus procerus*)
- origin: skin over nose;
- insertion: skin of forehead;
- innervation: facial;
- action: draws eyebrows down.

Rectus capitis anterior muscle (*musculus rectus capitis anterior*)
- origin: lateral mass of atlas;
- insertion: basilar process of occipital bone;
- innervation: first and second cervical;
- action: flexes, supports head.

Rectus capitis lateralis muscle (*musculus rectus capitis lateralis*)
- origin: upper surface of transverse process of atlas;
- insertion: jugular process of occipital bone;
- innervation: first and second cervical;
- action: flexes, supports head.

Rectus capitis posterior major muscle (*musculus rectus capitis posterior major*)
- origin: spinous process of axis;
- insertion: occipital bone;
- innervation: suboccipital and greater occipital;
- action: extends head.

Rectus capitis posterior minor muscle (*musculus rectus capitis posterior minor*)
- origin: tubercle on dorsal arch of atlas;
- insertion: occipital bone;
- innervation: suboccipital and greater occipital;
- action: extends head.

Rectus inferior bulbi muscle (*musculus rectus inferior bulbi*)
    origin: circumference of optic foramen;
    insertion: under side of sclera;
    innervation: oculomotor;
    action: adducts, rotates eyeball downward and medially.

Rectus lateralis bulbi muscle (*musculus rectus lateralis bulbi*)
    origin: lateral margin of optic foramen, margin of superior orbital fissure;
    insertion: lateral side of sclera;
    innervation: abducens;
    action: abducts eyeball.

Rectus medialis bulbi muscle (*musculus rectus medialis bulbi*)
    origin: circumference of optic foramen;
    insertion: medial side of sclera;
    innervation: oculomotor;
    action: adducts eyeball.

Rectus superior bulbi muscle (*musculus rectus superior bulbi*)
    origin: upper border of optic foramen;
    insertion: upper aspect of sclera;
    innervation: oculomotor;
    action: adducts, rotates eyeball upward and medially.

Semispinal muscle of head (*musculus semispinalis capitis*)
    origin: transverse processes of five or six upper thoracic and four lower cervical vertebrae;
    insertion: occipital bone;
    innervation: suboccipital, greater occipital, and branches of cervical;
    action: extends head.

Splenius muscle of head (*musculus splenius capitis*)
    origin: lower half of nuchal ligament, spines of seventh cervical and three upper thoracic vertebrae;
    insertion: occipital bone;
    innervation: middle and lower cervical;
    action: extends, rotates head.

Stapedius muscle (*musculus stapedius*)
    origin: interior of pyramid of tympanic cavity;
    insertion: posterior surface of neck of stapes;
    innervation: stapedial branch of facial;
    action: dampens stapedial movement.

Sternocleidomastoid muscle (*musculus sternocleidomastoideus*)
- origin: sternum and clavicle;
- insertion: mastoid process and superior nuchal line of occipital bone;
- innervation: accessory nerve and cervical plexus;
- action: flexes vertebral column, rotates head.

Styloglossus muscle (*musculus styloglossus*)
- origin: styloid process;
- insertion: margin of tongue;
- innervation: hypoglossal;
- action: raises and retracts tongue.

Stylohyoid muscle (*musculus stylohyoideus*)
- origin: styloid process;
- insertion: body of hyoid bone;
- innervation: facial;
- action: draws hyoid and tongue upward.

Stylopharyngeal muscle (*musculus stylopharyngeus*)
- origin: styloid process;
- insertion: thyroid cartilage and pharyngeal constrictors;
- innervation: pharyngeal plexus, glossopharyngeal;
- action: raises and dilates pharynx.

Superciliary corrugator muscle (*musculus corrugator supercilii*)
- origin: medial end of superciliary arch;
- insertion: skin of eyebrow;
- innervation: facial;
- action: draws eyebrow downward and medially.

Superior constrictor muscle of pharynx (*musculus constrictor pharyngis superior*)
- origin: medial pterygoid plate, pterygomandibular raphe, mylohyoid ridge of mandible, and mucous membrane of floor of mouth;
- insertion: median raphe of posterior wall of pharynx;
- innervation: pharyngeal plexus of vagus;
- action: constricts pharynx.

Superior oblique muscle of eyeball (*musculus obliquus superior bulbi*)
- origin: lesser wing of sphenoid above optic foramen;
- insertion: sclera;
- innervation: trochlear;
- action: rotates eyeball downward and outward.

Temporal muscle (*musculus temporalis*)
    origin: temporal fossa and fascia;
    insertion: coronoid process of mandible;
    innervation: mandibular division of trigeminal;
    action: closes jaws.

Tensor muscle of tympanic membrane (*musculus tensor tympani*)
    origin: cartilaginous portion of auditory tube;
    insertion: manubrium of malleus;
    innervation: mandibular division of trigeminal;
    action: tenses tympanic membrane.

Tensor muscle of velum palatini (*musculus tensor veli palatini*)
    origin: scaphoid fossa of sphenoid, wall of auditory tube;
    insertion: aponeurosis of soft palate, horizontal part of palatine bone;
    innervation: mandibular division of trigeminal;
    action: tenses soft palate, opens auditory tube.

Trapezius muscle (*musculus trapezius*)
    origin: occipital bone, nuchal ligament, spinous processes of seventh cervical and all thoracic vertebrae;
    insertion: clavicle, acromion, spine of scapula;
    innervation: accessory nerve and cervical plexus;
    action: rotates scapula to raise shoulder in abduction of arm, draws scapula backward.

Uvular muscle (*musculus uvulae*)
    origin: posterior nasal spine of palatine bone and aponeurosis of soft palate.
    insertion: uvula;
    innervation: pharyngeal plexus of vagus;
    action: raises uvula.

BIBLIOGRAPHY

Anson, B. J. and Donaldson, J. A. *Surgical Anatomy of the Temporal Bone and Ear,* 2nd ed. Philadelphia, London, Toronto, W. B. Saunders Company, 1973.

Clemente, C. D. *Anatomy. A Regional Atlas of the Human Body.* Philadelphia, Lea and Febiger, 1975.

Dorland's Illustrated Medical Dictionary. 25th ed. Philadelphia, London, Toronto, W. B. Saunders Company, 1974.

Etter, L. E. *Atlas of Roentgen Anatomy of the Skull.* Springfield, Ill., Charles C Thomas, 1964.

Grant, J. C. B. *Grant's Atlas of Anatomy.* 6th ed. Baltimore, The Williams & Wilkins Co., 1972.

Gray, H. *Gray's Anatomy: Anatomy of the Human Body,* 29th American ed. Edited by C. M. Goss, Philadelphia, Lea & Febiger, 1973.

McMinn, R. M. H. and Hutchings, R. T. *Color Atlas of Human Anatomy.* Chicago Year Book Medical Publishers, Inc., 1977.

Nomina Anatomica, 3rd ed. Amsterdam, Princeton, London, Excerpta Medica, 1972.

Pernkopf, E. *Atlas of Topographical and Applied Human Anatomy,* Vols. I and II. Edited by H. Ferner, Philadelphia and London, W. B. Saunders Company, 1963 and 1964.

Sobotta, J. *Atlas of Human Anatomy,* 9th English ed., Vols. I and III. Edited by F. H. J. Figge, New York, Hafner Press. A division of MacMillan Publishing Co., Inc., 1974.

Taveras, J. M. and Morello, F. *Normal Neuroradiology.* Chicago, London, Year Book Medical Publishers, Inc., 1979.

INDEX

If the presence of a structure is incidental, no reference is made to it in the index.

The Glossary and Table of Muscles are not part of this index.

Acoustic meatus
   external, 23, 28, 82, 122, 142, 145
   internal, 37, 127-131, 136, 137
Air cells, 48, 59
   ethmoidal, 67, 74, 104, 105
   mastoid, 133, 136, 144
Ala
   cristae galli, 60, 61, 67, 69, 104, 105
   of vomer, 61, 70-72, 100
Alveolar process, 92-94
Angle of mandible, 11, 16, 87
Anterior nasal spine, 5, 11, 59, 73, 92, 95
Aperture, posterior nasal, 51
Apex
   of orbit, 50
   of petrous part of temporal, 126, 127, 131, 134
Aqueduct of vestibule, 137
Arch
   superciliary, 11, 102
   zygomatic, 11, 19, 23, 80, 82, 83, 96
Arcuate eminence, 129
Articulation
   incudomalleolar, 132
   incudostapedial, 132
   temporomandibular, 82, 83
Asterion, 11
Auditory
   ossicles, 132, 138-145
      ligaments of, 140
      muscles of, 140, 141
   tube, 28, 134, 135
      opening of, 134-136
      sulcus for, 109
Auricular point, 11

Basioccipital, 31, 38, 40, 41
Basion, 23
Body
   of mandible, 84, 90
   of sphenoid, 61, 106, 109
Bone(s)
   cranial, 101-123
   ethmoid, 3, 9, 19, 104, 105
   facial, 3, 84-100
   frontal, 3, 9, 21, 32, 101-103
   infant skull, 146, 147
   inferior nasal concha, 3, 19, 59, 100
   lacrimal, 3, 9, 52, 53, 58, 59, 100
   mandible, 3, 9, 15, 19, 82-90
   maxilla, 3, 9, 15, 21, 91-95
   nasal, 3, 9, 66, 100
   occipital, 3, 9, 15, 19, 21, 32, 114-119
   ossicles, 132, 138-145
   palatine, 3, 15, 19, 21, 68, 69, 71, 79, 98, 99
   parietal, 3, 9, 15, 19, 32, 120-122
   sphenoid, 3, 9, 15, 19, 21, 61-65, 67-69, 78, 79, 106-113, 118, 119
   sutural, 32, 116, 118-120
   temporal, 3, 9, 15, 19, 21, 123-137
   vomer, 3, 15, 19, 21, 59, 62-64, 100
   zygomatic, 3, 9, 19, 21, 52, 53, 77, 96, 97
Bregma, 11

Calvarium, 32, 33
Canal(s)
   carotid, 29, 31, 80, 83, 124, 126, 128, 129, 131, 133, 136
   cochlear spiral, 131, 136, 137
   condylar, 22, 26, 29, 114, 116, 117
   facial, 133-135, 137, 142, 143
      genu of, 133
      hiatus, 129, 136
   for chorda tympani nerve, 144
   hypoglossal, 26, 27, 29, 114, 117, 119
   incisive, 56, 59, 65, 93
   infraorbital, 74, 93
   musculotubal, 131, 133
   nasolacrimal, 51, 58, 66, 95
   optic, 40, 43, 44, 47, 50, 65, 69, 76, 106, 107
   palatine
      great, 22, 72, 74, 77, 79, 92, 93, 98
      lesser, 22, 79, 93
   pharyngeal, 74, 75, 77, 78
   pterygoid, 74-78, 106, 107, 110, 112
   semi, 126, 136
   semicircular,
      lateral, 131, 142, 145
      posterior, 131, 145
      superior, 130, 131, 133, 136, 137, 145
Canaliculus
   inferior tympanic, 126
   of cochlea, 126, 128
Canine teeth, 23, 84, 85, 90, 91
Caroticoclinoid foramen, 41
Cavity(ies)
   nasal, 5, 73, 76, 94
      from behind, 70
      from below, 66
      from front, 60, 68
      midsagittal section, 56, 57
      oblique view, 63
      sphenopalatine foramen, 72
   orbital, 42-54
      bones, 42
      floor, 52
      foramina, 42, 43
      from front, 42
      from side, 54
      margins, 43
      nasolacrimal canal, 51
      rim, 46
      roof, 46
      walls, 49, 52
   oral, 94
   tympanic, 136-138, 141-145
      promontory of, 143
Choana, 23, 62, 65, 68-71
Clinoid processes
   anterior, 38-41, 49, 108, 109, 113, 119
   middle, 38, 108
   posterior, 38-41, 109, 113
Clivus, 38, 41, 119

Cochlea
   canaliculus, 126, 128
   cupola, 137
   modiolus, 137
   osseous spiral lamina, 136, 137
Cochleariform process, 132, 144
Cochlear spiral canal, 131, 136, 137
Cochlear window
   fossula of, 125, 135
Color code
   cranial bones, XII
   facial bones, XII
Concha
   inferior nasal, 56, 59, 60, 66, 70, 94, 95
   middle nasal, 57, 61, 66, 67, 69, 104, 105
   of sphenoid, 112
   superior nasal, 58, 104, 105
Conchal crest
   of maxilla, 93
   of palatine, 58
Condylar
   canal, 22, 26, 114, 116, 117
   fossa, 117
Condyle
   occipital, 17, 19, 23, 26, 27, 70, 117
   of mandible, 82-85, 87, 89
Coronal suture, 4, 10, 32, 146
Coronoid process, 82-85, 87
Cranial
   bones, 101-136
   cavity,
      base of, 37
      fossa(e), 37
      middle, 41
Crest
   conchal of maxilla, 93, 94
   conchal of palatine, 58, 98, 99
   ethmoidal of maxilla, 94
   ethmoidal of palatine, 58, 59, 98
   external occipital, 19, 116
   frontal, 103
   infratemporal, 62, 80, 112, 113
   nasal, 98, 99
   orbital, 112
   sphenoidal, 62, 108, 118
Cribriform plate, 56, 66, 67, 105
Crista galli, 41, 49, 53, 55, 56, 60, 61, 104, 105
   ala of, 60, 61, 67, 69, 104, 105
Cupola of cochlea, 137

Diploë, 35
Dorsum sellae, 38-41, 71, 113, 119

Eminence
   arcuate, 129
   canine, 92
   frontal, 11, 102, 146, 147
   parietal, 17, 32, 120, 146, 147
   pyramidal, 143
Epitympanic recess, 132, 137, 142, 145
Epitympanum, 142, 145
Ethmoid
   ala cristae galli, 60, 61, 67, 69, 104, 105
   air cells, 69, 74, 104, 105
   bone, 3, 9, 19, 104, 105
   olfactory groove, 105
   orbital plate, 49, 53, 55, 69, 73, 104, 105
   perpendicular plate, 56, 60, 61, 66, 67, 73, 104, 105
   uncinate process, 61, 66, 76, 104, 105
Ethmoidal
   crest of palatine, 58, 59, 98
   foramina, 44, 76, 104
   labyrinth, 104, 105
   notch, 48
External
   acoustic meatus, 23, 28, 82, 122, 142, 145
   occipital crest, 19, 116
   occipital protuberance, 16

Facial
   artery, groove for, 87
   bones, 84-100
   canal, 133-135, 137, 142, 143
      hiatus of, 129, 136
   nerve, 142-145
Fissure(s)
   of aqueduct of vestibule, 128
   inferior orbital, 29, 42-44, 51, 52, 55, 72, 74, 75, 80-82
   petrosquamosal, 123, 147
   petrotympanic, 81, 124, 135, 144
   pterygoid of sphenoid, 112
   pterygomaxillary, 77
   pterygopalatine, 69, 80, 81
   superior orbital, 5, 42-44, 47, 74, 75, 78, 106, 107, 109, 110
   tympanomastoid, 124, 125, 144
Fontanelle, 146, 147
   anterior, 146, 147
   mastoid, 146
   posterior, 147
   sphenoidal, 146
Foramen(ina)
   accessory, 109
   caroticoclinoid, 41
   carotid, 26, 28, 83
   cecum, 37, 56, 103
   ethmoid
      anterior, 44, 76, 104
      posterior, 44, 76, 104
   external carotid, 134, 135
   for emissary vein, 17, 109, 116
   incisive, 22, 95
   infraorbital, 5, 44
   jugular, 27, 29, 114
   lacerum, 22
      anterior wall of, 108, 109
   magnum, 23, 27, 29, 31, 114, 115, 118
   mandibular, 17, 85, 87, 89
   mastoid, of temporal bone, 17, 124, 128, 129
   mental, 5, 84
   olfactory, 105
   optic, 63, 75-77, 81, 112
   oval, 22, 26-29, 67, 71, 72, 78, 81, 106-111, 123
   palatine
      greater, 95
      lesser, 95
   parietal, 17, 32, 121

rotundum, 47, 50, 61, 62, 74-79, 106, 107, 109
singular, 127
sphenopalatine, 62, 66, 70, 72, 74, 77, 79
spinous, 22, 26-29, 66, 71, 72, 77, 80, 81, 83, 106-111, 130
stylomastoid, 22, 26, 28, 81, 124, 133-135

Fossa
canine, 92
cranial,
anterior, 37
middle, 37, 123
posterior, 37
for digastric muscle, 89
for lacrimal gland, 48
for lacrimal sac, 5, 51, 53, 73, 76
for sublingual gland, 12, 89
for submandibular gland, 89
hypophyseal, 38-41, 49, 101, 107, 113, 118
incisive, 29
infratemporal, 80, 81, 106, 112
jugular, 26, 28, 119, 126, 128, 133, 134
mandibular, 26, 28, 80, 81, 122, 124-126, 135
pterygoid, 110-112
pterygopalatine, 74, 77, 78, 106
diagrammatic, 74
from front, 74
from side, 74, 77
wall,
medial, 77
posterior, 78
scaphoid, 110, 111
subarcuate, 128
temporal, 11, 29, 55, 83, 96, 102

Fossula
of cochlear window, 125, 135, 143, 144
of vestibular window, 125, 135

Foveolae, granular, 33, 121

Frontal
bone, 3, 9, 21, 47, 101-103
crest of, 103
eminence, 11, 102, 146, 147
nasal spine, 101, 102
notch, ethmoidal, 48, 103
orbital plate, 47
parietal margin, 48, 101
sinuses, 55
squama, 8, 103
suture, 4, 146, 147
temporal line, 101
zygomatic process, 48
process, 53, 58, 65, 92, 93, 97

Frontolacrimal suture, 4, 43
Frontomaxillary suture, 4
Frontonasal suture, 4, 51
Frontozygomatic suture, 4, 10

Glabella, 5, 11, 30
Granular foveolae, 33, 121
Groove(s)
brain stem, 115
carotid, 109
facial artery, 87
for anterior ethmoidal nerve, 100
for bulb of jugular vein, 130, 131, 138
for greater petrosal nerve, 131
for inferior petrosal sinus, 35, 115, 128
for internal carotid artery, 35, 40
for meningeal vessels, 35, 103, 108, 121, 123, 129, 130
for occipital artery, 17, 116, 126, 134
for occipital sinus, 35, 114
for sigmoid sinus, 118, 128, 129, 131
for superior petrosal nerve, 131
for superior petrosal sinus, 35, 123, 127, 128
for superior sagittal sinus, 103, 114, 121
for transverse sinus, 35, 114, 115, 121, 130, 131
for tympanic part of glossopharyngeal nerve, 135
infraorbital, 42, 52, 75
lacrimal, 100
mylohyoid, 87
nasolacrimal, 92, 93
nasopalatine, 63, 100
olfactory, 105
optic, 38, 40
pterygoid, 112

Hamulus
lacrimal, 58, 100
pterygoid, 63, 64, 68, 110-113
Head of mandible, 19, 90
Hiatus of facial canal, 129, 136
Horizontal plate of palatine, 60, 71, 72, 95
Hypoglossal canal, 26, 27, 114, 117, 119
Hypophyseal fossa, 38-41, 49, 101, 107, 113, 118

Impression, trigeminal of temporal, 129
Incisive
canal, 56, 59, 65, 93
foramen, 22, 95
fossa, 92
Incisor teeth, 23, 84, 85, 90, 91
Incudomalleolar articulation, 132
Incudostapedial articulation, 132
Incus, 132, 138-140, 142, 143, 145
body, 132, 139
lenticular process, 139
long process, 132, 139, 141, 142, 145
short process, 139, 141-143
Inferior alveolar process, 11
Inferior nasal conchae, 3, 19, 59, 100
Inferior orbital fissure, 42-45, 51, 52, 55, 72, 74, 80, 82
Infraorbital
canal, 42, 74, 93
foramen, 5, 42, 44
groove, 42, 52, 75
margin of orbit, 43, 55
point, 5, 30
Infratemporal
crest, 62, 80
fossa, 80, 81, 106, 112, 113
surface of maxilla, 53, 73, 92
Intermaxillary suture, 4
Internal acoustic meatus, 37, 123, 127-131, 136, 137
Internasal suture, 4
Intrajugular process, 115, 129

Joint(s)
incudomalleolar, 132

incudostapedial, 132
temporomandibular, 82, 83
Jugular
   foramen, 27, 29, 114
   fossa, 26, 28, 119, 126, 128, 133, 134
   notch, 114, 115, 117, 128
   process, 26, 114, 115, 117-119
Jugum of sphenoid, 109
Labyrinth, ethmoidal, 104, 105
Lacrimal
   bone, 3, 9, 100
   foramen, 52
   gland, fossa for, 48
   groove, 100
   hamulus, 58, 100
   process, 100
   sac, fossa for, 5, 51, 53, 73, 76
Lacrimomaxillary suture, 76
Lambdoidal suture, 10, 16, 146, 147
Lateral pterygoid plate, 9, 15, 62, 63, 67-69, 71, 79, 110-113
Lenticular process of incus, 139
Ligaments of auditory ossicles, 140
Line(s)
   acanthiomeatal, 30
   auricular, 30
   base, 30
   glabelloalveolar, 30
   glabellomeatal, 30
   infraorbital, 30
   interorbital, 30
   mylohyoid, 89
   nuchal,
      inferior, 17, 116, 117
      highest, 17, 116
      superior, 17, 116
   oblique, 85, 87
   orbitomeatal, 30
   superior horizontal, 30
   temporal, 101, 102, 124
      inferior, 11, 120
      superior, 11, 120
   transverse,
      anterior, 31
      posterior, 31
Lingula
   of mandible, 89
   of sphenoid, 109-111, 123

Malleus, 138, 141-143
   anterior process, 138
   handle of, 132, 139, 141
   head of, 132, 138, 139, 145
   lateral process, 138, 141
   neck, 139
Mandible, 3, 9, 15, 19, 84-90
   angle, 11, 16, 87
   body, 84, 90
   condyle, 82-85, 87, 89
   coronoid process, 82-85, 87, 90
   head of, 19, 90
   landmarks, 87, 89
   lingula, 89
   mental foramen, 84, 87
   mental protuberance, 84

   mental spine, 85, 89
   muscle attachments, 87, 89
   mylohyoid line, 89
   neck, 87, 90
   oblique line, 85, 87
   ramus, 82-84
   symphysis, 84
   tubercle, 89
Mandibular
   foramen, 17, 85, 87, 89
   fossa, 26, 28, 80, 81, 89, 122, 124-126, 135
   notch, 82, 83, 85, 87
Margin(s)
   infraorbital, 43, 55
   of infraorbital fissure, 53
   orbital, 102
   supraorbital, 42
Mastoid
   air cells, 133, 136, 144
   foramen, 17, 124, 128, 129
   notch, 17, 126
   process, 17, 19, 26, 122, 124, 126, 128, 134, 144
Maxilla, 3, 9, 15, 19, 21, 43, 72, 91-94
   alveolar process, 92-94
   alveolar process removed, 91
   conchal crest, 93, 94
   diagrammatic, 94
   ethmoidal crest, 94
   frontal crest, 58
   frontal process, 53, 58, 65, 92, 93
   incisive canal, 56, 93
   infratemporal surface, 53, 73, 92
   orbital surface, 73, 94, 97
   palatine process, 93, 94
   spine, 53, 60, 66, 73, 95
   tubercle, 93
   tuberosity, 70, 92
   zygomatic process, 93-95
Maxillary sinus, 52, 55, 58, 65, 74, 92-95
Meatus
   external acoustic, 23, 28, 122, 125, 142, 145
   internal acoustic, 37, 123 127-131, 136, 137
      transverse crest, 127
   nasal,
      inferior, 58, 59, 70, 94, 98
      middle, 57, 59, 66, 70, 93, 94, 98
      superior, 59, 98
Medial pterygoid plate, 62, 64, 68, 78, 110-112
Membrane, tympanic, 132, 138, 142, 145
Mental
   foramen, 5, 84
   point, 5, 11
   protuberance, 84
   spine, 85, 89
   tubercle, 89
Modiolus, 137
Molar teeth, 23, 84, 85, 90, 91, 94
   unerupted, 53, 92
Muscle attachments (see also Table of Muscles, 000)
   buccinator, 7, 13, 87, 93
   depressor muscle of
      angle of mouth, 13, 87
      lower lip, 13, 87
      nasal septum, 93

digastric, 7, 13, 25, 89, 125
genioglossus, 89
geniohyoid, 89
greater zygomatic, 7, 13, 97
inferior oblique of eyeball, 44
lateral pterygoid, 25, 87, 112
lesser zygomatic, 7, 13
levator of
    angle of mouth, 7, 13, 93
    upper lip, 7, 13, 93, 97
    upper lip and ala of nose, 7, 13, 93
    velum palatini, 25, 126
longissimus of head, 13
long of head, 25, 117, 125
masseter, 7, 13, 25, 87, 93, 97, 125, 126
medial pterygoid, 25, 87, 99, 112
mental, 7, 13, 87
mylohyoid, 87
nasal, 7, 13, 93
occipitofrontal, 13, 25, 116, 125
orbicular of
    eye, 7, 13, 93, 100, 102
    mouth, 7, 13
platysma, 13, 87
posterior auricular, 13, 125
procerus, 7, 13
rectus capitis
    anterior, 25, 117
    lateralis, 25, 117
    posterior major, 25, 117
    posterior minor, 25, 117
semispinal of head, 13, 25, 116
splenius of head, 13, 25, 116, 125
stapedius, 140, 145
sternocleidomastoid, 13, 25, 116, 125
styloglossus, 13, 125
stylohyoid, 13, 125
stylopharyngeal, 126
superciliary corrugator, 7, 11, 102
superior constrictor of pharynx, 25, 87, 117
superior oblique of eyeball, 117
temporal, 7, 13, 87, 102, 112, 120, 125
tensor of
    tympanic membrane, 25, 131, 133, 145
    velum palatini, 25, 112, 131, 133
trapezius, 13, 25, 116
uvular, 25, 99
Musculotubal canal, 51, 66, 131, 133
Mylohyoid
    groove, 87
    line, 89

Nasal
    aperture, 51
    bone, 3, 9, 66, 100
    cavity, 5, 73, 76, 94
        floor of, 55
    concha,
        inferior, 56, 59, 60, 66, 70, 94, 95
        middle, 57, 61, 66, 67, 69, 104, 105
        superior, 58, 104, 105
    crest, 98, 99
    meatus,
        inferior, 58, 59, 70, 94, 98
        middle, 57, 59, 70, 93, 94, 98
        superior, 59, 98
    notch, 92
    septum, 61, 66, 72
    spine,
        anterior, 5, 48, 59, 60, 73, 92
        of frontal, 101
        posterior, 70, 71, 95, 98, 99
Nasion, 5, 11, 30
Nasolacrimal
    canal, 51, 58, 66, 95
    duct, opening of, 51
    groove, 92, 93
Nasopalatine groove, 63, 64
Neck of mandible, 87, 90
Notch
    ethmoidal, 48, 103
    jugular, 114, 115, 117, 128
    mandibular, 82, 83, 85, 87
    mastoid, 17, 126
    nasal, 92
    parietal, 122, 124
    pterygoid, 110
    sphenopalatine, 98, 99
    supraorbital, 43, 44, 48, 102

Oblique line, 85, 87
Occipital
    basilar part, 115, 117
    bone, 9, 15, 19, 21, 114-119
    condylar canal, 22, 26, 29, 114, 116, 117
    condylar fossa, 117
    condyle, 17, 19, 23, 26, 27, 70, 117
    crest, external, 19, 116
    inferior angle, 114
    jugular
        notch of, 114, 115, 117, 128
        process of, 114, 115, 117-119
    lateral angle, 115
    nuchal lines, 17, 116, 117
    pharyngeal tubercle, 23, 117
    protuberance
        external, 116
        internal, 114
    superior angle, 114
    tubercle, 115
Occipitomastoid suture, 10
Olfactory
    foramina, 105
    groove, 105
Opisthion, 23
Optic
    canal, 38, 40, 43, 44, 47, 50, 65, 69, 76, 106, 107
    foramen, 63, 75-77, 81, 112
    groove, 38, 40
Oral cavity, 94
Orbit
    apex, 50
    floor, 96
    roof, 47
    walls, 49, 54, 55
Orbital
    cavity, 42-54
    crest, 112
    fissures, 44, 47, 51, 72, 74, 75, 78, 80, 82,
        106, 107, 109, 110

margin, 102
plate, 48, 49, 53, 73, 104
process of palatine, 43, 45, 47, 65, 69, 79, 98, 99
surface of
  maxilla, 73, 94, 97
  sphenoid, 106, 107
Osseous spiral lamina of cochlea, 131, 136
Ossicles, auditory, 132, 138-145
Oval foramen, 22, 26-28, 66, 71, 72, 78, 81, 106-111, 123

Palatine
  bone(s), 15, 19, 21, 74, 98, 99
  canals, 22, 72, 74, 77, 79, 92, 93, 98
  conchal crest, 58, 98, 99
  ethmoidal crest, 58, 59, 98
  maxillary process, 98-100
  median, suture, 21, 95
  nasal crest, 98, 99
  plate,
    horizontal, 50, 60, 62, 65, 70-72, 95, 98, 99
    vertical, 63, 70, 98, 99
  sphenoidal process, 62, 65, 98, 99
  sphenopalatine notch, 98, 99
  transverse suture, 95
Parietal
  bone, 3, 9, 15, 19, 120-122
  eminence, 17, 32, 120, 146, 147
  foramen, 17, 95, 121
  frontal angle, 120
  lines, 11, 120
  mastoid angle, 120
  occipital angle, 120
  sphenoidal angle, 120
  squama, 120
Parietomastoid suture, 10
Perpendicular plate, 56, 71, 73, 104, 105
Petrosquamosal fissure, 123, 147
Petrotympanic fissure, 81, 124, 135, 144
Pharyngeal
  canal, 74, 75, 77, 78
  raphe, 25
  tubercle, 23
Plate(s)
  cribriform, 56, 66, 67, 105
  horizontal, of palatine, 50, 60, 62, 65, 70-72, 95, 98, 99
  orbital, 48, 49, 53, 55, 73, 104
  perpendicular of
    ethmoid, 56, 73, 104, 105
    vomer, 71
  pterygoid,
    lateral, 9, 15, 62, 63, 67-69, 71, 79, 110-113
    medial, 62, 64, 71, 110-112
  tympanic, 26
  vertical, of palatine, 63, 70, 98, 99
Point(s)
  anterior nasal spine, 5, 11
  asterion, 11, 16
  auricular, 11
  base of skull, 23
  basion, 23
  bregma, 11, 32
  external occipital protuberance, 16
  glabella, 5, 11, 30
  inferior alveolar, 5, 11
  infraorbital, 5, 30
  lambda, 16, 32
  mental, 5, 11
  nasion, 5, 11, 30
  opisthion, 23
  pterion, 11
  stephanion, 11
  superior alveolar, 5, 11
  supraorbital, 30
  vertex, 5, 16
Ponticulus, 143
Premolar teeth, 23, 84, 85
Process(es)
  alveolar, 92-94
  clinoid,
    anterior, 38-41, 49, 108, 109, 113, 119
    middle, 38, 108
    posterior, 38-41, 109, 113
  cochleariform, 132, 144
  coronoid, 82-85, 90
  frontal of
    maxilla, 53, 58, 65, 92
    zygomatic, 97
  jugular, 26, 114, 115, 117-119
  lacrimal, 100
  mastoid, 9, 19, 26, 122, 124, 126, 128, 134, 144
  maxillary of
    palatine, 98-100
    zygomatic, 97
  orbital of
    maxilla, 92
    palatine, 43, 53, 69, 79, 98, 99
  palatine, of maxilla, 93, 94
  pterygospinous, 106
  pyramidal, of palatine, 53, 59, 70, 73, 77, 79, 98, 99, 101
  sphenoidal, of palatine, 62, 98, 99
  styloid, 9, 15, 22, 26, 30, 70, 80, 81, 124, 126
  temporal, of zygomatic, 51, 53, 97, 122, 124, 126, 129, 134
  uncinate, 60, 61, 76, 104, 105
  vaginal, 106, 109, 110, 126
  zygomatic of
    frontal, 48, 92, 102, 103
    maxilla, 93-95
Protuberance
  external occipital, 116
  internal occipital, 114
  mental, 84
Pterion, 11
Pterygoid
  canal, 74-78, 106, 107, 110, 112
  fissure of sphenoid, 112
  fossa, 110-112
  groove, 112
  hamulus, 63, 64, 68, 110-113
  notch, 110
  plates
    lateral, 9, 15, 62, 64, 67, 68, 77, 79, 110-113
    medial, 62, 64, 68, 78, 110-112
Pterygomaxillary fissure, 77
Pterygopalatine
  fissure, 69, 80, 81

fossa, 74, 77, 78, 106
sulcus, 106, 112
Pterygospinous process, 106
Pyramidal
   eminence, 143
   process of palatine, 53, 59, 63, 70, 73, 77, 79, 98, 99

Ramus, 82-84
Raphe, pharyngeal, 25
Recess, epitympanic, 132, 137, 142, 145
Roof of tympanum, 26, 131
Rostrum of sphenoid, 63, 65, 67, 78, 107, 111, 112
Sagittal suture, 16, 32, 147
Scaphoid fossa, 110, 111
Semicanals, 126, 136
Semicircular canals, 130, 131, 133, 136, 137, 142, 144, 145
Sheath of styloid process, 9, 81, 144
Sigmoid sinus, 144
Sinus
   frontal, 55, 57
   maxillary, 52, 55, 58, 65, 74, 92-95
   sigmoid, 144
   sphenoid, 56, 74
   tympanic, 125, 143
Skull
   base of internal surface, 34-37
   bones, 3, 9, 15, 19, 21, 32, 33, 38
   from above, 32, 33
   from behind, 14-17
   from below, 18-26
   from front, 2-6
   from side, 8-12
   infant, 146, 147
   lines, 30, 31
Sphenoid
   body of, 61, 106, 109
   bone, 3, 9, 15, 19, 21, 61-65, 67-69, 78, 79, 106-113, 118, 119
   clinoid processes, 38-41, 49, 108, 109
   clivus, 38, 41, 119
   concha, 112
   greater wing, 35, 42, 49, 64, 74, 107-109, 113
      orbital surface, 47
      temporal surface, 63, 67
   infratemporal crest, 62, 80, 112, 113
   jugum, 109
   lesser wing, 35, 40, 42, 49, 63, 64, 74, 107-109, 118
   lingula, 109-111, 123
   orbital
      crest, 112
      surface, 106, 107
   pterygoid
      hamulus, 63, 64, 68, 71, 110-113
      notch, 110
      plates, 62, 64, 67-71, 77, 78, 101, 110-112
   pterygospinous process, 106
   rostrum, 63, 65, 67, 78, 107, 111, 112
   scaphoid fossa, 110, 111
   sinus, 56, 74
   spine, 26, 27, 63, 67, 69, 71, 83, 101, 108, 109, 112, 113, 123
   tubercle, 50, 112
   vaginal process, 106, 109, 110
Sphenoidal
   crest, 62, 108, 118
   notch, 64
   process of palatine, 62, 98, 99
   sinuses, 56, 74
Sphenofrontal suture, 4, 10, 47, 101
Sphenooccipital synchondrosis, 38
Sphenopalatine foramen, 62, 66, 70, 72, 74, 77, 79
Sphenosquamous suture, 10, 123
Sphenozygomatic suture, 4, 10
Spine
   anterior nasal, 5, 11, 48, 59, 60, 66, 92, 95
   mental, 85, 89
   of frontal, 102
   of maxilla, 53
   posterior nasal, 70, 71, 95, 98, 99
   sphenoid, 26, 27, 63, 67, 71, 83, 101, 108, 109, 112, 113,
Spinous foramen, 22, 26, 27, 66, 71, 72, 77, 80, 81, 83, 106-111, 131
Spiral canal of cochlea, 131, 136
Squama
   frontal, 48, 103
   parietal, 120
   temporal, 35, 122-124, 126, 128, 134
Squamosoparietal suture, 10
Squamous suture, 4, 122
Stapes, 138, 139, 141, 143, 144
   anterior limb, 139, 141, 143
   base, 138, 139, 141
   head, 139, 142, 145
   neck, 139
   posterior limb, 139, 141, 143
Stephanion, 11
Styloid process, 9, 22, 26, 70, 80, 81, 124, 126, 144
   sheath of, 9, 81, 144
Stylomastoid foramen, 22, 26, 28, 81, 124, 133-135
Subarcuate fossa, 128
Sublingual gland, fossa for, 85
Sulcus
   for auditory tube, 109
   pterygopalatine, 106, 112
Superciliary arch, 11, 102
Superior alveolar point, 5, 11
Superior orbital fissure, 5, 42-44, 47, 74, 75, 78, 106, 107, 109, 110
Superior temporal line, 101, 102
Supraorbital
   foramen, 5
   margin, 42
   notch, 43, 44, 48, 102
   point, 5, 30
Sutural bone, 32, 116, 118-120
Suture(s)
   coronal, 4, 10, 32, 146, 147
   frontal, 4, 146, 147
   frontolacrimal, 4, 43
   frontomaxillary, 4
   frontonasal, 4, 51
   frontozygomatic, 4, 10
   intermaxillary, 4

internasal, 4
lacrimomaxillary, 76
lambdoidal, 10, 16, 146, 147
median palatine, 21, 95
occipitomastoid, 10, 16, 21
parietomastoid, 10, 16
sagittal, 16, 32, 147
sphenofrontal, 4, 10, 47, 101
sphenosquamous, 10, 123
sphenozygomatic, 4, 10
squamosoparietal, 10, 16
squamous, 4, 122
temporozygomatic, 10
transverse palatine, 21, 95
vomeromaxillary, 56
zygomaticomaxillary, 4, 51
zygomaticotemporal, 82

Symphysis
  menti, 146
  of mandible, 84

Synchondrosis
  sphenobasilar, 118
  sphenooccipital, 38

Table
  inner, 35
  outer, 35

Teeth, 23, 84, 85, 90, 91

Temporal
  bone, 3, 9, 15, 19, 82, 123-136
    apex, 126, 127, 131, 134
    arcuate eminence, 129
    auricular tubercle, 124
    external acoustic meatus, 23, 28, 82, 122, 125
    facial canal, 132, 133
    infant, 147
    internal acoustic meatus, 123, 127-131, 136 137
    intrajugular process, 115, 129
    line
      inferior, 11, 124
      superior, 11, 101, 102, 124
    mandibular fossa, 26, 28, 80, 81, 89, 122, 124-126
    mastoid
      foramen, 17, 124, 128, 129
      notch, 17, 126
      process, 9, 19, 26, 122, 124, 126, 128
    parietal notch, 122-124
    petrous portion, 35, 123, 126-134
    semicircular canals, 132, 133, 136
    subarcuate fossa, 128
    tubercle, 82
    tympanic part, 122, 124, 147
    vaginal process, 126
    zygomatic process, 122, 124, 126, 129, 134

Temporal fossa, 11, 29, 55, 83, 96, 102
Temporal process of zygomatic, 5, 97
Temporal squama, 35, 122-124, 126, 134
Temporozygomatic suture, 10
Trigeminal impression, 128, 129
Tube, auditory, 28, 134-136
Tubercle
  of mandible, 89
  of temporal bone, 82
  pharyngeal, 23
Tuberculum sellae, 38, 40, 108
Tuberosity of maxilla, 70, 92
Tympanic
  cavity, 136-138, 141-145
  membrane, 132, 138, 141, 142, 145
  part, 122, 124, 147
  plate, 26
  ring, 141, 142
  sinus, 125, 143
Tympanomastoid fissure, 124, 125, 144
Tympanum, roof of, 26, 132

Umbo, 141
Uncinate process, 60, 61, 66, 76, 104, 105

Vaginal process, 106, 109, 110, 126
Vertex, 5
Vestibular window, 144
  fossula of, 125, 135
Vestibule, aqueduct of, 137
Vomer, 3, 15, 19, 21, 66, 70-72, 81, 100
  ala, 61, 70-72, 100
  nasopalatine groove, 63, 64, 100
  perpendicular plate, 71
  posterior border, 100
  septum, 61, 62, 72
Vomeromaxillary suture, 56

Zygomatic
  arch, 19, 80, 82, 83, 96
  bone, 3, 9, 19, 21, 96, 97
    maxillary process, 97
    temporal process, 97
  process of
    frontal, 48, 92, 102, 103
    maxilla, 93-95
    temporal, 122, 124, 126, 129, 134
Zygomaticofacial foramen, 44, 97
Zygomaticomaxillary suture, 4, 51
Zygomaticoorbital foramen, 51, 52, 97
Zygomaticotemporal suture, 82